Molecular Pathology in Cancer Research

Sunil R. Lakhani • Stephen B. Fox
Editors

Molecular Pathology in Cancer Research

Springer

Editors
Sunil R. Lakhani
UQ Centre for Clinical Research
The Royal Brisbane and Women's Hospital
University of Queensland and Pathology Queensland
Brisbane, QLD, Australia

Stephen B. Fox
Director of Pathology
Peter MacCallum Cancer Centre
Professorial Fellow
University of Melbourne
Melbourne, VIC, Australia

ISBN 978-1-4939-6641-7 ISBN 978-1-4939-6643-1 (eBook)
DOI 10.1007/978-1-4939-6643-1

Library of Congress Control Number: 2016958971

Printed on acid-free paper

This Springer imprint is published by Springer Nature
The registered company is Springer Science+Business Media LLC
The registered company address is: 233 Spring Street, New York, NY 10013, U.S.A.

Introduction: Setting the Scene, Morphology to Molecular

Revolution and/or evolution is a constant theme in pathology, but never has it been more pertinent than today with the adoption by diagnostic laboratories of a number of new key technologies. Borne out of the research environment, these have driven significant changes in laboratory practice and changes in workforce especially in the field of cancer diagnostics. The increase in our understanding underlying the biology of cancer has spawned a new generation of tests using previously unused "tissue" sources with levels of detection that only a few years ago would have seen inconceivable. Thus, we are now able to identify DNA sequence mutations using whole-genome approaches in single cells isolated from peripheral blood, assess the underlying aetiology of tumours from their mutational signatures and perform methylation and expression analysis with profiles that give prognostic and/or predictive information.

Much of our increased knowledge have been underpinned by platform enablers such as autopsy and tissue banking which have become disciplines of their own, with the emergence of speciality expertise that is needed to provide appropriately collected, processed and stored samples for molecular interrogation in cancer research.

The changes in the use of new methods and applications in the pathology of cancer are in many ways analogous to the introduction of the microscope in Paris in the 1840s when morphology challenged conventional wisdom and the use of clinical classification of tumours. Thus, the applications from the technologies and techniques outlined in this book in a similar manner raise fundamental questions on tumour classification, biology and therapeutics. Indeed for the classification of tumours, there is a similar discourse around whether tumours should be genomically "binned" or whether conventional morphology should continue to be used. There is emerging data to demonstrate close similarities between tumours of vastly different origins but look, behave and respond to therapeutic targets in the same manner. This discussion will continue over the coming decade, but it is likely that a combination of conventional and innovative technologies will be used to go from a standard classification system to more of an ontology definition of tumours with all the ancillary information that this provides. A highly thought-provoking and provocative "future scoping" further explores how pathology and the revolution in technology and platform enablers may change the face of pathology across not only the neoplastic diseases but throughout all pathology.

Melbourne, VIC, Australia Stephen B. Fox
Brisbane, QLD, Australia Sunil R. Lakhani

Contents

Molecular Diagnostics: Translation from Discovery to Clinical Practice

Fares Al-Ejeh and Andrew V. Biankin

Over the past decade, there has been broad publicity and discussions over the potential of "personalized medicine" to transform clinical practice in oncology. In its broad definition, personalized medicine in oncology refers to the use of biomarkers to make decisions such as the type of therapies, prognosis, and extent of monitoring of disease progression. As such, oncologists have been practicing personalized medicine throughout modern medicine where patients are treated according to clinical staging and the current understanding of specific cancer behaviors. It may be argued that even chemotherapy is personalized, not only in terms of using different chemotherapeutics for different cancer types but also for the concept of using anti-proliferation cytotoxic drugs against highly, uncontrolled proliferative cancers.

Recent examples of personalized targeted therapies include trastuzumab and crizotinib. Diagnostic tests for *Erbb2* amplification for Herceptin in breast cancer and *ALK*-gene fusion for crizotinib in non-small cell lung cancer are required to identify the patients who would benefit from these treatments. More recently, genome-wide molecular profiling has accelerated deeper understanding of the architecture of cancers in general and their heterogeneity in their response to therapies specifically. Transcriptome and genome profiling, and proteome profiling to some extent, have expounded the heterogeneity of cancers even of similar origins which has been recognized by clinicians and pathologists for decades. The rapid growth of molecular profiling has been paralleled with an exponential growth in the use of the term "personalized medicine" (Fig. 1). The promise made is that such

F. Al-Ejeh (✉)
Personalised Medicine, QIMR Berghofer Medical Research Institute, Bancroft Building, 300 Herston Road, Herston, QLD 4006, Australia
e-mail: Fares.Al-Ejeh@qimrberghofer.edu.au

A.V. Biankin (✉)
Institute of Cancer Sciences, Wolfson Wohl Cancer Research Centre, Garscube Estate, Switchback Road, Bearsden, Glasgow, Scotland G61 1BD, UK
e-mail: Andrew.Biankin@glasgow.ac.uk

S.R. Lakhani, S.B. Fox (eds.), *Molecular Pathology in Cancer Research*,
DOI 10.1007/978-1-4939-6643-1_1

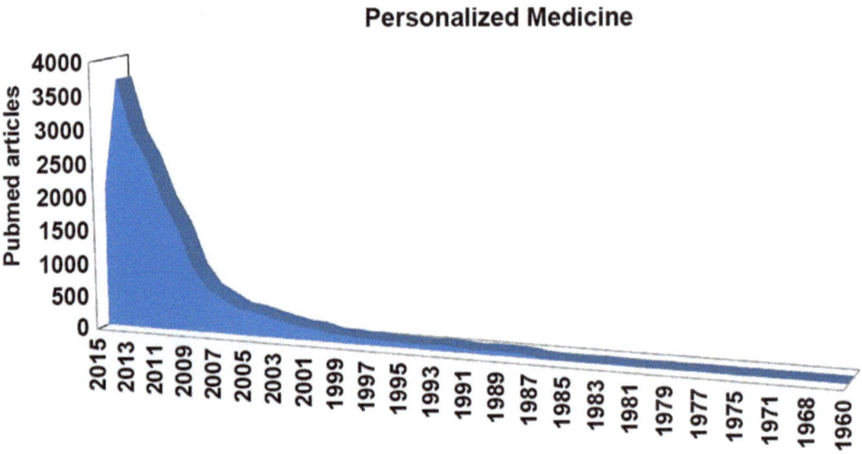

Fig. 1 PubMed search of "personalized medicine" (http://www.ncbi.nlm.nih.gov/pubmed/?term=personalized+medicine)

advances in the field would deliver sophisticated diagnostic tests and precise therapies beyond the "one marker-one drug" model of the past. Accordingly, it may be more suitable to refer to the new model as "precision medicine" or "precision oncology" to distinguish from personalized medicine although both terms may be related.

In this chapter, we will discuss the advances of "omics" technologies particularly their translation into multiplexed molecular pathological tests, which are central for personalized/precision oncology. The pathways towards clinical development of such tests will also be discussed within the current and future regulatory landscapes, and the perspective of clinical utility and impact on patient management and benefit.

New Technologies and Their Promise

Cancer molecular diagnostics are based on the analysis of biomarkers such DNA, RNA, or protein to identify risk or incidence of disease, determine disease progression (prognostic tests), determine therapy, and/or predict response (predictive or companion diagnostic tests). Microarrays enable high throughput measurements of DNA, RNA, or protein and have contributed vastly to our current research practice. One of the earliest descriptions of the use of "DNA microarrays" or "DNA chips" is probably the study by Augenlicht et al. in 1987 which measured the relative expression of each of 4000 complementary DNA (cDNA) sequences from biopsies of human colonic tissue and in colonic carcinoma cells [1]. A follow-up study focused on 30 cDNA clones in an attempt to compare the expression profiles between two genetic groups from patients at high risk for developing colorectal cancer and normal colonic mucosa in low-risk individuals [2]. The earliest precursors of current gene-expression microarrays were reported in 1995 and 1996 by Schena et al. [3, 4]. Genomic sequencing arrays, "sequencing by hybridization" (SBH), were first reported in 1992 by Drmanac et al. [5]. Shortly after, Drmanac et al. in 1994 [6] and 1996 [7] reported on newly developed methods for large-scale production of cDNA and genomic DNA microarrays. One particularly pivotal development was the Affymetrix gene-expression array by Lockhart et al. in 1996 [8, 9]. Comparative genome hybridization arrays (array—CGH or aCGH) were also developed in 1992 to enable high throughput cytogenic analysis in cancer [10] and further developed in 1999 for higher resolution and applied in breast cancer [11]. Protein microarrays, based on reverse-phase method where proteins are spotted onto membranes and detected by antibodies, were not developed until 1999 by Lueking et al. [12]. Alternative to the solid surface chips described so far, bead-based microarrays (BeadArray™ technology) were a later addition to the field which were based on the invention of David Walt and colleagues reported in 1998 [13] and developed by Illumina Inc. (founded in 2001). Illumina Inc. was the new competitor of the then microarray-market dominant Affymetrix Inc. (founded in 1992). BeadArrays™ were initially used for single nucleotide polymorphism (SNP) profiling but later developed for gene-expression profiling in 2004 [14]. Since the 1990s, microarrays have developed and expanded considerably to enable high throughput, genome-wide profiling of DNA mutations, copy number variations (CNVs), gene expression (mRNA), proteins, methylation, and microRNA (miRNA). Next-generation sequencing (NGS) methods developed in the late 1990s are now providing the next levels of accuracy, speed, and low cost to employ in molecular profiling of cancer. NGS applies to genome sequencing, transcriptome profiling (RNA-Seq), DNA–protein interactions (ChIP-sequencing), and epigenome characterization.

Early examples of differential molecular profiling include those carried out in melanoma in 1996 [15], and prostate [16], renal [17], and breast [18, 19] cancers in 1999. Such studies have attracted several reviews, commentaries, and views on the paradigm shift in research where "the hypothesis is there is no hypothesis" [20] and early recognition of the potential of microarrays in drug discovery and response to therapies [21, 22]. The feasibility of molecular classification of cancer based solely on gene expression was first demonstrated in leukemia in 1999 based on class discovery/prediction methods [23]. Perhaps, the seminal studies by Perou et al. in 2000 [24] and Sørlie et al. in 2001 [25] were the first to demonstrate the utility of molecular profiling to explain heterogeneity and more importantly the discovery of the association between the distinct molecular profiles and clinical outcomes. The utility of gene-expression profiling in predicting clinical outcomes was further demonstrated in the study by van 't Veer in 2002 which identified gene-expression signature that is strongly predictive of a short interval to distant metastases in lymph node-negative breast cancer [26]. The importance behind these three fundamental studies in breast cancer [24–26] is their translation into three commercial diagnostic products: the Oncotype DX® breast cancer assay, the MammaPrint® 70-gene breast cancer recurrence assay, and the Prosigna® Breast Cancer Prognostic Gene Signature Assay. Oncotype DX® was developed from 250 candidate genes selected from three studies by Perou et al. [24], Sørlie et al. [25], and van 't Veer [26] and 16 cancer-related genes and 5 reference genes (21-genes) were selected for RT-PCR based test to calculate a recurrence score [27–29]. MammaPrint® is based on a 70-gene signature developed from the study by van 't Veer [26]. Prosigna® is based on the PAM50 breast cancer subtype predictor [30] which was developed from the intrinsic gene-expression profiles [24, 25, 31]. These success stories in breast cancer support the burst in molecular profiling over the past decade. This eruption may be beyond any review capability, however, may be appropriately illustrated by two major international projects, The Cancer Genome Atlas (TCGA) [32] and the International Cancer Genome Consortium (ICGC) [33] projects. The TCGA and ICGC projects continue to expand our understanding of several cancer types at multiple levels of molecular architecture by genome-scale profiling of DNA, RNA, and proteins as well as the integration of such profiles to provide more comprehensive portraits of complex regulatory interactions in cancers. Notwithstanding the clear benefit of biological insights delivered by such "big data" studies, one persistent question remains: *how can we translate our findings to benefit patients?*

Currently Approved Molecular Diagnostics in Pathology

In Europe, CE marking for an in vitro diagnostic (IVD) product is required before it can be launched in the market (Directive 98/79/EC). In the USA, pathological tests are regulated by the Food and Drug Administration (FDA) as in vitro diagnostic medical devices (IVDMD) under two main types of applications: 510(k) and the more comprehensive Premarket Approval (PMA). These applications govern pathological tests and related instrumentation used to carry out testing when used to assist in clinical diagnosis/patient management. We reviewed all the FDA 510(k)[1]-(Table 1) and PMA[2]-approved (Table 2) IVDMD in the field of pathology with decision made in the past 25 years (1990–2015).

Apart from clinical pathology instruments and single biomarker tests, the FDA-approved IVDMD lists to date contain very few examples of multi-biomarker tests (gene expression of mutation panels). This is in contrast to the growing consensus envisioned by genome-wide profiling (e.g., TCGA and ICGC) that multiple biomarkers are a better reflection of the multidimensional nature of cancer heterogeneity and association with clinical outcomes. Multi-biomarker tests are limited to the following:

- *Cologuard*® (Exact Sciences Corp.) is a DNA test that includes quantitative molecular assays for *KRAS* mutations, aberrant *NDRG4* and *BMP3* methylation, and β-actin as reference control, plus a hemoglobin immunoassay [34].
- *GeneSearch BLN Assay* (Veridex, LLC) is a stand-alone intraoperative RT-PCR molecular test for sentinel node staging in breast cancer which measures the

[1] http://www.accessdata.fda.gov/scripts/cdrh/cfdocs/cfpmn/pmn.cfm [Pathology "Panel"].

[2] http://www.accessdata.fda.gov/scripts/cdrh/cfdocs/cfPMA/pma.cfm [Pathology "Advisory Committee"].

Table 1 FDA 510(k)-approved in vitro diagnostic in medical devices in pathology

Company	Product	510(k) Number	Year
Abbott Molecular Inc.	Vysis D7S486/CEP 7 FISH Probe Kit	K131508	2013
	Vysis EGR1 FISH Probe Kit	K091960	2011
	Vysis CLL FISH Probe Kit (TP53/ATM)	K100015	2011
Agendia	MammaPrint FFPE	K141142	2015
	MammaPrint	K101454	2011
	Modification to MammaPrint	K081092	2009
	Modification to MammaPrint	K080252	2008
	MammaPrint	K070675	2007
American Fluoroseal Corp.	Kapton peel pouch	K933228	1994
Aperio Technologies	ScanScope XT system	K080564	2009
	ScanScope XT system	K080254	2008
	ScanScope XT system	K073677	2008
	ScanScope XT system	K071671	2007
	ScanScope XT system	K071128	2007
Applied Imaging Corp.	Ariol HER-2/neu FISH	K043519	2005
	Ariol	K033200	2004
	Ariol HER-2/neu IHC	K031715	2004
Applied Spectral Imaging	ScanView System	K110345	2011
	GenASIs HiPath IHC family	K140957	2015
	GenASIs ScanView System	K122554	2013
	ScanView HER2/neu FISH system	K101291	2010
	FISHView	K050236	2005
	BandView system	K012103	2001
Asuragen Inc.	RNA Retain	K113420	2012
AsymmetRx	Prostate-63 cancer diagnostic test	K050063	2005
AutoGenomics Inc.	INFINITI System	K060564	2007
BioGenex Laboratories	Anti-Progesterone Receptor (InSite® PR)	K012960	2002
	Anti-Estrogen Receptor (InSite® ER)	K013148	2002
BioImagene Inc.	PATHIAM™ with iScan for p53 and Ki67	K092333	2010
	PATHIAM™ system for HER2/neu IHC reagents & kits	K080910	2009
	PATHIAM™ imaging software for HER2/neu	K062756	2007
BioView Ltd.	BioView Duet™ System (Automated ALK FISH Scanning of Lung Cancer)	K130775	2014
Celera Diagnostics	Cystic Fibrosis Genotyping Assay 6L20-01	K062028	2007
Cell Analysis Inc.	QCA (version 3.1)	K031363	2004
Cepheid	Xpert HemosIL Factor II and Factor V assay	K082118	2009
ChromaVision Medical Systems, Inc.	ACIS (automated cellular imaging system)	K032113	2003
	Modification to ACIS (automated cellular imaging system)	K012138	2002

(continued)

Table 1 (continued)

Company	Product	510(k) Number	Year
Clinical Micro Sensors, Inc.	eSensor Cystic Fibrosis Carrier test (eSensor 4800 DNA detection system)	K051435, K060543	2006
Dako Corp.	FLEX Monoclonal Mouse Anti-Human Progesterone Receptor, Clone PgR 636	K020023	2002
	Monoclonal Mouse Anti-Human Estrogen Receptor b1 (Clone PPG5/10)	K993957	2000
	FLEX Monoclonal Mouse Anti-Human Progesterone Receptor, Clone PgR 636	K130861	2013
	Monoclonal Rabbit Anti-Human Estrogen Receptor α Clone EP1	K120663	2013
	Monoclonal Rabbit Anti-Human Estrogen Receptor α Clone SP1 model M3634	K081286	2009
	ER/PR pharmDx™ Kits	K042884	2005
DiagnoCure, Inc.	ImmunoCyt/uCyt+	K994356	2000
Hologic Inc.	Invader Factor II	K100943	2011
	Invader Factor V	K100980	2011
	Invader MTHFR 677	K100987	2011
	Invader MTHFR 1298	K100496	2011
Ikonisys Inc.	Ikoniscope oncoFISH HER2 test system model 2000	K080909	2008
Illumina Inc.	Illumina MiSeqDx Cystic Fibrosis 139-Variant Assay	K124006	2013
	Illumina MiSeqDx Cystic Fibrosis clinical sequencing assay	K132750	2013
	VeraCode Genotyping Test for Factor V (Leiden) and Factor II	K093129	2010
Immunicon Corp.	CELLTRACKS ANALYZER II® System	K060110	2006
	CELLTRACKS ANALYZER II® System	K050145	2005
Incstar Corp.	Incstar Herpes Simplex Virus I/II IgG "fast" ELISA assay	K955362	1996
Instrumentation Laboratory Co.	HemosiL F11 & FV DNA Control	K093737	2010
International Remote Imaging Systems	iQ200 System	K022774	2002
	IRIS 939 Udx urine pathology system	K000373	2000
IRIS International Inc.	iQ® 200 System and iQ Lamina Cradle	K093861	2010
Lab Vision Corp.	NeoMarkers Rabbit Monoclonal Anti-Human Estrogen Receptor (Clone SP1)	K061360	2006
	NeoMarkers Rabbit Monoclonal anti-Human Progesterone Receptor (Clone SP2)	K060462	2006
Leica Biosystems Inc.	Aperio ePathology eIHC IVD system	K141109	2014
	Vision biosystems estrogen receptor clone 6F11	K122556	2014

(continued)

Table 1 (continued)

Company	Product	510(k) Number	Year
Luminex Corp.	The FLEXMAP 3D	K133302	2014
	xTAG Cystic Fibrosis 60 (CF60) Kit v2	K083845	2009
	xTAG Cystic Fibrosis 39 (CF39) Kit v2	K083846	2009
Maine Molecular Quality Controls Inc.	INTROL CF Panel I Control (model: G106)	K083171	2008
MetaSystems GMBH	Ikaros Karyotyping System	K940240	1995
Nanosphere Inc.	Verigene CFTR and Verigene CFTR PolyT Nucleic Acid Tests	K083294	2009
NanoString Technologies Inc.	Prosigna breast cancer prognostic gene signature assay	K141771	2014
Olympus Inc./ Scientific Equipment Group	Virtual Slide System Olympus system	K111914	2012
Omnyx LLC	Omnyx IDP for HER2 manual application	K131140	2014
Osmetech Molecular Diagnostics	eSenor FII-FV-MTHFR genotyping test	K093974	2010
	eSensor cystic fibrosis (CF) genotyping test models XT-8	K090901	2009
Pathwork Diagnostics Inc.	Pathwork Tissue of Origin Test Kit-FFPE	K092967	2010
	Pathwork Tissue of Origin Test Kit-FFPE	K120489	2012
	Pathwork Tissue of Origin Test Kit-FFPE	K080896	2008
Philips Medical Systems Nederland B.V.	Philips HercepTest digital score	K130021	2013
QC Sciences LLC	QCS HER2 immunocontrols (product no. C010)	K023335	2003
Roche Diagnostics Corp.	Factor II (Prothrombin) G20210A Kit	K033612	2003
Sequenom Inc.	IMPACT Dx Factor V Leiden and Factor II Genotyping Test	K132978	2014
Sysmex Inc.	Sysmex UF-500i automated urine particle analyzer	K083002	2009
	Sysmex UF-1000i automated urine particle analyzer with software	K080887	2008
Tecan Ltd.	ProfiBlot	K933996	1994
Third Wave Technologies Inc.	InPlex CF Molecular Test	K063787	2008
Tm Bioscience Corp.	Tag-It Cystic Fibrosis Kit	K060627	2006
TriPath Imaging Inc.	Ventana image analysis system—pathway HER2 (4b5)	K061613	2007
	Ventana image analysis system (VIAS)	K062428	2006
	Ventana image analysis system Ki-67	K053520	2006
	Ventana image analysis system—HER2/neu	K051282	2005
	Ventana image analysis system	K050012	2005

(continued)

Table 1 (continued)

Company	Product	510(k) Number	Year
Ventana Inc.	Ventana ER primary antibody (clone 6F11)	K984567	1999
Ventana Medical Systems Inc.	Virtuoso system for IHC ER (SP1), HER2 (4B5), PR (1E2), Ki67 (30-9), p53 (DO-7), and PgR (1A6)	K140465, K130515, K121516, K122143, K121033 K110215, K121350, K111872, K111869, K111755, K111543, K990618, K103818	1999–2014
Veridex LLC	Celltracks analyzer II system, Autoprep system and kits	K130794, K122821, K110406, K113181, K103502, K073338, K071729, K062013, K052191, K050245	2005–2013
Vision BioSystems Inc.	Vision BioSystems Progesterone Receptor PGR Clone 16	K062615	2007
	Vision BioSystems Estrogen Receptor Clone 6F11	K060227	2006
Vysis Inc.	AneuVysion Multicolor DNA Probe Kits	K010288, K972200, K954214, K962873, K953591	1996–2001

expression of two genes: mammaglobin (*SCGB2A2*) and cytokeratin 19 (*KRT19*) in lymph nodes [35].

- *ResponseDX Tissue of Origin Test* (previously Pathwork® Tissue of Origin Test Kit-FFPE) is an FDA-cleared test marketed Response Genetics Inc. and performed on FFPE tissue at their CLIA (Clinical Laboratory Improvement Amendments) certified laboratory. The test measures the expression of 2000 genes using microarrays to assist pathologists and oncologists in determining the primary sites of tumors, thus help guiding appropriate therapy [36–38].
- *BRACAnalysis CDx* (Myriad Genetics, Inc.) is a companion diagnostic (CDx), CLIA-certified laboratory test for *BRCA1* and *BRCA2* mutations to aid in treatment decision of ovarian cancer with the PARP inhibitor Lynparza™ (olaparib, AstraZeneca Ltd.). Myriad Genetics Inc. also markets several other multigene

Table 2 FDA PMA approved in vitro diagnostic in medical devices in pathology

Company	Product	PMA Number(s)	Year
Abbott Molecular Inc. Vysis	Vysis ALK break apart FISH probe kit	P110012, P110012/S001-P110012/S008	2011–2014
	UroVysion bladder cancer kit	P030052, P030052/S001-P030052/S016	2005–2014
	PATH Vysion HER-2 DNA probe kit	P980024, P980024/S001-P980024/S013	1998–2014
BD Diagnostic Systems TriPath Imaging Inc.	BD PrepStain System	P970018, P970018/S001-P970018/S031	1999–2015
	BD FocalPoint Slide Profiler	P950009, P950009/S001-P950009/S018	1995–2014
	PapNet(R) testing system	P940029, P940029/S001-P940029/S003	1995–1997
BioGenex Laboratories Ltd.	InSite Her-2/neu kit	P040030	2004
bioMerieux Inc.	THxID-BRAF kit	P120014, P120014/S005-P120014/S005	2013–2015
Cytyc Corp. Hologic Inc.	ThinPrep Imaging System	P020002, P020002/S001-P020002/S010	2003–2015
	ThinPrep Processors	P950039, 950039/S001-950039/S032	1996–2014
Dako Corp.	HercepTest	P980018, P980018/S001-P980018/S019	1998–2015
	HER2 CISH pharmDx Kit	P100024, P100024/S001-P100024/S006	2011–2014
	TOP2A FISH pharmDx Kit	P050045, P050045/S001-P050045/S004	2008–2012
	HER2 IQFISH pharmDx	P040005, P040005/S001-P040005/S010	2005–2013
	c-Kit pharmDx Kits	P040011, P040011/S001-P040011/S002	2005–2012
	EGFR pharmDx Kits	P030044, P030044/S001-P030044/S002	2004–2006
Exact Sciences Corp.	Cologuard	P130017, P130017/S001	2014–2015
Gen-Probe Inc. Hologic Inc.	Progensa PCA3 assay	P100033, P100033/S001-P100033/S003	2012–2015
Hologic Inc. MonoGen Inc.	MonoPrep Pap Test	P040052-P040052/S008	2006–2008
Janssen Diagnostics Veridex LLC	GeneSearch Breast Lymph Node Assay	P060017, P060017/S001-P060017/S004	2007–2009
Leica Biosystems Ltd.	Leica Bond Oracle HER2 IHC System	P090015, P090015/S001-P090015/S003	2012–2015
Life Technologies Inc. Invitrogen Inc.	SPOT-Light HER2 CISH Kit	P050040, P050040/S001-P050040/S003	2008–2012

(continued)

Table 2 (continued)

Company	Product	PMA Number(s)	Year
Myriad Genetics Laboratories	BRACAnalysis CDx	P140020	2014
QIAGEN Manchester Ltd.	Therascreen EGFR RGQ PCR Kit	P120022, P120022/S002-P120022/S009	2013–2015
	Therascreen KRAS RGQ PCR Kit	P110030, P110027, P110027/S002-P110027/S007	2012–2015
Roche Molecular Diagnostics Inc.	Cobas 4800 BRAF V600 Mutation Test	P110020, P110020/S001-P110020/S013	2011–2015
	Cobas EGFR Mutation Test	P120019, P120019/S001-P120019/S006	2013–2015
	Cobas KRAS Mutation Test	P140023	2015
Ventana Medical Systems Inc.	INFORM HER2 Dual ISH DNA Probe Cocktail	P100027, P100027/S001-P100027/S021	2011–2015
	PATHWAY anti-c-KIT (9.7) Primary Antibody	P020055, P020055/S001-P020055/S016	2004–2014
	PATHWAY anti-HER-2/neu (4B5) Antibody	P990081, P990081/S001-P990081/S031	2000–2015
	Inform HER-2/neu breast cancer test	P940004, P940004/S001	1997, 2000
	VENTANA ALK (D5F3) CDx Assay	P140025	2015
Zeus Technology	Fluoro-Cep Estrogen assay reagent	P860015/S001	1992

kits (not FDA approved) including: hereditary cancer tests which detect gene mutations (myRisk™ Hereditary Cancer, BRACAnalysis®, COLARIS®, COLARIS AP®, MELARIS®, and PANEXIA®) and prognostic tests which measure gene-expression panels (myPlan® lung cancer, myPath® melanoma, and Prolaris® for prostate cancer).

- *Oncotype Dx®* (Genomic Health Inc.) discussed earlier is another CLIA-certified laboratory test and is supported by level of evidence (LOI) II [39, 40], and may be reaching LOI IB in accordance to a proposed refined LOI guidelines for prognostic/predictive biomarkers using archival specimens [41, 42]. ASCO (American Society of Clinical Oncology) and NCCN (National Comprehensive Cancer Network) have included the Oncotype DX assay in their guidelines as an option to aid decision whether patients with node-negative (N0), estrogen-receptor-positive (ER^+) breast cancer will benefit from chemotherapy. Two large prospective clinical trials are underway to support Oncotype DX by level I of confidence and hopefully can address whether the test prediction is not driven by inclusion of non-luminal breast cancer or low ER^+ and HER2-positive ($HER2^+$) cases [43–46]. Other products by Genomic Health Inc. (not approved yet) include: Oncotype DX Genomic

Prostate Score which measures the expression of 12 genes and 5 reference genes using RT-PCR to aid clinical decisions by assessing the aggressiveness of early stage prostate cancer [47]; Oncotype DX® Colon Cancer Assay is another RT-PCR based test that measures the expression of 7 genes and 5 reference genes for predicting recurrence risk in patients with stage II and III colon cancer [48].

- *MammaPrint®* (Agendia Inc.) is supported by LOI III evidence [49–51] and has FDA 510(k) approval. Similar to Oncotype DX, several clinical trials are undergoing with MammaPrint® to obtain higher LOI for the utility in $ER^+/HER2^-$ breast cancer.[3] Agendia is also running a prospective study assessing recurrence risk in stage II colon cancer patients (NCT00903565) using the ColoPrint® 18-Gene Colon Cancer Recurrence Assay [52].
- *Prosigna®* (NanoString Technologies Inc.) is another FDA 510(k)-approved multigene expression test for use in postmenopausal women with hormone receptor-positive (HR^+) N0 stage I or II or N1 stage II node-positive (Stage II) breast cancer to be treated with adjuvant endocrine therapy. Prosigna® is supported by level I of evidence based on prospective clinical trials [53, 54] and additional indications for influencing treatment decisions [55] and predicting response to neoadjuvant chemotherapy in $HR^+/HER2^-$ patients [56].

In line with the purpose of regulating IVDMDs that assist in clinical diagnosis/patient management, the FDA-approved molecular pathology tests invariably serve as companion diagnostic (CDx) tests for currently approved drugs (Table 3).[4] Similarly, the multi-biomarker molecular diagnostics outlined above also operate as CDx (BRACAnalysis CDx for Lynparza™; Oncotype DX®, MammaPrint®, and Prosigna® CDx for the benefit of chemotherapy in $ER^+/HER2^-$ breast cancer) or aid clinical diagnosis (Cologuard® for diagnosis of colorectal cancer, GeneSearch BLN Assay for diagnosis of sentinel lymph node involvement in breast cancer, and ResponseDX Tissue of Origin Test to diagnose the origin of cancers of unknown origins). This marriage between companion diagnostics and therapies clearly illustrates the practice of personalized medicine and brings clinical and commercial benefits; however, one question arises: *Do the molecular profiles and prognostics gene-expression/mutation signature from "omics" have any clinical/commercial potential?*

Translation of 'Omics to Molecular Diagnostics

Molecular profiling studies, best illustrated by the TCGA and ICGC projects, are cohort studies primarily focused on gaining biological insights of the molecular architecture of human cancers. These studies are not controlled clinical trials, observatory or intervention trials, thus any findings from the 'omics studies require rigorous validation. Such validations are demonstrated by the studies conducted to

[3] http://www.agendia.com/healthcare-professionals/breast-cancer/current-clinical-trials/.

[4] http://www.fda.gov/MedicalDevices/ProductsandMedicalProcedures/InVitroDiagnostics/ucm301431.htm.

Table 3 FDA approved in vitro diagnostic tests as companion diagnostics

Drug trade name	NDA/BLA	Device trade name	PMA	Device manufacturer	Intended use (IU)/indications for use (IFU)
Lynparza™ (olaparib)	NDA 206162	BRACAnalysis CDx™	P140020	Myriad Genetic Laboratories, Inc.	Qualitative detection and classification of variants in the protein coding regions and intron/exon boundaries of the BRCA1 and BRCA2 genes using genomic DNA obtained from whole blood specimens collected in EDTA. An aid in identifying ovarian cancer patients with deleterious or suspected deleterious germline BRCA variants eligible for treatment with Lynparza™ (olaparib)
Xalkori (crizotinib)	NDA 202570	VENTANA ALK (D5F3) CDx Assay	P140025	Ventana Medical Systems, Inc.	Qualitative detection of the anaplastic lymphoma kinase (ALK) protein in FFPE of NSCLC tissue as an aid in identifying patients eligible for treatment with XALKORI® (crizotinib)
Gleevec/Glivec (imatinib mesylate)	NDA 021335; NDA 021588	DAKO C-KIT PharmDx	P040011S001-S002	Dako North America, Inc.	Qualitative IHC for the identification of c-kit expression in normal and neoplastic FFPE tissues as an aid in the differential diagnosis of gastrointestinal stromal tumors (GIST) and identifying those patients eligible for treatment with Gleevec/Glivec (imatinib mesylate)
Mekinist (tramatenib); Tafinlar (dabrafenib)	NDA 204114; NDA 202806	THxID™ BRAF Kit	P120014	bioMérieux Inc.	RT-PCR qualitative detection of the BRAF V600E and V600K mutations in DNA of FFPE melanoma tissue as an aid in selecting melanoma patients whose tumors carry the BRAF V600E mutation for dabrafenib treatment and tumors that carry the BRAF V600E or V600K mutation for treatment with trametinib
Xalkori (crizotinib)	NDA 202570	VYSIS ALK Break Apart FISH Probe Kit	P110012S001-S003	Abbott Molecular Inc.	Qualitative test to detect rearrangements involving the *ALK* gene via FISH in FFPE of NSCLC tissue specimens to aid in identifying patients eligible for treatment with Xalkori (crizotinib)

Zelboraf (vemurafenib)	NDA 202429	COBAS 4800 BRAF V600 Mutation Test	P110020S001-S010	Roche Molecular Systems, Inc.	RT-PCR qualitative detection of the BRAF V600E mutation in DNA from FFPE of melanoma as an aid in selecting melanoma patients whose tumors carry the BRAF V600E mutation for vemurafenib treatment
Erbitux (cetuximab); Vectibix (panitumumab)	BLA 125084; BLA 125147	The cobas® KRAS Mutation Test	P140023	Roche Molecular Systems, Inc.	For use with the cobas® 4800 System. RT-PCR test for seven somatic mutations in codons 12 and 13 of the *KRAS* gene in DNA derived from FFPE human colorectal cancer (CRC) tumor tissue. CDx for CRC patients for whom treatment with Erbitux® (cetuximab) or with Vectibix® (panitumumab) may be indicated based on a no mutation detected result
Erbitux (cetuximab); Vectibix (panitumumab)	BLA 125084; BLA 125147	therascreen KRAS RGQ PCR Kit	P110030, P110027	Qiagen Manchester, Ltd.	RT-PCR assay for the detection of seven somatic mutations in the human *KRAS* oncogene in DNA derived from FFPE CRC tissue to aid the identification of CRC patients for treatment with Erbitux (cetuximab) and Vectibix (panitumumab) based on a KRAS no mutation detected test result
Erbitux (cetuximab)	BLA 125084	DAKO EGFR PharmDx Kit	P030044S001-S002	Dako North America, Inc.	IHC kit system to identify EGFR expression in normal and neoplastic tissues routinely fixed for histological evaluation. Aid in identifying colorectal cancer patients for treatment with Erbitux (cetuximab) or Vectibix (panitumumab)
Vectibix (panitumumab)	BLA 125147				
Tarceva (erlotinib)	NDA 021743	cobas EGFR Mutation Test	P120019S001-S004	Roche Molecular Systems, Inc.	RT-PCR test for the qualitative detection of exon 19 deletions and exon 21 (L858R) substitution mutations of *EGFR* gene in DNA from FFPE of non-small cell lung cancer (NSCLC) tumor tissue. Aids in selecting patients with metastatic NSCLC for Tarceva® (erlotinib) treatment
Gilotrif (afatinib)	NDA 201292	therascreen EGFR RGQ PCR Kit	P120022	Qiagen Manchester, Ltd.	RT-PCR test for the qualitative detection of exon 19 deletions and exon 21 (L858R) substitution mutations *EGFR* gene in DNA from FFPE NSCLC tissue. Select patients with nsclc for whom gilotrif (afatinib) or iressa (gefitinib), egfr tyrosine kinase inhibitors (tkis), is indicated

(continued)

Table 3 (continued)

Drug trade name	NDA/BLA	Device trade name	PMA	Device manufacturer	Intended use (IU)/indications for use (IFU)
Herceptin (trastuzumab)	BLA 103792	INFORM Her-2/ Neu	P940004S001	Ventana Medical Systems, Inc.	FISH DNA probe assay for *Her-2/Neu* gene amplification on FFPE breast tissue to stratify breast cancer patients according to risk for recurrence or disease-related death
	BLA 103792	PathVysion HER-2 DNA Probe Kit	P980024S001-S012	Abbott Molecular Inc.	Detect amplification of the *HER-2/neu* gene via FISH in FFPE breast tissue. Prognostic an aid in the assessment of patients for whom Herceptin (trastuzumab) treatment is being considered
	BLA 103792	PATHWAY anti-HER-2/neu (4B5) Rabbit mAb	P990081S001-S028	Ventana Medical Systems, Inc.	Semiquantitative detection of HER2 in FFPE normal and neoplastic tissue as an aid in the assessment of breast cancer patients for Herceptin treatment
	BLA 103792	InSite HER2/neu KIT	P040030	BioGenex Labs Inc.	IHC assays Her-2/neu in FFPE normal and neoplastic tissue sections. An aid in the assessment of breast cancer patients for Herceptin (Trastuzumab) therapy
	BLA 103792	SPOT-Light HER2 CISH Kit	P050040S001-S003	Life Technologies, Inc.	Quantitatively determine *HER2* gene amplification in FFPE breast carcinoma tissue sections using CISH. An aid in the assessment of patients for whom Herceptin (trastuzumab) treatment is being considered
	BLA 103792	Bond Oracle HER2 IHC System	P090015S001	Leica Biosystems	IHC assay to determine HER2 in FFPE breast cancer tissue An aid in the assessment of patients for Herceptin (trastuzumab)
	BLA 103792	HER2 CISH PharmDx Kit	P100024S001-S005	Dako Denmark A/S	Dual-color CISH probes targeting the *HER2* gene and centromeric region of chromosome 17 to use in FFPE as an aid in the assessment of patients for whom Herceptin (trastuzumab) treatment is being considered. Prognosis in stage II, node-positive breast cancer patients
	BLA 103792	INFORM HER2 DUAL ISH DNA Probe Cocktail	P100027S001-S017	Ventana Medical Systems, Inc.	Dual-color CISH probes targeting the *HER2* gene and centromeric region of chromosome 17 to use in FFPE as an aid in the assessment of patients for whom Herceptin (trastuzumab) treatment is being considered

Herceptin (trastuzumab); Perjeta (pertuzumab); and Kadcyla (ado-trastuzumab emtansine)	BLA 103792; BLA 125409	HercepTest	P980018S001-S018	Dako Denmark A/S	IHC for HER2 in breast cancer, metastatic gastric or gastroesophageal junction adenocarcinoma. An aid in the assessment of breast and gastric cancer patients for Herceptin (trastuzumab) and for breast cancer patients for PERJETA (pertuzumab) or KADCYLA (ado-trastuzumab emtansine) treatment
	BLA 103792; BLA 125409	HER2 FISH PharmDx Kit	P040005S001-S010	Dako Denmark A/S	FISH assay to quantitatively determine *HER2* gene amplification in FFPE breast cancer tissue, metastatic gastric or gastroesophageal junction adenocarcinoma. Assessment of breast and gastric cancer patients for Herceptin and for breast cancer patients for PERJETA or KADCYLA treatments

support the currently approved gene signatures in ER^+ breast cancer (Oncotype DX, MammaPrint, and Prosigna). Earlier, we raised the question *how can we translate our 'omics findings to benefit patients?* A prior question may be *which findings would have clinical utility if translated?* More specifically is the question whether *the molecular profiles and prognostics gene-expression/mutation signature from "omics" have any clinical/commercial potential?*

Scientific publications using whole genome arrays and describing molecular profiles that identify cancer molecular subtypes and/or predict clinical outcomes are beyond any review capacity. Some studies take such defined profiles to the next step and validate and/or develop more accurate mathematical models to predict outcomes using retrospective cohorts. However, it is disappointing that less than handful of genomic tests have made it to FDA-approved clinical practice (ResponseDX Tissue of Origin, Oncotype DX, MammaPrint, and Prosigna) and not many more are being marketed as commercial products without an approved clinical decision utility (e.g., Myriad Genetics Inc. products mentioned earlier). This has been discussed as a "diagnostics pipeline problem" with a key issue being clinically relevant biomarkers that have clear utility to improve clinical outcomes [57]. Other major issues behind the "pipeline problem" include regulation and development models, money, and lack of sufficient clinical specimens for rigorous validations [57].

In 2012, a comprehensive review of the translation of 'omics-based tests to clinical trials/practice was conducted by a committee of experts convened by the US Institute of Medicine (IOM). The committee defined an 'omics-based test as "*an assay composed of or derived from multiple molecular measurements and interpreted by a fully specified computational model to produce a clinically actionable result.*" Importantly, the committee developed a 30-item checklist as a guideline to the development of such tests for clinical use [58, 59] and has been adopted by the National Cancer Institute (NCI) [60]. The roadmap to follow during the development of new 'omics-based tumor biomarker tests comprises of three stages: (1) discovery, (2) test development, and (3) evaluation of clinical utility and use [61]. The 30-item checklist identified five key issues and the corresponding checklist attempts to guide and facilitate appropriate developments of 'omics-based tests. The issues included: (1) specimen issues, (2) assay issues, (3) model development, specification, and preliminary performance evaluation, (4) clinical trial design, and (5) ethical, legal, and regulatory issues. This effort mirrors the previous NCI guideline for REporting recommendations for tumor MARKer prognostic studies (REMARK) in 2005 [62–64]. In 2013, the FDA published a document "*Paving the Way for Personalized Medicine: FDA's role*"[5] describing how the FDA is responding to regulate and drive the rapid developments in personalized medicine. In 2014, the FDA notified the US Congress regarding laboratory developed tests (LDTs; such as CLIA-cleared tests)[6] with a draft guidance of framework for regulatory oversight of

[5] http://www.fda.gov/scienceresearch/specialtopics/personalizedmedicine/default.htm.

[6] http://www.fda.gov/medicaldevices/productsandmedicalprocedures/ucm407296.htm.

such tests,[7,8] which are becoming more complex and with higher risk in the new era of 'omics. More recently, the Precision Medicine Initiative (PMI) was announced by the US President [65, 66].

Examples of efforts to support personalized medicine in Europe include the 2012 Declaration in Rome [67] and The European Alliance for Personalised Medicine (EAPM)[9] founded in 2012, an advocate initiative with the objective to accelerating the development and uptake of personalized medicine and diagnostics. In 2013, the European Commission published a report "*Use of 'omics technologies in the development of personalised medicine*"[10] to evaluate the progress made in personalized medicine, and the opportunities and challenges it presents for healthcare systems. The report recognized the potential of personalized medicine as well as the challenges (e.g., in research). Collectively, the new era of personalized/precision medicine is currently being recognized and efforts to support its development and provide guidelines and regulations are taking place. While these policy changes are underway, we, the scientific/medical community, have the responsibility to address the major issue: *what is the clinical utility of 'omics?*

During the first stage of development, discovery, biologically and clinically interesting 'omics-based biomarkers need to demonstrate the intended clinical use. The vast majority of 'omics-based molecular profiles arise from cancer patient cohort studies where a statistically significant separation of clinical outcomes (e.g., cancer-specific survival) based on a "genomic-profile" brings claims and suggestions of clinical utility. Some studies, although limited, take the next leap to illustrate the ability of the 'omics-based profiles as an assay, which may or may not have analytical validity, to stratify some clinical outcomes using independent retrospective cohorts. Even with this effort, this remains to be in the "discovery" phase and should be labeled as clinical validity rather than utility. Clinical utility is achieved if high levels of evidence have been generated to consistently demonstrate that the 'omics-based tests can improve clinical outcomes for the patient when compared to not using the assay. Such evidence comes either from prospectively directed clinical trials [68, 69] or from "prospective-retrospective" studies exploiting archived specimens from previous clinical trials [41]. Thus, the validation of new 'omics-based tests requires financial investment which is merited by the intellectual property position as well as the return on investment projected based on the utility of the tests.

Genomic studies, such as the TCGA and ICGC projects, are not clinical trials although clinical outcomes such as overall survival or recurrence/metastasis-free survival are recorded for the patients who are receiving the standard treatments for the given cancer. As such, 'omics-profiles or derived assays remain to be limited to prognosis or at best serving as predictive biomarkers for standard therapies (surgery,

---

[7] http://www.regulations.gov/#!documentDetail;D=FDA-2011-D-0360-0002.

[8] http://www.regulations.gov/#!documentDetail;D=FDA-2011-D-0357-0002.

[9] http://euapm.eu/.

[10] http://ec.europa.eu/health/human-use/personalised-medicine/index_en.htm.

radiotherapy, or chemotherapy). The genomic tests currently approved for breast cancer are very clear illustration of this nature. Oncotype DX, MammaPrint, and Prosigna were all discovered and developed based on gene-expression profiling of retrospective breast cancer cohorts. Estrogen receptor (ER) status is a major player in the molecular stratification of breast cancer reported according to the PAM50 and the 70-gene signature; for example, the group identified by the 70-gene signature with high metastatic incidence is almost entirely composed of high grade (Grade 3) tumors with lymphocytic infiltrate and are negative for ER by IHC (Figure 1 in [26]). It may be argued that the prognostication by these signatures being limited to ER^+ breast cancer is merely driven by the identification, molecularly, of ER^+-annotated breast cancer cases with gene-expression profiles that are closer to ER^- breast cancer with the worse prognosis [43–46, 70]. Without the demonstration of the clinical utility in aiding the decision for inclusion of chemotherapy in the treatment planning of ER^+/$HER2^-$ patients, these breast cancer kits would have remained as prognostic signatures without any clinical success.

Researcher: "This prognostic test identifies aggressive tumors"
Clinician: "so what can we offer these patients?"

'Omics provide panels of DNA, RNA, or protein and even integration of these biomarkers to reflect the biological complexity of cancers. Such information needs to translate into an action in order to make an impact on patient management. Prognostication that lacks insight, and more importantly evidence, for how to treat or not-treat the "poor prognosis" patient will be just bad news. Additionally, developing prognostic 'omics-tests would be such an expensive practice particularly if the clinician can reach to similar prognosis using current, simpler clinicopathological indictors. To this end, there may be two models for translating 'omics-based tests in the future; one model has been previously proposed "fit-for-purpose" [71] where the 'omics-based tests are developed along with the therapies—*prospective CDx/drug development model*. This model is probably limited to pharmaceutical companies where biomarker development is integrated in their drug development program. Indeed, there has been a marked increase in co-development and co-commercialization agreements between pharmaceutical companies and diagnostics/sequencing companies. Currently, it is not clear whether these joint development efforts are focusing on whole genome profiling or limited to predefined cancer panels (gene mutation panels or cancer transcriptome profiles). The latter may be less fruitful than genome-wide profiling which do not make prior assumptions but may require investment in bioinformatics teams (in-house or by contract). The second model is more appropriate to the already existing 'omics knowledge or new knowledge generated from genome-wide studies in cohort studies—*prospective development of retrospective signatures*. Molecular profiles/signatures discovered in 'omics studies reflect biological characteristics that often associate with clinical outcomes and behavior, but may also hold companion diagnostic ability to certain classes of drugs. Biological and pathway understanding of these prognostic signatures may enable some prediction of the types of drugs that may be effective against these molecular profiles. Alternatively, preclinical studies using cancer cell lines and more appropriately patient-derived xenografts may aid in determining if prognostic signatures have CDx capabilities [71–73]. These "preclinical trials" would support further discovery studies to be conducted on human cancer specimens from clinical trials to test such CDx capabilities of 'omics signatures. A simpler path may be possible if whole genome profiling arrays conducted by pharmaceutical companies during drug development are published. These data could be mined by researchers similar to the extensive mining of scientific publications of all genome studies.

Future Outlook

The IVDs market, although a fraction of the pharmaceutical market today, is projected for strong annual growth owing to the rising promise of new and novel personalized/precision medicine approaches. Factors affecting the growth of the IVDs market are a topic for specialist commercial discussions [74–77]. These factors include regulatory landscape, clinician endorsement/use, and uptake by payers (consumers/health insurers' reimbursement). As mentioned earlier, policies and guidelines are changing to standardize how 'omics-based tests are developed and validated to obtain the appropriate evidence. Evidence should include clinical validation to gain regulatory approvals as well as evidence to demonstrate clinical utility to gain coverage by payers (health systems/health insurers).

For personalized/precision medicine to have a significant impact on the increasing cost of health care, approaches to stratify patients effectively to determine who will require or be spared additional therapy and to decide on the types of additional therapies need to be robust. This precision model will need to outperform the "blockbuster" model which favors treatments and pharmaceuticals that may be prescribed to "all comers" and are less affected by the heterogeneous molecular architectures of tumors. An integrative model to modernize clinical management of cancer patients would translate the accumulating biological knowledge from 'omics studies. Routine clinical and pathological workups have been standardized over decades of clinical practice and research to provide actionable information for clinical decisions in oncology (e.g., TNM staging and clinicopathology). Prognostic 'omics-tests may play an important role in decisions about patients who can be spared more extensive treatments than patients who may require additional therapies due to poorer prognosis (Fig. 2). This is already practiced in oncology using single biomarkers; for example, early stage ER^+ node-negative breast cancers are spared from chemotherapy whereas chemotherapy is included for ER^+ node-positive breast cancer patients. Indeed, the risk prediction based on gene expression based tests in breast cancer such as Oncotype DX is now aiding clinical decisions to manage the fact that some early ER^+ node-negative breast cancer patients have disease recurrence or metastatic spread while others do not. More value would be added if such prognostic 'omics signature can make validated predictions for the type of

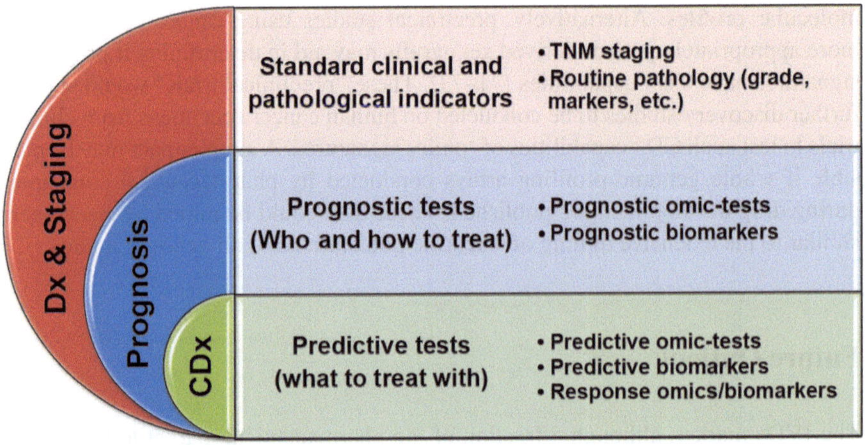

Fig. 2 Integrative model to exploit 'omics as molecular diagnostics in precision medicine

treatments that would be effective against the tumors which these tests classify as high risk. For example, the gene-expression based tests that are prognostic in ER^+ breast cancer contain genes involved in proliferation, or more accurately regulating chromosome segregation and aneuploidy [78, 79], thus anti-aneuploidy drugs such as Aurora kinase inhibitors may benefit the high-risk patients [79]. Following the stratification of patients according to risk of cancer progression, high-risk patients would benefit from predictive tests (companion diagnostics, CDx such as single marker CDx in Table 3) to determine their eligibility for targeted therapies (Fig. 2). Development/validation of 'omics-based CDx would also contribute to patient selection for targeted therapies and could provide deeper understanding of complex disease than single biomarker CDx. Moreover, 'omics-based CDx may enable expansion of drug labels to additional indications and drug repositioning if similar molecular profiles are observed across cancers of different origins. Future development of single biomarker and/or 'omics-based CDx tests to determine the response to a given therapy may add to this integrated model (Fig. 2) and could provide mechanistic explanation for the failure of therapy to allow consequent adjustment or modification (e.g., stop the treatment). Finally, while the use of NGS in clinical diagnostics is currently undergoing debates [80, 81], some advances have been achieved already such as the FDA clearance of the first NGS sequencer, Illumina's MiSeqDx, in 2013 [82]. Notwithstanding the reform in regulatory aspects, success of NGS in clinical diagnostics will depend on clinical relevance and actionability of sequencing information. Better understanding of genotype–phenotype and genotype–drug response relationships and ethical and effective bioinformatics management of genomic are required for the transition of NGS to clinical use [83]. It is important to emphasize here that NGS is not limited to genotype (DNA variation) but also to transcriptome (RNA-Seq) which provides another layer of biological regulation of cancer behavior and should not be ignored for clinical relevance as genomic mutation profiles, while informative, are a limited view of the complex cancer biology. Over the next 5–10 years, NGS may provide deep understanding of complex genetic profiles of human cancers and enable informed selection of effective treatments tailored to individual patients given that the relationships between genotype–phenotype–drug responses are well-defined.

References

1. Augenlicht LH, Wahrman MZ, Halsey H, Anderson L, Taylor J, Lipkin M (1987) Expression of cloned sequences in biopsies of human colonic tissue and in colonic carcinoma cells induced to differentiate in vitro. Cancer Res 47(22):6017–6021
2. Augenlicht LH, Taylor J, Anderson L, Lipkin M (1991) Patterns of gene expression that characterize the colonic mucosa in patients at genetic risk for colonic cancer. Proc Natl Acad Sci U S A 88(8):3286–3289
3. Schena M, Shalon D, Davis RW, Brown PO (1995) Quantitative monitoring of gene expression patterns with a complementary DNA microarray. Science 270(5235):467–470
4. Schena M, Shalon D, Heller R, Chai A, Brown PO, Davis RW (1996) Parallel human genome analysis: microarray-based expression monitoring of 1000 genes. Proc Natl Acad Sci U S A 93(20):10614–10619
5. Drmanac R, Drmanac S, Labat I, Crkvenjakov R, Vicentic A, Gemmell A (1992) Sequencing by hybridization: towards an automated sequencing of one million M13 clones arrayed on membranes. Electrophoresis 13(8):566–573
6. Drmanac S, Drmanac R (1994) Processing of cDNA and genomic kilobase-size clones for massive screening, mapping and sequencing by hybridization. Biotechniques 17(2):328–329, 332–326
7. Drmanac S, Stavropoulos NA, Labat I, Vonau J, Hauser B, Soares MB, Drmanac R (1996) Gene-representing cDNA clusters defined by hybridization of 57,419 clones from infant brain libraries with short oligonucleotide probes. Genomics 37(1):29–40. doi:10.1006/geno.1996.0517
8. Lockhart DJ, Dong H, Byrne MC, Follettie MT, Gallo MV, Chee MS, Mittmann M, Wang C, Kobayashi M, Horton H, Brown EL (1996) Expression monitoring by hybridization to high-density oligonucleotide arrays. Nat Biotechnol 14(13):1675–1680. doi:10.1038/nbt1296-1675
9. Editorial (1996) To affinity … and beyond! Nat Genet 14(4):367–370. doi:10.1038/ng1296-367
10. Kallioniemi A, Kallioniemi OP, Sudar D, Rutovitz D, Gray JW, Waldman F, Pinkel D (1992) Comparative genomic hybridization for molecular cytogenetic analysis of solid tumors. Science 258(5083):818–821
11. Pollack JR, Perou CM, Alizadeh AA, Eisen MB, Pergamenschikov A, Williams CF, Jeffrey SS, Botstein D, Brown PO (1999) Genome-wide analysis of DNA copy-number changes using cDNA microarrays. Nat Genet 23(1):41–46. doi:10.1038/12640
12. Lueking A, Horn M, Eickhoff H, Bussow K, Lehrach H, Walter G (1999) Protein microarrays for gene expression and antibody screening. Anal Biochem 270(1):103–111. doi:10.1006/abio.1999.4063
13. Michael KL, Taylor LC, Schultz SL, Walt DR (1998) Randomly ordered addressable high-density optical sensor arrays. Anal Chem 70(7):1242–1248
14. Kuhn K, Baker SC, Chudin E, Lieu MH, Oeser S, Bennett H, Rigault P, Barker D, McDaniel TK, Chee MS (2004) A novel, high-performance random array platform for quantitative gene expression profiling. Genome Res 14(11):2347–2356. doi:10.1101/gr.2739104
15. DeRisi J, Penland L, Brown PO, Bittner ML, Meltzer PS, Ray M, Chen Y, Su YA, Trent JM (1996) Use of a cDNA microarray to analyse gene expression patterns in human cancer. Nat Genet 14(4):457–460. doi:10.1038/ng1296-457

16. Bubendorf L, Kolmer M, Kononen J, Koivisto P, Mousses S, Chen Y, Mahlamaki E, Schraml P, Moch H, Willi N, Elkahloun AG, Pretlow TG, Gasser TC, Mihatsch MJ, Sauter G, Kallioniemi OP (1999) Hormone therapy failure in human prostate cancer: analysis by complementary DNA and tissue microarrays. J Natl Cancer Inst 91(20):1758–1764
17. Moch H, Schraml P, Bubendorf L, Mirlacher M, Kononen J, Gasser T, Mihatsch MJ, Kallioniemi OP, Sauter G (1999) High-throughput tissue microarray analysis to evaluate genes uncovered by cDNA microarray screening in renal cell carcinoma. Am J Pathol 154(4):981–986. doi:10.1016/S0002-9440(10)65349-7
18. Perou CM, Jeffrey SS, van de Rijn M, Rees CA, Eisen MB, Ross DT, Pergamenschikov A, Williams CF, Zhu SX, Lee JC, Lashkari D, Shalon D, Brown PO, Botstein D (1999) Distinctive gene expression patterns in human mammary epithelial cells and breast cancers. Proc Natl Acad Sci U S A 96(16):9212–9217
19. Sgroi DC, Teng S, Robinson G, LeVangie R, Hudson JR Jr, Elkahloun AG (1999) In vivo gene expression profile analysis of human breast cancer progression. Cancer Res 59(22):5656–5661
20. Mir KU (2000) The hypothesis is there is no hypothesis. The Microarray Meeting, Scottsdale, Arizona, USA, 22–25 September 1999. Trends Genet 16(2):63–64
21. Debouck C, Goodfellow PN (1999) DNA microarrays in drug discovery and development. Nat Genet 21(1 Suppl):48–50. doi:10.1038/4475
22. Gray JW, Collins C (2000) Genome changes and gene expression in human solid tumors. Carcinogenesis 21(3):443–452
23. Golub TR, Slonim DK, Tamayo P, Huard C, Gaasenbeek M, Mesirov JP, Coller H, Loh ML, Downing JR, Caligiuri MA, Bloomfield CD, Lander ES (1999) Molecular classification of cancer: class discovery and class prediction by gene expression monitoring. Science 286(5439):531–537
24. Perou CM, Sorlie T, Eisen MB, van de Rijn M, Jeffrey SS, Rees CA, Pollack JR, Ross DT, Johnsen H, Akslen LA, Fluge O, Pergamenschikov A, Williams C, Zhu SX, Lonning PE, Borresen-Dale AL, Brown PO, Botstein D (2000) Molecular portraits of human breast tumours. Nature 406(6797):747–752. doi:10.1038/35021093
25. Sorlie T, Perou CM, Tibshirani R, Aas T, Geisler S, Johnsen H, Hastie T, Eisen MB, van de Rijn M, Jeffrey SS, Thorsen T, Quist H, Matese JC, Brown PO, Botstein D, Lonning PE, Borresen-Dale AL (2001) Gene expression patterns of breast carcinomas distinguish tumor subclasses with clinical implications. Proc Natl Acad Sci U S A 98(19):10869–10874. doi:10.1073/pnas.191367098
26. van 't Veer LJ, Dai H, van de Vijver MJ, He YD, Hart AA, Mao M, Peterse HL, van der Kooy K, Marton MJ, Witteveen AT, Schreiber GJ, Kerkhoven RM, Roberts C, Linsley PS, Bernards R, Friend SH (2002) Gene expression profiling predicts clinical outcome of breast cancer. Nature 415(6871):530–536. doi:10.1038/415530a
27. Paik S, Shak S, Tang G, Kim C, Baker J, Cronin M, Baehner FL, Walker MG, Watson D, Park T, Hiller W, Fisher ER, Wickerham DL, Bryant J, Wolmark N (2004) A multigene assay to predict recurrence of tamoxifen-treated, node-negative breast cancer. N Engl J Med 351(27):2817–2826. doi:10.1056/NEJMoa041588
28. Cronin M, Pho M, Dutta D, Stephans JC, Shak S, Kiefer MC, Esteban JM, Baker JB (2004) Measurement of gene expression in archival paraffin-embedded tissues: development and performance of a 92-gene reverse transcriptase-polymerase chain reaction assay. Am J Pathol 164(1):35–42. doi:10.1016/S0002-9440(10)63093-3
29. Cobleigh MA, Tabesh B, Bitterman P, Baker J, Cronin M, Liu ML, Borchik R, Mosquera JM, Walker MG, Shak S (2005) Tumor gene expression and prognosis in breast cancer patients with 10 or more positive lymph nodes. Clin Cancer Res 11(24 Pt 1):8623–8631. doi:10.1158/1078-0432.CCR-05-0735
30. Parker JS, Mullins M, Cheang MC, Leung S, Voduc D, Vickery T, Davies S, Fauron C, He X, Hu Z, Quackenbush JF, Stijleman IJ, Palazzo J, Marron JS, Nobel AB, Mardis E, Nielsen TO, Ellis MJ, Perou CM, Bernard PS (2009) Supervised risk predictor of breast cancer based on intrinsic subtypes. J Clin Oncol 27(8):1160–1167. doi:10.1200/JCO.2008.18.1370

31. Perreard L, Fan C, Quackenbush JF, Mullins M, Gauthier NP, Nelson E, Mone M, Hansen H, Buys SS, Rasmussen K, Orrico AR, Dreher D, Walters R, Parker J, Hu Z, He X, Palazzo JP, Olopade OI, Szabo A, Perou CM, Bernard PS (2006) Classification and risk stratification of invasive breast carcinomas using a real-time quantitative RT-PCR assay. Breast Cancer Res 8(2):R23. doi:10.1186/bcr1399
32. Cancer Genome Atlas Research Network (2008) Comprehensive genomic characterization defines human glioblastoma genes and core pathways. Nature 455(7216):1061–1068. doi:10.1038/nature07385
33. International Cancer Genome Consortium (2010) International network of cancer genome projects. Nature 464(7291):993–998. doi:10.1038/nature08987
34. Imperiale TF, Ransohoff DF, Itzkowitz SH, Levin TR, Lavin P, Lidgard GP, Ahlquist DA, Berger BM (2014) Multitarget stool DNA testing for colorectal-cancer screening. N Engl J Med 370(14):1287–1297. doi:10.1056/NEJMoa1311194
35. Blumencranz P, Whitworth PW, Deck K, Rosenberg A, Reintgen D, Beitsch P, Chagpar A, Julian T, Saha S, Mamounas E, Giuliano A, Simmons R (2007) Scientific Impact Recognition Award. Sentinel node staging for breast cancer: intraoperative molecular pathology overcomes conventional histologic sampling errors. Am J Surg 194(4):426–432. doi:10.1016/j.amjsurg.2007.07.008
36. Pillai R, Deeter R, Rigl CT, Nystrom JS, Miller MH, Buturovic L, Henner WD (2011) Validation and reproducibility of a microarray-based gene expression test for tumor identification in formalin-fixed, paraffin-embedded specimens. J Mol Diagn 13(1):48–56. doi:10.1016/j.jmoldx.2010.11.001
37. Handorf CR, Kulkarni A, Grenert JP, Weiss LM, Rogers WM, Kim OS, Monzon FA, Halks-Miller M, Anderson GG, Walker MG, Pillai R, Henner WD (2013) A multicenter study directly comparing the diagnostic accuracy of gene expression profiling and immunohistochemistry for primary site identification in metastatic tumors. Am J Surg Pathol 37(7):1067–1075. doi:10.1097/PAS.0b013e31828309c4
38. Nystrom SJ, Hornberger JC, Varadhachary GR, Hornberger RJ, Gutierrez HR, Henner DW, Becker SH, Amin MB, Walker MG (2012) Clinical utility of gene-expression profiling for tumor-site origin in patients with metastatic or poorly differentiated cancer: impact on diagnosis, treatment, and survival. Oncotarget 3(6):620–628
39. Lo SS, Mumby PB, Norton J, Rychlik K, Smerage J, Kash J, Chew HK, Gaynor ER, Hayes DF, Epstein A, Albain KS (2010) Prospective multicenter study of the impact of the 21-gene recurrence score assay on medical oncologist and patient adjuvant breast cancer treatment selection. J Clin Oncol 28(10):1671–1676. doi:10.1200/JCO.2008.20.2119
40. Mamounas EP, Tang G, Fisher B, Paik S, Shak S, Costantino JP, Watson D, Geyer CE Jr, Wickerham DL, Wolmark N (2010) Association between the 21-gene recurrence score assay and risk of locoregional recurrence in node-negative, estrogen receptor-positive breast cancer: results from NSABP B-14 and NSABP B-20. J Clin Oncol 28(10):1677–1683. doi:10.1200/JCO.2009.23.7610
41. Simon RM, Paik S, Hayes DF (2009) Use of archived specimens in evaluation of prognostic and predictive biomarkers. J Natl Cancer Inst 101(21):1446–1452. doi:10.1093/jnci/djp335
42. Ahern TP, Hankinson SE (2011) Re: Use of archived specimens in evaluation of prognostic and predictive biomarkers. J Natl Cancer Inst 103(20):1558–1559. doi:10.1093/jnci/djr327, author reply 1559–1560
43. Ingoldsby H, Webber M, Wall D, Scarrott C, Newell J, Callagy G (2013) Prediction of oncotype DX and TAILORx risk categories using histopathological and immunohistochemical markers by classification and regression tree (CART) analysis. Breast 22(5):879–886. doi:10.1016/j.breast.2013.04.008
44. Milburn M, Rosman M, Mylander C, Tafra L (2013) Is oncotype DX recurrence score (RS) of prognostic value once HER2-positive and. low-ER expression patients are removed? Breast J 19(4):357–364. doi:10.1111/tbj.12126

45. Gage MM, Rosman M, Mylander WC, Giblin E, Kim HS, Cope L, Umbricht C, Wolff AC, Tafra L (2015) A validated model for identifying patients unlikely to benefit from the 21-gene recurrence score assay. Clin Breast Cancer. doi:10.1016/j.clbc.2015.04.006
46. Sun Z, Prat A, Cheang MC, Gelber RD, Perou CM (2015) Chemotherapy benefit for 'ER-positive' breast cancer and contamination of nonluminal subtypes-waiting for TAILORx and RxPONDER. Ann Oncol 26(1):70–74. doi:10.1093/annonc/mdu493
47. Knezevic D, Goddard AD, Natraj N, Cherbavaz DB, Clark-Langone KM, Snable J, Watson D, Falzarano SM, Magi-Galluzzi C, Klein EA, Quale C (2013) Analytical validation of the Oncotype DX prostate cancer assay – a clinical RT-PCR assay optimized for prostate needle biopsies. BMC Genomics 14:690. doi:10.1186/1471-2164-14-690
48. You YN, Rustin RB, Sullivan JD (2015) Oncotype DX colon cancer assay for prediction of recurrence risk in patients with stage II and III colon cancer: a review of the evidence. Surg Oncol 24(2):61–66. doi:10.1016/j.suronc.2015.02.001
49. Nguyen B, Cusumano PG, Deck K, Kerlin D, Garcia AA, Barone JL, Rivera E, Yao K, de Snoo FA, van den Akker J, Stork-Sloots L, Generali D (2012) Comparison of molecular subtyping with BluePrint, MammaPrint, and TargetPrint to local clinical subtyping in breast cancer patients. Ann Surg Oncol 19(10):3257–3263. doi:10.1245/s10434-012-2561-6
50. Cusumano PG, Generali D, Ciruelos E, Manso L, Ghanem I, Lifrange E, Jerusalem G, Klaase J, de Snoo F, Stork-Sloots L, Dekker-Vroling L, Lutke Holzik M (2014) European inter-institutional impact study of MammaPrint. Breast 23(4):423–428. doi:10.1016/j.breast.2014.02.011
51. Exner R, Bago-Horvath Z, Bartsch R, Mittlboeck M, Retel VP, Fitzal F, Rudas M, Singer C, Pfeiler G, Gnant M, Jakesz R, Dubsky P (2014) The multigene signature MammaPrint impacts on multidisciplinary team decisions in ER+, HER2- early breast cancer. Br J Cancer 111(5):837–842. doi:10.1038/bjc.2014.339
52. Maak M, Simon I, Nitsche U, Roepman P, Snel M, Glas AM, Schuster T, Keller G, Zeestraten E, Goossens I, Janssen KP, Friess H, Rosenberg R (2013) Independent validation of a prognostic genomic signature (ColoPrint) for patients with stage II colon cancer. Ann Surg 257(6):1053–1058. doi:10.1097/SLA.0b013e31827c1180
53. Dowsett M, Sestak I, Lopez-Knowles E, Sidhu K, Dunbier AK, Cowens JW, Ferree S, Storhoff J, Schaper C, Cuzick J (2013) Comparison of PAM50 risk of recurrence score with oncotype DX and IHC4 for predicting risk of distant recurrence after endocrine therapy. J Clin Oncol 31(22):2783–2790. doi:10.1200/JCO.2012.46.1558
54. Gnant M, Filipits M, Greil R, Stoeger H, Rudas M, Bago-Horvath Z, Mlineritsch B, Kwasny W, Knauer M, Singer C, Jakesz R, Dubsky P, Fitzal F, Bartsch R, Steger G, Balic M, Ressler S, Cowens JW, Storhoff J, Ferree S, Schaper C, Liu S, Fesl C, Nielsen TO, Austrian B, Colorectal Cancer Study Group (2014) Predicting distant recurrence in receptor-positive breast cancer patients with limited clinicopathological risk: using the PAM50 risk of recurrence score in 1478 postmenopausal patients of the ABCSG-8 trial treated with adjuvant endocrine therapy alone. Ann Oncol 25(2):339–345. doi:10.1093/annonc/mdt494
55. Martin M, Gonzalez-Rivera M, Morales S, de la Haba-Rodriguez J, Gonzalez-Cortijo L, Manso L, Albanell J, Gonzalez-Martin A, Gonzalez S, Arcusa A, de la Cruz-Merino L, Rojo F, Vidal M, Galvan P, Aguirre E, Morales C, Ferree S, Pompilio K, Casas M, Caballero R, Goicoechea U, Carrasco E, Michalopoulos S, Hornberger J, Prat A (2015) Prospective study of the impact of the Prosigna assay on adjuvant clinical decision-making in unselected patients with estrogen receptor positive, human epidermal growth factor receptor negative, node negative early-stage breast cancer. Curr Med Res Opin 31(6):1129–1137. doi:10.1185/03007995.2015.1037730
56. Prat A, Galvan P, Jimenez B, Buckingham W, Jeiranian HA, Schaper C, Vidal M, Alvarez M, Diaz S, Ellis C, Nuciforo P, Ferree S, Ribelles N, Adamo B, Ramon YCS, Peg V, Alba E (2015) Prediction of response to neoadjuvant chemotherapy using core needle biopsy samples with the Prosigna assay. Clin Cancer Res. doi:10.1158/1078-0432.CCR-15-0630
57. Phillips KA, Van Bebber S, Issa AM (2006) Diagnostics and biomarker development: priming the pipeline. Nat Rev Drug Discov 5(6):463–469. doi:10.1038/nrd2033

58. McShane LM, Cavenagh MM, Lively TG, Eberhard DA, Bigbee WL, Williams PM, Mesirov JP, Polley MY, Kim KY, Tricoli JV, Taylor JM, Shuman DJ, Simon RM, Doroshow JH, Conley BA (2013) Criteria for the use of omics-based predictors in clinical trials. Nature 502(7471):317–320. doi:10.1038/nature12564
59. McShane LM, Cavenagh MM, Lively TG, Eberhard DA, Bigbee WL, Williams PM, Mesirov JP, Polley MY, Kim KY, Tricoli JV, Taylor JM, Shuman DJ, Simon RM, Doroshow JH, Conley BA (2013) Criteria for the use of omics-based predictors in clinical trials: explanation and elaboration. BMC Med 11:220. doi:10.1186/1741-7015-11-220
60. Editorial (2014) NCI issues omics checklist for tests. Cancer Discov 4(2):OF6. doi:10.1158/2159-8290.CD-NB2013-157
61. Hayes DF (2013) omics-based personalized oncology: if it is worth doing, it is worth doing well! BMC Med 11:221. doi:10.1186/1741-7015-11-221
62. McShane LM, Altman DG, Sauerbrei W, Taube SE, Gion M, Clark GM, Statistics Subcommittee of the NCI-EORTC Working Group on Cancer Diagnostics (2005) Reporting recommendations for tumor marker prognostic studies (REMARK). J Natl Cancer Inst 97(16):1180–1184. doi:10.1093/jnci/dji237
63. McShane LM, Altman DG, Sauerbrei W, Taube SE, Gion M, Clark GM, Statistics Subcommittee of the NCI-EORTC Working Group on Cancer Diagnostics (2005) REporting recommendations for tumor MARKer prognostic studies (REMARK). Nat Clin Pract Oncol 2(8):416–422
64. Altman DG, McShane LM, Sauerbrei W, Taube SE (2012) Reporting Recommendations for Tumor Marker Prognostic Studies (REMARK): explanation and elaboration. PLoS Med 9(5), e1001216. doi:10.1371/journal.pmed.1001216
65. Jaffe S (2015) Planning for US Precision Medicine Initiative underway. Lancet 385(9986):2448–2449. doi:10.1016/S0140-6736(15)61124-2
66. Ashley EA (2015) The precision medicine initiative: a new national effort. JAMA 313(21):2119–2120. doi:10.1001/jama.2015.3595
67. Brand A, Lal JA; Public Health Genomics European Network (2012) European Best Practice Guidelines for quality assurance, provision and use of genome-based information and technologies: the 2012 declaration of Rome. Drug Metabol Drug Interact 27(3):177–182. doi:10.1515/dmdi-2012-0026
68. Sargent DJ, Conley BA, Allegra C, Collette L (2005) Clinical trial designs for predictive marker validation in cancer treatment trials. J Clin Oncol 23(9):2020–2027. doi:10.1200/JCO.2005.01.112
69. Freidlin B, McShane LM, Polley MY, Korn EL (2012) Randomized phase II trial designs with biomarkers. J Clin Oncol 30(26):3304–3309. doi:10.1200/JCO.2012.43.3946
70. Cheang MC, Martin M, Nielsen TO, Prat A, Voduc D, Rodriguez-Lescure A, Ruiz A, Chia S, Shepherd L, Ruiz-Borrego M, Calvo L, Alba E, Carrasco E, Caballero R, Tu D, Pritchard KI, Levine MN, Bramwell VH, Parker J, Bernard PS, Ellis MJ, Perou CM, Di Leo A, Carey LA (2015) Defining breast cancer intrinsic subtypes by quantitative receptor expression. Oncologist 20(5):474–482. doi:10.1634/theoncologist.2014-0372
71. de Gramont A, Watson S, Ellis LM, Rodon J, Tabernero J, de Gramont A, Hamilton SR (2015) Pragmatic issues in biomarker evaluation for targeted therapies in cancer. Nat Rev Clin Oncol 12(4):197–212. doi:10.1038/nrclinonc.2014.202
72. Lieu CH, Tan AC, Leong S, Diamond JR, Eckhardt SG (2013) From bench to bedside: lessons learned in translating preclinical studies in cancer drug development. J Natl Cancer Inst 105(19):1441–1456. doi:10.1093/jnci/djt209
73. Bertotti A, Trusolino L (2013) From bench to bedside: does preclinical practice in translational oncology need some rebuilding? J Natl Cancer Inst 105(19):1426–1427. doi:10.1093/jnci/djt253
74. Biophoenix (2008) The convergence of biomarkers and diagnostics: therapy area analyses, key products and future trend. Business Insights Ltd, London
75. Falkingbridge S (2009) Expanding applications of personalized medicine: use of biomarkers in prognostic, predictive and pharmacogenetic tests in a targeted approach. Business Insights Ltd, London

76. Aldridge S (2009) Innovations in oncology diagnostics: technological advances, growth opportunities and future market outlook. Business Insights Ltd, London
77. Nolen BM, Lokshin AE (2013) Biomarker testing for ovarian cancer: clinical utility of multiplex assays. Mol Diagn Ther 17(3):139–146
78. Cheng WY, Ou Yang TH, Anastassiou D (2013) Biomolecular events in cancer revealed by attractor metagenes. PLoS Comput Biol 9(2), e1002920. doi:10.1371/journal.pcbi.1002920
79. Al-Ejeh F, Simpson PT, Sanus JM, Klein K, Kalimutho M, Shi W, Miranda M, Kutasovic J, Raghavendra A, Madore J, Reid L, Krause L, Chenevix-Trench G, Lakhani SR, Khanna KK (2014) Meta-analysis of the global gene expression profile of triple-negative breast cancer identifies genes for the prognostication and treatment of aggressive breast cancer. Oncogenesis 3, e100. doi:10.1038/oncsis.2014.14
80. Evans BJ, Burke W, Jarvik GP (2015) The FDA and genomic tests—getting regulation right. N Engl J Med 372(23):2258–2264. doi:10.1056/NEJMsr1501194
81. Litwack ED, Mansfield E, Shuren J (2015) The FDA and genetic testing. N Engl J Med 372(23):2273–2274. doi:10.1056/NEJMc1504604
82. Collins FS, Hamburg MA (2013) First FDA authorization for next-generation sequencer. N Engl J Med 369(25):2369–2371. doi:10.1056/NEJMp1314561
83. Lu JT, Campeau PM, Lee BH (2014) Genotype-phenotype correlation—promiscuity in the era of next-generation sequencing. N Engl J Med 371(7):593–596. doi:10.1056/NEJMp1400788

Biobanking in Cancer Research

Lisa Devereux, Heather Thorne, and Stephen B. Fox

Introduction

The availability of a biological resource such as human tissue and its derivatives for research that is fit for purpose and linked to well-annotated clinical data under approved ethical protocols is an essential facility for biomedical research, especially in the present era of personalized, translational medicine. The importance of these facilities have been recognized in the popular media with Time Magazine (2009) identifying biobanks as one of the ten tools of significance in recent times that have contributed to health and well-being [1]. Recent investments to upgrade the health department's databases held by government and institutional registries, with electronic data mining and linkage tools, now means it is possible to perform data linkage to a specific disease, such as a cancer diagnosis and the related treatments but in addition, to have access to the other non-cancer related conditions and treatments so the effect of co-morbidities can be researched and the overall influence of the treatments determined. This important data linkage can be routinely performed by a biobank with the participant's informed consent whilst still protecting the privacy

L. Devereux (✉)
Peter MacCallum Cancer Centre, Victorian Comprehensive Cancer Centre Building,
305 Grattan Street, Melbourne, VIC 3000, Australia
e-mail: lisa.devereux@petermac.org

H. Thorne (✉)
Peter MacCallum Cancer Centre, Victorian Comprehensive Cancer Centre Building,
305 Grattan Street, Melbourne, VIC 3000, Australia
e-mail: heather.thorne@petermac.org

S.B. Fox
Department of Pathology, Peter MacCallum Cancer Centre, 305 Grattan Street,
Victorian Comprehensive Cancer Centre Building, Melbourne, VIC 3000, Australia
e-mail: Stephen.Fox@petermac.org; http://www.petermac.org

S.R. Lakhani, S.B. Fox (eds.), *Molecular Pathology in Cancer Research*,
DOI 10.1007/978-1-4939-6643-1_2

and security of all personal information [2]. Access to the national health department's clinical databases also provides practical and great economies to a biobank whose routine task is to perform clinical follow-up on all recruited participants. The reason being, the national health records database provides the additional clinical history and treatment regimen information that a biobank cannot currently obtain, as it is impractical for the biobank team to know about, or even try to cover, all hospital and/or general practitioner interactions that a biobank participant may have.

Whilst biobanks have been established over many centuries, over the past 15 years as the current facilities have matured, there has been recognition that harmonization and professionalism for all work activities needed to be established globally. This approach has provided benefits and is now allowing researchers and clinicians to advance our knowledge in disease prevention and treatment and translate these research findings to provide better health outcomes for patients. Just as critical when establishing and operating a biobank is an understanding of the public's perception and acceptance of the decision-making process in deciding to be a participant and contribute to a biobank. This requires extra consideration, specifically when it involves a biobank for genomic research [2–4].

One essential activity of a biobank manager is the constant review of the facilities strategic plan to ensure they are always relevant to the market's needs. Incorporated into the strategic planning is the constant engagement and collaboration with academic and commercial researchers and clinicians accessing their resource so that the management team can modify the collection protocol regimes, if indicated. That may include the type of biological sample being collected and the processing, storage conditions of the samples so they are suitable for new technologies being used and the type of linked clinical data. A willingness to frequently review and modify a strategic plan and collection protocols at different time points is essential, as there is no point in spending allocated funds to collect biological samples and data that are never used. In regards to funding, the identification of as many avenues as possible to fund the biobank facility is prudent and this may involve applications and engagement with government agencies, not for profit agencies and philanthropic donations. All forward budgets need to include a plan for sustainability to protect the infrastructure for long term storage, all within the framework of ethics approvals, the legal jurisdictions of where the biobank is located and observance to social acceptance and the research and community needs.

Biobanks have been broadly defined into three major types [5]:

1. Population biobanks where biological samples and data are used to determine markers of susceptibility and population identity, representative of a country or ethnic cohort.
2. Disease focus biobanks (or cohorts) for epidemiology and genomic analysis where the research focus is on exposure and modifier influences using DNA and the large collection of specifically collected baseline and clinical follow-up data. All generated results are frequently compared to a population based healthy control group.

3. Disease focussed general biobanks whose participants provide biological samples and limited clinical data to be used to identify markers for disease. The data in most cases is minimal and baseline and is not collected for a specific research project but is a more speculative collection in nature.

There are many hundreds if not thousands of combinations of the three types of biobanks worldwide and beyond the numerous references in the scientific literature [6–11].

Our understanding of what is required to facilitate translational research and the role a biobank may play for improved patient outcomes is becoming more streamlined and specialist in nature due to experience. In a recent British "gap-analysis" publication of the critical tools needed for translational research to advance the understanding and successful treatment of breast cancer, the authors outlined ten points to be addressed, one of the points being the vital need for the development of collaborative infrastructure (biobanks) that contains clinically annotated and longitudinal biological sampling in patients with this disease, was repeatedly identified [12]. The southern Swedish malignant melanoma research initiatives have also highlighted the gap between the bio-analytical and clinical translation and have developed the required biobank infrastructure with a multi-discipline group membership to provide best practice protocols and procedures for an integrated platform and work flow to advance the understanding and improved diagnosis and treatment in this disease stream [13].

Biobanks can be established as individual entities; however, it is increasingly common for biobanks to contribute their collected resources as part of a national or international network. This enables biospecimens and data to be provided in sufficient numbers for large scale analysis that generates the required power calculations and adequate sample size for biomarker studies, analysis of rare diseases, or small subtypes of common diseases [14–16]. The models available for the establishment of a biobank are also numerous with all models acceptable and best framed around the most suitable conditions related to the geography, social and political landscape. There are two common forms of biobank structures, the first one is a network that is known as a Centralized model where samples and data are collected from staff at potentially multiple sites and transported to a central laboratory and data centre for processing, storage, value adding and distributed to approve research projects. Examples of this are BancoADN, Spain (http://www.bancoadn.org/en/presentacion.htm); kConFab, Australia (http://www.kconfab.org); the Singapore Tissue Network, now the Singapore Biobank (http://www.stn.org.sg); and the UK Biobank (http://www.ukbiobank.ac.uk). The second model is known as the Federated model and is where samples are stored at numerous collection sites and the collections are combined in a virtual sense by transferring sample information to a central database. This allows researchers to identify collections or series of samples of interest and access them from multiple collection sites. Examples include the Australian Prostate Cancer BioResource (http://www.apccbioresource.org.au), the Canadian Tumour Repository Network (https://www.ctrnet.ca), the Wales Cancer Bank (http://www.walescancerbank.com), Tubafrost (http://www.tubafrost.org), as well as the P[3]G catalog of large epidemiology based cohorts (http://www.p3g.org). The two models have different advantages and disadvantages. The Centralized model allows for the storage and processing of samples to be easily controlled and processed and managed in a standardized manner, which is an extremely important benefit. The disadvantage is all health professional contributors need to support having the one site as the overall custodian of the biobank collection. The Federated model is potentially more acceptable to stakeholders at the numerous collection sites because they remain more involved throughout the biobanking and decision-making processes for the samples and data being collected. The disadvantage is the difficulty in processing the biological samples and data in a standardized manner and, delivering the material in a timely, organized, coordinated manner from multiple sites to approved research projects [17]. While a clinical pathology department's primary role is not as a biobank, the clinical diagnostic samples and linked data retained by these facilities are a valuable resource for research. Many of these facilities have been actively involved with biobanking activity and are integral to the operations of the two biobanking models detailed above. This retained resource is an economical use of an established infrastructure that can be made available to researchers for analysis where appropriate approvals are obtained.

Ethical Considerations

A biobank, as the "trusted third party" in any setting involving patients, researchers and clinicians, provides an essential mechanism for separating consent to clinical care from consent to use donated biospecimens and data in research.

In most jurisdictions around the world, oversight of the use of human biospecimens for research is the responsibility of a committee charged with reflecting community norms and ethically defensible opinion on appropriate research conduct. These independent ethics committees, known as Institutional Review Boards (IRBs) or Human Research Ethics Committees (HRECs), commonly have responsibility for monitoring the activity of a biobank at their respective institutions. Very large, national biobanks such as the UK Biobank in the United Kingdom and CaHUB in the USA have established an independent entity to oversee the governance and ethical oversight of their large biobank facilities. In Australia, the National Health and Medical Research Council (NH&MRC) provides guidance to HRECs through the National Statement and has established a registry of committees [18].

Consent

A vital consideration and key to the usefulness of a biobank for future research is the explanation and definition of what the term "consent" means when approaching donors. The goal of informed consent is to ensure that subjects are fully aware of the risks and potential benefits of the research to be performed and make a voluntary decision about participating in the research. The dilemma facing all biobank facilities is that due to the speed of advances in technology, the future use of the stored biospecimens and data are sometimes unpredictable and not fully articulated at the time of the consenting process. Previously collected biospecimens and data distributed to researchers, sometimes years after the collection, made it difficult, if not impossible, in many cases to describe in detail at the time of the consenting process, what the exact future research of the samples and linked data will entail, or the significance and impact of possible findings. Therefore practically, it is important to recognize that in many cases the legal and ethical requirements of informed consent for all future uses cannot be satisfied at the time the biospecimens and data are collected. Even so, research into the community attitudes about this issue indicate that most participants would agree that stored specimens and data are a valuable resource and should be used to advance research if appropriate protections are in place.

Basically, the type of participant consent gained falls broadly into three categories:

1. Specific consent for use in a defined and finite project;
2. Extended consent for use of biospecimens in research that is related to, or a direct extension of, the original project for which consent was given;
3. Unspecified, broad consent for use in future research. Such consent is usually underpinned by the knowledge that any future research will be conducted under oversight of the relevant ethical oversight committee.

Many ethical oversight committees now consider, in some circumstances, a waiver of donor consent. This option is included on the majority of IRB/HREC application forms in recent times. For a waiver to be granted, in general the following conditions need to be determined:

1. The research involves no more than minimal risk to the subjects;
2. The waiver or alteration will not adversely affect the rights and welfare of the subjects;
3. The research could not practicably be carried out without the waiver or alteration; and
4. Whenever appropriate, the subjects will be provided with additional pertinent information after participation.

In regards to point 1, when a waiver is requested by the researcher, there needs to be consideration by the IRB/HREC about whether research using stored biospecimens and/or data meets the criteria of involving no more than minimal risk. Some have argued that because the risks are primarily informational, as long as adequate privacy protections have been adopted the research should be considered minimal risk. However, research shows that IRB/HREC chairs are not always in agreement of what constitutes minimal risk and tend in the main to be conservative in their judgments. A standardized, global definition of minimal risk in this context would aide researchers and IRB/HRECs in their determinations of whether a particular research project using identifiable specimens or data can go forward with a waiver of informed consent. With point 2 and 3, as long as appropriate security measures are in place and the research does not involve traits or conditions that would be viewed by the subject or the community to be highly sensitive or stigmatizing, a waiver of consent should not adversely affect the rights and welfare of consented participants. In addition, in many cases granting a waiver to consent is a practical decision as there can be logistical difficulties in re-contacting participants, some might have changed address therefore potentially difficult or impossible to locate or they may be deceased with no next-of-kin details available to obtain a proxy consent. For this reason it is understood that some valid research facilitated by a biobank could not practicably be carried out without a waiver of consent on occasion [19–23].

There has been discussion and some resistance to such a waiver in recent times. The University of California, San Francisco initiated a blog calling for public comment on issues of consent to raise awareness and generate community discussion about research ethics and the use of de-identified biospecimens in genome sequencing activity [24]. Oliver et al. recently conducted a randomized trial of three consent types affording varying levels of control over data release decisions on participants recently recruited into one of six on-going genetic research studies that covered a broad spectrum of diseases and traits. Follow-up interviews were held to assess their attitudes towards genetic research, privacy and data sharing. The results found that participants were more restrictive in their reported data sharing preferences than in their actual data sharing decisions as they saw both benefits and risks associated with sharing their genomic data. Risks were seen as less concrete or happening in the future, and were largely outweighed by asserted benefits [25]. In the discussions around waiver of consent, these studies highlight the ethical conduct considerations when it specifically involves genomc research and that proposed policy changes should carefully consider the research participants perspectives, including privacy concerns [26, 27].

Another major consideration for biobank governance and management that has been widely discussed in the literature and tested in law is the question and observance of ownership and secondary use of the biospecimens and data samples [28–30]. In Australia, community and industry views have been important in shaping guidelines for the ethical use of human tissue in research and jurisdictions internationally have also developed positions relevant to local law [31]. Training of biobank management in this area is essential so there is adherence to the legal requirements and statutory guidelines to maintain the public's trust and respect of these facilities. Therefore, when preparing a Participant Information and Consent Form (PICF), a number of broad questions should be considered and addressed:

- Will the biobank collect for future unspecified research?
- Will the biobank supply de-identified, or potentially re-identifiable biospecimens for research?
- Will the biobank return genetic research results of clinical significance to the donor, the donor's nominated family member(s) and treating doctors?
- What is the scope of applications considered by the biobank for access to resources, i.e. would the biobank be open to receive applications from researchers in both the academic and commercial setting?

- What other records or information will be needed. Does the biobank plan to acquire clinical or other data to annotate specimens through linkage with other, external national or international databases?
- Will the collected de-identified data potentially be downloaded and pooled with other similar external research groups?

Including the relevant clauses in the PICF at the establishment phase of the biobank operations will deliver the most efficient and cost effective ethical framework under which to operate the biobank and be ethically compliant.

Consideration must also be given to the legal and ethical framework of the jurisdiction in which the biobank operates as we are increasingly moving towards international collaborations so it is essential that the elements of consent under which biospecimens and data are collected are robust to allow sharing of this material across international borders. The guidelines and policies established by The International Cancer Genome Consortium (ICGC) on informed consent and access policies were specifically drafted to address the requirements of an international genome wide sequencing project. The elements of informed consent described in this document are an excellent guide to produce a robust and enduring consent document for biobanks in the genomic age [32].

Translational work is being performed by cancer genetic cohort research studies that collect blood for germ-line mutation identification. When personally relevant genetic information is discovered, established protocols are in place to notify participants of these clinically significant, research generated mutation test results. With the best intent by the researchers and clinicians involved in these cohorts, and a confirmed indication from the participants when they were recruited into the genetic research study that they wanted to be notified of clinically significant mutation test results, it remains to be a challenge to effectively notify participants from high risk cancer families, and increase the proportion whose risk is managed clinically, particularly for males and individuals unaffected by cancer. Improving notification of at-risk cancer individuals is an important goal in both the research and clinical environment. Further investigation of the potential barriers to communication between genetics research groups, family cancer clinicians, at-risk individuals and their family members is urgently needed. The ethical implications of these types of studies are also important, and highlight issues for further discussion in the genetics community. A key ethical question does remain unanswered: "If research studies are obliged to notify participants when new genetic information becomes available, to what lengths should they go to meet these obligations?" This question also raises important financial and logistical considerations regarding how many resources research studies should (and can) use to notify their at-risk research participants involved with a biobank or cohort study [33].

The other important issue to be considered and addressed when developing the PICF is when researchers accessing the resource are conducting genetic research that has the potential for finding a heritable genetic alteration that is incidental to the purpose of the analysis or research, and may be considered by a research participant, a clinician or the researcher to be of significance to the health or reproductive decision making of the research participant or their family. This is particularly true of research involving whole genome sequencing (WGS) or any high throughput research techniques that have the potential to generate incidental findings of heritable genetic alterations. Incidental findings (IFs) are made in the course of conducting research, but are, by definition, beyond the aims of the research project. IFs may or may not be anticipated and researchers and others may disagree about their validity, reliability, significance and need for reporting back to the participant.

Whether researchers have a moral obligation to provide findings of this nature back to research participants is a vexed issue with an international consensus just starting to be defined [34, 35]. The matter is complicated by the fact that, unlike clinically validated genetic tests, WGS and other high throughput research techniques are research tools, that is they are not designed for clinical diagnosis and they produce results that are of questionable clinical utility and difficult to validate. Results from these analyses may identify genetic variants that are related to increases in disease risk, but the increased risk may be particularly small. Further, genetic associations that are found may well be non-replicable or difficult to interpret. Nevertheless, findings that are currently uncertain may in future become clinically relevant. Accordingly, the identification of an IF raises questions regarding the potential need for evaluation of the finding and for communication to the participant's clinician or to the participant. One recent study examined the attitudes of individuals diagnosed with sarcoma and their family members towards genetics, genomic research and incidental information arising as a result of participating in genetic research. The results demonstrated that no matter whether they were individuals affected with cancer or their family members, they were generally positive about new genetic discoveries and genetic testing. Possibly not surprisingly, age and gender were factors that influenced how people thought about genetic discoveries and genetic testing. Although intention to receive results did not necessarily translate into action by attending a clinic to obtain their personal genetic test results, the research team believe that if genetic testing for sarcoma becomes available in the foreseeable future, it is likely that family members may demonstrate more reservation towards such testing than the cancer affected and their spouses and this should be taken into consideration. Finally, the majority of sarcoma participants believe people would like to be informed about incidental information arising as a result of research [36].

For these reasons, developing management pathways between the researchers, biobank management and the local IRB/RHEC to determine whether or what information should be feedback to research participants, their families, or clinicians involved in their care and who should be responsible for feeding back these results is necessary. More research in this area about decision-aids to notify participants and evidence based research on attitudes and what is understood by the participants who are signing consent forms to be engaged in this area of researcher is essential. The PICF or other information designed for presentation to research participants should be designed, as a minimum, to clarify that, in the course of the research, information may arise suggesting the presence of mutations that are unrelated to the specific disease or trait being investigated [37].

Sustainability

Whether a biobank, cohort or registry is established with a defined participant recruitment criteria to address a specific research question, or is a speculative collection with the view for future use, the issue of sustainability for the estimated life of the resource that includes procurement, sample processing, data linkage, value adding, clinical follow-up, maintenance, infrastructure and the supply chain for distribution needs to be addressed with all management and operations scoped before a facility is started. Establishing such facilities, even small ones, is a very expensive operation [6, 10, 38–40] (Table 1). Funding for such facilities can be sourced from one or a combination of organizations: not-for profit granting agencies, universities and research facilities, government, private foundations, commercial biotechnology and pharmaceutical companies. In addition to the financial commitment, sustainability may also require long term support and commitment from the donor participants to provide updates about clinical information and biological samples as new diagnosis and treatments are made. The on-going interaction and communication strategies between the biobank management and donors are an important demonstration that the facility is adding to our knowledge base about the population that has been recruited and is beneficial and rewarding to all parties. Such interaction

Table 1 Costs in establishing a bio-bank facility

Facility name	Funding source	Funding received	Years funds awarded	Facility size
Australian (Oncology only) enabling grants	Federal Government (NH&MRC)	Aus$ 22 million	2004–2014	Medium, 12 independent networks
The pan-European Biobanking and Biomolecular Resources Research Infrastructure—European Research Infrastructure Consortium (BBMRI-ERIC)	National governments	€135 million	2013	Large, 10 networked counties
caHUB, NCI, USA	NCI	USA $23 million	2009	Large, numerous networks
Israel National Biorepository	National government and philanthropy	$10 million	2008–2013	Small, numerous networks
Stellenbosch Biobank H3Africa—H3Africa Consortium	National and International government	$74 million	2012–2018	Large, numerous networks
UK Biobank	Numerous national and regional government and not for profit organizations	£87 million	2006–2016	Large, numerous networks

can be invaluable in highlighting a biobanks aim and purpose to funding agencies. When establishing a facility it is important to recognize in the budget projections that most biological samples linked to data become more valuable for research purposes after 5–10 years of clinical follow-up, often with repeat sampling of biological material and data. Exploring all options to ensure that the collected material is used to facilitate research is essential, for example, biological samples and data may have been collected originally for a specific project with the samples embargoed until that research project is completed. At the completion of that project the remaining samples, excess to the original project's needs, will be free to be used by other groups, potentially with the value added data generated from the original projects findings. Therefore, when determining the overall costs for sustainability it is important to capture all expenses that include the costs for the initial set up and the on-going operational costs, on average 15–20 years, for management and laboratory staff, laboratory facilities, equipment and maintenance, databases, supply of material to researchers and data linkage to external agencies such as cancer agencies, medical records and health departments, as the facility matures.

Government and not for profit granting agencies will frequently provide funds for the establishment of a biobank, cohort or registry but unfortunately, it is frequently the case that the agencies then fail to recognize that on-going funding, even at a small percentage of the original grant total, is still required to fund the existing infrastructure to maintain the facilities operations, even when co-funding from other organizations and cost recovery schemes are in place. A common occurrence appears to be that at best, biobank facilities can obtain 5–10 years of funding before being informed that they need to be 100 % self-sufficient by their funding agencies. For biobank facilities who are relying on predominately academic researchers funded through peer review government awarded grants, being 100 % self-sufficient by the implementation of a cost recovery scheme isn't an achievable goal as the researchers grants are usually lean in value and committed to other aspects of their research project. Academic based researchers do not have grant funding anywhere near the levels required for a reasonable cost recovery linked to the true collection and supply cost. In addition, the demand to supply biological samples and data held by a biobank and linked to a cost recovery fee can vary year to year, sometimes dramatically. This makes budget predictions based on a cost recovery scheme alone very difficult and not sustainable.

Linked to the issue of biospecimen and data usage and sustainability, it is important at the facilities establishment stage to understand what the market requirements are and the potential demand so to optimize the use of the collected resource. The strategy of collecting all surgical material in a speculative manner in the hope that it might be used at some stage has led, in the majority of cases, to a very small percentage of usage of the collected material. A more strategic plan can be seen where the biobank is embedded within the clinical pathology department as the biological samples and associated pathology and treatment data have already been collected for clinical diagnostic purposes and eventually will be available for research purposes when deemed to be in excess to diagnostic purposes [40, 41]. The other productive model is associated with the cohort biorepositories and registries where there is often a rich collection of biological samples such as multiple sampling of tissue, primary and metastatic, and blood from the recruited participants. The biological samples and data in a cohort facility have usually been collected to support a series of specific research questions. The associated data may have been collected for 10 years or more linked to comprehensive epidemiological and clinical follow-up data. The cohorts traditionally demonstrate flexibility in its expansion of sample and data collection as the specific clinical and scientific aims of the cohort study evolve [42–44] with a large percentage of the biological samples and associated data used by research approved groups multiple times.

A recent review of 636 biobanks in the USA characterized their origins, specimens collected, market context and the issue of sustainability. Importantly, linked to the issue of sustainability, the researchers found that having a biobank embedded in a larger organization, such as a hospital or research institution, was essential to the biobanks financial structure and survival. The majority were associated with an academic institution 78 %, hospitals or research institutions 27 % and 15 %, respectively.

When the biobanks were asked about how competitive they were in the market place, only 14 % answered in the affirmative. Significantly, only 4/57 (7 %) of the biobanks stated that there was a "great deal" of competition for their resource, 51 % stated a modest amount and 42 % indicated that they had very little competition. They found that *for profit* biobanks were significantly more likely to be competitive (61 % vs. 12 %) and it appeared that most of the biobanks surveyed filled a specific niche within their organization and were not concerned about holding the share of their "market". In response to a question about the demand for their collected facility products, 51 % reported that demand for their biobanks products had increased over the past 2 years, 6 % found demand for material had decreased and 45 % answered that it had remained about the same. In the current period of financial restraint, it was a surprise that only 13 % of biobanks had a major concern about the under-utilization of their resource, 28 %, respectively, had moderate or minor concerns while for 31 % of biobank facilities it was not a concern at all [7].

As on-going funding for such facilities is a recognized challenge internationally, options for sustainability have led to many funding avenues being explored and adopted, depending on the local funding landscape. By far the best and potentially easiest option is to have the host institute fund the facility 100 %, and for the facility to become part of the health care structure of the institute. From a practical sense this model is best suited to a smaller, single site type of facility as the political logistics of funding sourced from a single site to support a multi-site network where numerous sites collect and pool their samples and data is problematic. In the expanded, multi-site collection model, ultimately one site from the group will have the responsibility for specimen and data collation and responsibility for the facilities management. The challenge for the management of a multi-site networked group is to convince a local hospital or research institutions management where the centralized facility is based that funds should not just support the local sites collection but that part of the funds are needed to support the broader networks activities to collate, value add and distribution the biobanks material.

Whilst there are many valid arguments about the merits for the consolidation of biobank activity across networks with standardization being the essential aim for best practice and efficiency, unfortunately, the funding required to support this type of centralized, broader network is rarely taken up by one sole institute due to that organizations own budget restraints, legal nuances and other competing interests within that groups health structure. In the current financial climate, largess is becoming harder to find in publically funded organizations. Fortunately there are a few good examples of regional and inter- and intra-country networks that have managed to resolves many of these problems and established medium to large scale facilities receiving a combination of government, not for profit and foundation funding [6, 10, 14, 38, 44, 45] to support their combined networks.

As previously mentioned, another popular model for sustainability is the leverage of a cost recovery or administration fee. Many facility managers strongly argue that even if a facility has 100 % funding coverage for all aspects associated with the operational, logistics and supply costs, a small fee labelled as an administration or cost recovery fee that is associated with the managers time and for a small portion of the costs associated with the collection and value added component of the biological material or data being supplied can be a useful tool as it makes researcher accessing the resource think responsibly about what material they are requesting. This fee also assists in avoiding inappropriate or over ordering, therefore, wastage of the valuable biological samples. In addition to an administration fee, the dilemma for all biobank facility managers is how to structure a cost recovery fee that takes into account the original cost to recruit and consent a participant, collect the relevant biological material and data, value add and supply to an approved research project, hopefully multiple times. Against this value, and already mentioned in this article, is the realistic cost that a researcher wishing to access the resource can afford to pay, especially if they are an academic researcher on a modest government awarded research grant that needs to cover multiple laboratory wages and consumables and where there has not been an allocated budget to access a biobanks resource due to a research project evolving. In addition, most biobanks in the not for profit arena need permission from the relevant government authority to charge a cost recovery fee and certainly are not allowed to make a profit from the supply of material to an approved research project [46]. When dealing with a private or industry partner, the regulations in some countries are even stricter in regards to the implementation of a cost recovery fee [6]. Cost recovery charges are usually reviewed annually and adjusted, if need be, to reflect the level of grant funding from other sources. At best, internationally most biobanks would only be recouping 10 % of their total budget costs related to recruitment and on-going overheads.

Though small in number, economic models have recently been developed around using centralized consolidated biobank resources to produce budget savings and efficiencies to aid the long term sustainability of these facilities. The prepared economic model provides a more accurate estimation of direct vs. indirect biobank costs and establishes the cost effectiveness and cost benefit evidence that is required to justify, usually to government, spending in this area. Variable and fixed costs, cost recovery schedules that incorporate internal and or external funding sources, access fees, administration infrastructure costs and potentially intellectual property considerations have all been built into these models. The authors hypothesize that these models will lead to analysis efficiencies, improved data accuracy and infrastructure costs, therefore, improvements in patient welfare and a higher professionalism within the work place and sustainability [47]. The authors also highlight the practical aspect of an efficient biobank, that is, it isn't the number of samples that are collected and stored but rather how many samples are out going to support research projects, making sure that you do have a "product" that the market wants, and ensuring that the biological material and data is of the highest quality possible. Linked to this is the aspect of a cost recovery fee that can be modified at any stage, that is, "stepped" or "graduated" as the collection matures, or funding streams become available or cease, hence a developing business model over time to ensure the sustainability and protection of a vital resource [48]. Another funding revenue source that has also been increasingly adapted by biobank facilities in recent times is a contracted service fee offered by research groups and pharmaceutical companies who want a patient group recruited with specific biological samples and data collected under strict collection protocols. There are economies for all involved by *piggy-backing* onto this type of customized service as a business model. The down side is that the biobanks main collection and recruitment may be reduced as the contracted service takes up the biobanks routine scheduled work time, but at least it secures another streaming of funding to keep the facility operational. Whilst the majority of biobanks are supporting academic research groups, facility managers should be aware of the demand for well-annotated biological samples linked to clinical, genomic data and treatments that can be used by pharmaceutical companies for drug discovery and validation analysis. In determining a cost recovery fee to be charged to academic researchers vs. commercial research groups, academic researchers may argue that a public biobank that has been funded in the main from a government grant whose revenue has been obtained from public taxes should not be charged all, whereas private commercial entities who have the ability to pay for access to the resource, due to the potential profit generated, should be required to pay a higher cost recovery user fee. This is not an easy issue to manage, especially against the background of lower government grants to fund biobank facilities in recent time. Overall, there does appear to be consensus that it is reasonable and fair that the cost recovery schedule for the supply of biological samples and data to an academic researcher is less in value than to a commercial company and charging all of the groups that access the resources, thereby, the established infrastructure in the public domain to advance their own interests promotes an equitable approach for the financing of public funded programs [44, 48].

Best Practice and Access Protocols

The protocols around sample collection are driven by a number of factors as not all biobanks have the same brief. A biobank may focus on specimens collected during routine clinical care for therapeutic or diagnostic purposes that are then in excess to clinical purposes, specimens collected for clinical trials, specimens collected as part of specific research projects or specimens collected as part of population based or cohort biobanks.

Collections for a specific research question will be built around the methodology developed by the researchers although whatever the biobanks design and function is, harmonization of biobank operational procedures and a recognition of international best practices is an essential requirement for the management and all staff. As the diverse types of biobanks in existence have different roles in their operations and translation of the generated research results, there is also a recognition that it impossible and potentially undesirable to harmonize completely all practices, policies and operations [49]. It is essential though that the biobank manager and staff recognize when complete adherence is or isn't required, for example, in the era of large scale global collaboration that requires the exchange and pooling of data and samples, exacting standardization of SOPs and harmonization ensures the effective interchange of valid information and samples from numerous groups to be pooled for analysis [50]. There are many excellent papers and documents listing extensive descriptions of appropriate collection protocols and SOPs developed by peak biobank groups over many years [8, 51–53]. As well as their documents being a practical resource for day to day use, these protocols also provide confidence to funders, participants and researchers accessing the resource. One dilemma that is currently being discussed and urgently needs to be addressed by all biobank facilities is the fact that researchers are highlighting that they are having trouble accessing sufficient numbers of samples from a biobank and that the available samples are not always suitable for their research purpose. From the biobank operational and usage milestones, they have 1000s upon 1000s of biospecimens and data stored that no one is using. This is one of the issues that the ESBB working party (http://www.esbb.org) is currently addressing (personal communication with Drs Dominic Allen and Christina Schröder) with the development of a register of all biobanks that list their specific features. This is to be followed by a greater dialogue between the biobanks, academic and commercial researchers to help identify and solve collection and supply issues between all parties. Linked to the problem of previously collected biological samples not being fit for purpose, or having an under developed matched data set, in recent times it has become common for the research departments of pharmaceutical companies to contract a biobank to collect the samples required for a defined research protocol. There is still a problem with this strategy as the challenge then lies in recruiting enough participants and collecting enough biospecimens linked to clinical follow-up data in a prompt and reasonable time frame, i.e. what was the response to a first line treatment regime that may be administered over a 6–8 month period.

The final governance issue for the biobank management to address is the development of a policy and procedures document that clearly outlines how to access the biobank resource to facilitate research in a transparent, effective and equitable manner. It is also important that these documents are visible and easily located so external researchers can see what biospecimens and data are available and what the formal application process is to access the material. Establishing such policies and procedures can be a challenge as obviously some of the stakeholders involved in establishing the biobank will have a legitimate and positive vested interest in establishing the infrastructure as it will bring a productive flow of material and data to their personal research work. The handling of potential conflict of interest (COI) does need to be addressed and worked through. Many groups have resolved the COI issue by establishing a working committee that includes multi-discipline professionals with a wide range of relevant specialties and that includes community representation. There also needs to be a recognition and a willingness by everyone in the organization and stated in the groups Terms of Reference (ToR) that members will be excused from discussions and decision making if a COI is evident or perceived. The primary aim for the whole organization is to provide access to the biobank resource to everyone, internal and external investigators, who apply and who has an IRB/HREC approved project linked to a peer reviewed research study that is funded. Operationally, it is important that all applications are reviewed formally and all decisions documented. This can be achieved by researchers submitting a research proposal to the biobank management that outlines the research aims, hypothesis, plan and conclusion, the overall number of participant numbers required and the specific amounts of biospecimens and data points and suggested timelines for the supply of material. Proposals can be reviewed and commented on by a sub-committee with expertise in the field of interest. Once all questions and concerns about the project application are addressed, the project can be approved. It is essential at this stage that the researcher accessing the resource is aware of the terms and conditions required to access the biobank resource. Practically this can occur in a formal project acceptance letter that may include requirements such as the submission of annual progress report, the length of the project approval period, depending on the circumstance the signing of either a Material Transfer Agreement (MTA), Memorandum of Understanding (MOU) and Data Transfer Agreement (DTA), acknowledging the biobank in all publications and the return of all generated research results back to the biobank after publication. This last step is an invaluable data adding contribution from the researcher to the overall data held by the biobank facility and has the added benefit of avoiding duplication of research efforts on the precious biological samples. The same requirements have been applied when a biobank or researcher accessing a biobank resource then contributes to another consortia with a new set of MOU, MTA and DTA agreements being signed to protect all stakeholder's interest. Whilst non-compliance to the ToRs by researchers accessing a biobank resource is thankfully uncommon, it is important for the biobanks policy and procedures document to state what penalties will be enforced should any access policy and procedure requirements be broken.

Quality Assurance and Quality Control

It is essential that a biobank incorporate a quality assurance (QA) and quality control (QC) programme into their routine work so the facility managers are observing the international best practice guidelines and the programme requirements are meet to achieve the highest standard possible for the supplied biological samples and data for research projects [54–58]. Robust QA and QC protocols will ensure that the biobanks management group is aware of what specimens and associated data has been received, the relevant transport conditions, the multiple samples that have been processed and stored and what has then been supplied to approved research projects. Researchers receiving the biological samples and data need to be confident that the materials they have received are of high quality and able to support their planned research projects. The process through which the product is obtained is referred to QA, whereas the product generated is part of the QC. For example, quality assurance is defined as "that part of quality management that focuses on providing confidence that quality requirements will be fulfilled" [59]. QA requires the systematic monitoring and evaluation of all aspects of the biobank processes; it covers the way in which the biobank is operated as well as the quality of the samples and data held. QC consists of specific tests defined by the QA programme to be performed to monitor procurement, processing, preservation and storage, specimen quality and test accuracy. These tests may include but are not limited to: performance evaluations, testing and controls used to determine the accuracy and reliability of the biobank equipment and operational procedures as well as monitoring of the supplies, reagents, equipment and facilities. Standard operating procedures (SOPs) are an essential part of quality assurance; a biobank will determine and document its ways of working to ensure that samples and data are collected and handled consistently. As the global biobank community matures, most of the different facilities have shared and published their SOPs to further standardize what is being done in QA and QC. This has been a benefit for researchers receiving samples from different biobanks as there is a degree of confidence that all of the samples will be of a similar quality and for facility managers who are starting to establish a biobank facility [8, 52, 60, 61].

As published by the NCI Best practices in 2007 and revised in 2011, and adopted by many biobank facilities internationally, QA and QC should address the following:

Facility Infrastructure

Equipment validation and change control, calibration, maintenance, repair procedures and environmental monitoring; e.g., temperature monitoring of freezers.

Supplier management programmes, including inspection and validation of reagents and other supplies.

Biospecimen Control and Documentation

Control of biospecimen collection, processing and tracking.

Documentation of biospecimen collection, processing and tracking, with detailed annotation of pre-analytical parameters.

Measurement and analysis of key process indicators to drive quality improvement.

System Security

Recordkeeping and document control.

Employment of a data quality management, assessment and reporting system.

Clinical Data Records

Accessibility of policies and procedures.

Documentation records, including audit reports, deviation reports and corrective action and preventive action reports.

External document monitoring to ensure that the facility remains up to date with relevant laws, standards and best practice publications.

Staff training records, including record of staff adherence to training schedules.

Data quality management (source documentation and electronic records), assessment of reporting system.

Supply Records

Internal audit of program and its policies, scheduled and unscheduled.

Audit for accuracy of all annotation data; e.g., the biospecimen and where it is purported to be, in the purported volume, with the appropriate labels and unique identifiers.

Audit for accuracy of patient data associated with biospecimens; e.g., age, gender, date of diagnosis and processing, etc.

Audit of the compliance of the biospecimen resource with institution policies; e.g., human subjects and privacy and confidentiality protections, prioritization of biospecimen use.

Audit of SOPs for all activities, processes and supply.

Each biospecimen resource ensures that SOPs are written, reviewed and are an appropriately approved process that exists for review and updating at designated time intervals.

In addition to the best practice guidelines published by the NCI in recent times, as many biobanks have matured they have been innovative in the era of molecular pathology and genomics to also include a rich collection of phenotypic data and comprehensive clinical follow-up data linked to each biological samples. This extra data value adding via these new technologies has led to the biobank facility sharing new genomic platforms either within or external to their work site for high throughput technology to derive data for bio-informatics analysis. A key requirement for data to be analysed when samples and data have been pooled from multiple sites depends on harmonization and standardization of SOPs have been developed and used by all of the contributing biobanks so uniform, combined data analysis can occur. In recent times there are good international examples where cancer consortia have agreed on harmonized SOPs so large scale analysis can occur to gain the required statistical power for analysis [62, 63]. In these multi-site global consortia, biological samples being supplied such as DNA or RNA undergo rigorous QA and QC using concentration determination methods best suited to the platform technology being used, i.e. picogreen or qubit readings for DNA concentration are often used in preference to nanodrop for Next Gen Sequencing or whole genome Copy Number Analysis. For the supply of fresh or archival tumour tissue, researchers will often request to know the % tumour, stroma, ducts and necrotic tumour in every specimen before deciding on the optimal case with the best cellular component for their analysis [64]. Using a scanning microscope to capture and catalogue the tissue image also aids external researchers in deciding the tissue they wish to access by having log in rights to view the scanned tissue. This process has provided a speedier supply of tissue for review, and eliminated the hazards and delays associated with the shipment of glass slides to external sites [65].

Whilst it is a challenge to implement flexibility and adaption for biospecimen SOPs to address the needs of emerging technologies, a lack of attention to SOPs and adjusting the SOPs for a project requirements in molecular pathology research can lead to misconception of molecular findings and discrepant results if the sample being tested isn't of a high quality, contains contaminants or has not been prepared under the appropriate protocol. Having the correct specimen characteristics, prepared under standardized SOPs that includes stringent QA and QC, is the recognized way to advance translational research.

Databases

In addition to the acquisition of biospecimens and data, the other important item in the SOPs is the purchase or development of a database so that the entry of all of the items associated with the biological samples linked to a data dictionary are entered for every participant recruited. With the evolving technology and laboratory findings translated into clinical practice, it is also important to consider when purchasing or developing a database that upgrades and modifications may need to be made within a few years of purchase due to the extension of the data dictionary to record additional data parameters or potential linkage with other groups due to data sharing. Awareness of the international standardization and published protocols for specimen and data collection is required so compatible data fields are implemented. Analysing data generated from biological samples and clinical data is becoming more complex as combined datasets become larger due to the new technologies and data sharing of multiple groups to gain statistical power. Therefore, databases are integral to performing the required large scale analysis to understand a complex disease such as cancer. Depending on the nature of the biobank, databases can be purchased off the shelf or purpose built. A database provides not just management, operations, governance and details of the researcher access approval, but they also enable the results that have been generated on the supplied samples to be downloaded back into the database as a value adding exercise for further analysis. Tracking of this information manually is not practical and in fact impossible to do, especially with the larger networked consortia, without a database. It also ensures that specific data searches can be performed to avoid duplication of work and locate a required sub-set of biological samples. There are many international examples of where supplied biological samples and/or data to a researcher are returned back to the facility after publication by the researcher [44, 66]. A good practical example of this activity is when a researcher might perform immunohistochemistry (IHC) using a panel of antibodies on supplied tissue. These IHC results can be returned to the biobank and entered into the facilities database and then made available to other researchers with the same research question.

All countries have laws and guidelines for the security of information entered into a database to protect privacy and confidentiality of the participants to protect and minimize accidental or intentional abuse. In most cases participant's identifiable information can't be entered into a biobank database without a participant's acknowledgement and approval although these requirements can vary between regions and countries [67–70]. A database containing information should have the information held on servers at a secure site and be password protected. This security aspect has become even more important now that many databases are web based and accessed via multiple sites. SOPs should include specific guidelines for frequent staff training about all aspects of best practice in regards to the information entered into the database with adherence to the privacy and confidentiality laws and guidelines for their location or country whilst facilitating the researchers' needs for appropriate biospecimens and data [71, 72].

Recently developed databases can generate a de-identified number that can be used when supplying material to researchers but that can be decoded, if required, by the biobank staff. Many cohort studies also have the ability to generate via their databases a unique family specific number when multiple family members are recruited into familial cancer research studies. The decoding of the de-identified number by the biobank staff is a practical function as it means in addition to the baseline data collected at the recruitment phase, clinical data from external facilities, such as the death and cancer registry, hospital discharge diagnosis data, general practitioner data; medication prescriptions, pathology reports, imaging reports, screening practices and health-related data can be linked at any stage which greatly adds to the depth of the data available for research purposes. The external number of all clinical procedures should also have been entered into the database, if possible, so contact with the clinical service can be made if extra treatment details are required.

Equipment and General Requirements

It is difficult to be to prescriptive about what equipment is required for a biobank as variation will occur depending on the scope of the facility. SOPs have been published by all of the major groups such as ABNA oncology, BBMRI, ISBER and the

NCI on general requirements. In brief, all groups agree that thorough planning and resource design is required so suitable space is available for equipment such as freezers that includes −20, −80 and liquid nitrogen banks, bio hazard hoods and centrifuges all with automated alarm systems in place to alert the staff of equipment failures. Linked to this is the need for maintenance, delivery, warranty, service contracts, lifespan, performance and efficiency cost savings, along with current and future service provision options. The depreciation for all capital equipment and replacement costs need to be factored into on-going budgets.

As part of the QA/QC protocols and to ensure the best conditions possible for the biological resource, all maintenance visits and routine staff checks for equipment should be scheduled for and be logged into the dedicated files and cover validation and change control, calibration, maintenance, repair procedures and environmental monitoring; e.g., temperature monitoring of freezers. Contingencies plan also needs to be in place and part of the SOPs for back up equipment should there be equipment failure, especially for freezers and liquid nitrogen vats.

Conclusion

After 15 years of a professional approach to the operations associated with biobanking, it has been demonstrated that these facilities have more than just the potential to be a major infrastructure to facilitate a range of benefits for improved health benefits for our community. Global efforts are already utilizing biobanks that are leading to translation of new research findings. Harmonization by biobanks is recognized as being crucial in order to make facilities more robust, targeted and economical that is associated with the important issue of sustainability. The efforts made by the various professional biobank groups have led to a high observance in the development of policies and procedures in the design and management of biobanks, the SOPs for sample handling linked to QA and QC, database entry and data cleaning, all within the national and international ethico-legal frameworks. As research funding for all activity becomes more difficult to secure, one of the biggest challenges for biobanks is to keep networking and forming strategic alliances between governmental bodies, funding agencies, public and private science enterprises and other stakeholders to keep the importance of our work on the agenda.

References

1. Parks A (2009) Biobanks. 10 Ideas changing the world right now. Time Magazine, March
2. Australian Government (2011) D.o.H. Factsheet Medicare Australia pdf
3. Etchegary H et al (2013) Community engagement with genetics: public perceptions and expectations about genetics research. Health Expect
4. Ahram M et al (2014) Factors influencing public participation in biobanking. Eur J Hum Genet 22:445–451

5. Riegman PH et al (2008) Biobanking for better healthcare. Mol Oncol 2(3):213–222
6. BBMRI (2013) Biobanking and Biomolecular Resources Research Infrastructure. November. www.bbmri.eu
7. Henderson GE et al (2013) Characterizing biobank organizations in the U.S.: results from a national survey. Genome Med 5(1):3
8. Network ABNA. http://www.abna.org.au
9. Network CTRnet. https://www.ctrnet.ca
10. UK BioBank http://www.ukbiobank.ac.uk
11. The Cancer Human Biobank – NCI. https://biospecimens.cancer.gov/programs/cahub
12. Eccles SA et al (2013) Critical research gaps and translational priorities for the successful prevention and treatment of breast cancer. Breast Cancer Res 15(5):R92
13. Welinder C et al (2013) Establishing a Southern Swedish Malignant Melanoma OMICS and biobank clinical capability. Clin Transl Med 2(1):7
14. Watson RW, Kay EW, Smith D (2010) Integrating biobanks: addressing the practical and ethical issues to deliver a valuable tool for cancer research. Nat Rev Cancer 10(9):646–651
15. Michailidou K et al (2013) Large-scale genotyping identifies 41 new loci associated with breast cancer risk. Nat Genet 45(4):353–361, 361e1-2
16. Curtis C et al (2012) The genomic and transcriptomic architecture of 2,000 breast tumours reveals novel subgroups. Nature 486(7403):346–352
17. Hewitt R, Hainaut P (2011) Biobanking in a fast moving world: an international perspective. J Natl Cancer Inst Monogr 2011(42):50–51
18. https://www.nhmrc.gov.au/health-ethics/
19. McGuire AL, Beskow LM (2010) Informed consent in genomics and genetic research. Annu Rev Genomics Hum Genet 11:361–381
20. Bathe OF, McGuire AL (2009) The ethical use of existing samples for genome research. Genet Med 11(10):712–715
21. White MT, Gamm J (2002) Informed consent for research on stored blood and tissue samples: a survey of institutional review board practices. Account Res 9(1):1–16
22. Wendler D et al (2005) Quantifying the federal minimal risk standard: implications for pediatric research without a prospect of direct benefit. JAMA 294(7):826–832
23. Shah S et al (2004) How do institutional review boards apply the federal risk and benefit standards for pediatric research? JAMA 291(4):476–482
24. Bernard Lo M. The Research Ethics Blog. http://accelerate.ucsf.edu/blogs/ethics/biobank-genomics-research-do-we-need-patient-consent
25. Oliver JM et al (2012) Balancing the risks and benefits of genomic data sharing: genome research participants' perspectives. Public Health Genomics 15(2):106–114
26. Hudson KL (2011) Genomics, health care, and society. N Engl J Med 365(11):1033–1041
27. Pullman D et al (2012) Personal privacy, public benefits, and biobanks: a conjoint analysis of policy priorities and public perceptions. Genet Med 14(2):229–235
28. Gaffney EF, Madden D, Thomas GA (2012) The human side of cancer biobanking. Methods Mol Biol 823:59–77
29. Allen MJ et al (2010) Human tissue ownership and use in research: what laboratorians and researchers should know. Clin Chem 56(11):1675–1682
30. Hakimian R, Korn D (2004) Ownership and use of tissue specimens for research. JAMA 292(20):2500–2505
31. Commission ALR (1996) Essentially yours: protection of human genetic information in Australia. http://www.alrc.gov.au/publications/essentially-yours-protection-human-genetic-information-australia-alrc-report-96
32. ICGC. Goals, structure, policies and guidelines. http://icgc.org/icgc/goals-structure-policies-guidelines
33. Wakefield CE et al (2013) Improving mutation notification when new genetic information is identified in research: a trial of two strategies in familial breast cancer. Genet Med 15(3):187–194

34. Appelbaum PS et al (2014) Informed consent for return of incidental findings in genomic research. Genet Med 16:367–373
35. Viberg J et al (2014) Incidental findings: the time is not yet ripe for a policy for biobanks. Eur J Hum Genet 22:437–441
36. Young MA et al (2013) The attitudes of people with sarcoma and their family towards genomics and incidental information arising from genetic research. Clin Sarcoma Res 3(1):11
37. Issues (2013) Anticipate and communicate. Ethical Management of Incidental and Secondary Findings in the Clinical, Research, and Direct-to-Consumer Contexts. December. http://www.bioethics.gov
38. National Health and Medical Research Council A (2009) Outcomes enabling grant funding rounds. https://www.nhmrc.gov.au/grants-funding/outcomes
39. Vaught J, Rogers J, Myers K, Compton CC (2011) An NCI perspective on creating sustainable biospecimen resources. JNCI Monogr (42):1–7
40. Watson PH, Wilson-McManus JE, Barnes RO, Giesz SC, Png A, Hegele RG, Brinkman JN, Mackenzie IR, Huntsman DG, Junker A, Gilks B, Skarsgard E, Burgess M, Aparicio S, McManus BM (2009) Evolutionary concepts in biobanking – the BC BioLibrary. J Transl Med 7:95
41. Kaye J (2012) Embedding biobanks as tools for personalised medicine. Norsk Epidemiologi, pp 169–175
42. Henderson GE, Cadigan RJ, Edwards TP, Conlon I, Nelson AG, Evans JP, Davis AM, Zimmer C, Weiner BJ (2013) Characterizing biobank organizations in the U.S.: results from a national survey. Genome Med 5:3
43. NCI (2013) Epidemiology and Genomics Research – cohort consortium. http://epi.grants.cancer.gov/Consortia/cohort.html
44. consortium k (1997–2013) kConFab: A national consortium for research into families at high risk of breast cancer. www.kconfab.org
45. ACC (2012) T.F.H.C.R.C.-T.A.C.C. http://www.fhcrc.org/en/labs/phs/projects/asia-cohort-consortium.html
46. Australia TVG (1982) Human Tissue Act, D.f. Health, Editor
47. Rogers J, Carolin T, Vaught J, Compton C (2011) Biobankonomics: a taxonomy for evaluating the economic benefits of standardized centralized human biobanking for translational research. J Natl Cancer Inst Monogr 2011(42):32–38
48. Vaught J, Rogers J, Carolin T, Compton C (2011) Biobankonomics: developing a sustainable business model approach for the formation of a human tissue biobank. J Natl Cancer Inst Monogr 2011(42):24–31
49. The Wellcome Trust (2011) Sharing research data to improve public health: full joint statement by funders of health research. The Wellcome Trust, UK, 10 January 2011
50. Fortier I, Doiron D, Burton P, Raina P (2011) Invited commentary: consolidating data harmonization—how to obtain quality and applicability? Am J Epidemiol 174:261–264
51. OECD Organization for Economic Cooperation and Development (2010) OECD guidelines on human biobanks and genetic research databases. Eur J Health Law 17:191–204
52. ISBER (2012) Best practices for repositories. Collection, storage, retrieval, and distribution of biological materials for research. Biopreserv Biobank 10:79–161
53. NCI (2011) Best practices for biospecimen resources. http://biospecimens.cancer.gov/bestpractices/2011
54. Betsou F, Gunter E, Clements J, DeSouza Y, Goddard KA, Guadagni F, Yan W, Skubitz A, Somiari S, Yeadon T, Chuaqui R (2013) Identification of evidence-based biospecimen quality-control tools: a report of the International Society for Biological and Environmental Repositories (ISBER) Biospecimen Science Working Group. J Mol Diagn 1:3–16
55. U.S. Department of Energy Washington, D.C., Implementation guide for quality assurance programs for basic and applied research
56. Riegman PH, Morente MM, Betsou F, de Blasio P, Geary P; Marble Arch International Working Group on Biobanking for Biomedical Research (2008) Biobanking for better healthcare. Mol Oncol 2:213–222

57. Carter A, Betsou F, Clark BJ (2011) Quality management and accreditation of research tissue banks. Virchows Arch 458(2):247–248, author reply 249–50
58. Betsou F et al (2009) Human biospecimen research: experimental protocol and quality control tools. Cancer Epidemiol Biomarkers Prev 18(4):1017–1025
59. ISO 9000:2005 (2011) Quality management systems – Fundamentals and vocabulary. http://www.iso.org/iso/catalogue_detail?csnumber=42180
60. Centers for Disease Control and Prevention (CDC) 24/7 (2013) Saving lives. Protecting people. Saving money through prevention. Clinical Laboratory Improvement Amendments (CLIA). http://wwwn.cdc.gov/clia/
61. Pukkala E (2011) Nordic biological specimen bank cohorts as basis for studies of cancer causes and control: quality control tools for study cohorts with more than two million sample donors and 130,000 prospective cancers. Methods Mol Biol 675:61–112
62. Harris JR et al (2012) Toward a roadmap in global biobanking for health. Eur J Hum Genet 20(11):1105–1111
63. Centre for Cancer Genetic Epidemiology School of Clinical Medicine, C.U. iCOGS. 2013. http://ccge.medschl.cam.ac.uk/research/consortia/icogs/
64. Webster JD et al (2011) Quantifying histological features of cancer biospecimens for biobanking quality assurance using automated morphometric pattern recognition image analysis algorithms. J Biomol Tech 22(3):108–118
65. Solutions Ae (2013) Transforming the practice of Pathology
66. AOCS Study (2013) http://www.aocstudy.org/gp_about.asp
67. Office of the Australian Information Commission https://www.oaic.gov.au/privacy-law
68. Commission TE (2013) Data protection
69. The Wellcome Trust https://www.wellcome.ac.uk/funding/managing-grant/policy-and-position-statements (2013)
70. U.S. Dept of Health and Human Services. https://www.hhs.gov/hipaa
71. OEDC (2009) Guidelines on human biobanks and genetic research databases. https://www.oecd.org/sti/biotech/44054609.pdf
72. Schroder C et al (2011) Safeguarding donors' personal rights and biobank autonomy in biobank networks: the CRIP privacy regime. Cell Tissue Bank 12(3):233–240

Cytogenetics: Methodologies

Chiyan Lau

Introduction

A range of clinical disorders can be caused by abnormal changes in the DNA of a patient's cells. These 'genetic disorders' can arise from large DNA changes which affect whole chromosomes, such as numerical loss or gain of entire chromosomes (i.e. aneuploidies, including trisomies and monosomies), or breaking and joining of parts of chromosomes (i.e. structural variations, such as translocations, inversions, insertions, duplications, and deletions). At the other end of the size spectrum, the DNA change may be quite small and only affect a single nucleotide base or a few bases (i.e. sequence variations). In between these extremes, some DNA changes involve gains or losses of a few thousand to a few million bases (commonly referred to as 'copy number changes', or CNCs), or expansions in the number of units of repetitive DNA (e.g. oligonucleotide repeat expansions, such as triplet repeats). These DNA changes may be present in every single cell of an individual (germline change, which can be passed on to the next generation because gonadal cells are also affected), or only in a subpopulation of cells (somatic change or mosaicism).

The methodology needed to detect these various types of genetic changes depends on both the size of the change and whether it is somatic or germline. For numerical and structural aberrations affecting whole chromosomes or a large part of a chromosome, cytogenetic methodologies are commonly used. Classical cytogenetics are based on visual analysis of chromosomal material by microscopy. For the detection of small DNA changes such as those at the nucleotide sequence level, molecular genetic methods are used which include such techniques as direct DNA sequencing, Southern blotting, multiplex ligation-dependent probe amplification (MLPA), and so on.

C. Lau (✉)
SEALS Genetics, NSW Health Pathology, Level 4, Campus Centre,
Prince of Wales Hospital, Randwick, NSW 2031, Australia
e-mail: chiyan.lau@health.nsw.gov.au

S.R. Lakhani, S.B. Fox (eds.), *Molecular Pathology in Cancer Research*,
DOI 10.1007/978-1-4939-6643-1_3

Traditionally, cytogenetic and molecular genetic testing were considered separate disciplines, but there are now some laboratory methods which bridge this separation and are termed 'molecular cytogenetic' methods. These include fluorescence in situ hybridisation (FISH) and microarray analyses. In addition, there are emerging methods which defy strict classification as cytogenetic or molecular genetic techniques, such as massively parallel sequencing (MPS) (so-called next generation sequencing) mostly used in research laboratories but which are now increasingly adopted for clinical use. The division between cytogenetics and molecular genetics is becoming blurred, and the critical issue for the practising clinician, scientist, or pathologist is the knowledge of the strengths and limitations of each technique, and to match the appropriate method to the particular disorder or patient presentation in question.

In this chapter, we will not only look at common cytogenetic and molecular cytogenetic laboratory techniques, but also discuss some commonly used molecular genetic techniques. We will focus on the principles underpinning these techniques as well as their strengths and weaknesses. In the following chapter, we will look at how these laboratory techniques are applied in some common clinical situations.

Classical Cytogenetics/Karyotyping

In classical cytogenetics, the aim of the analysis is to determine if the cells in a specimen have the correct number of chromosomes and whether the structure of each individual chromosome is normal. This analysis is called 'karyotyping', where individual chromosomes in a cell are visualised by microscopy and arranged in a conventional order and format (essentially in order of size from largest to smallest) to facilitate analysis and comparison [1].

In order to perform this analysis, first a suitable specimen is obtained from the patient. For diagnosis of constitutional genetic disorders in the postnatal period, the most common sample type is a peripheral blood specimen collected in lithium heparin anticoagulant. For haematological malignancies, bone marrow aspirate collected in lithium heparin is usually suitable. The specimen should be transported at ambient temperature or in a cool container (not frozen), to arrive in the cytogenetics laboratory within 48 h of collection.

The critical thing about specimens for karyotyping is that the cells should be viable and can be induced to divide in cell culture. For obvious reasons the specimen should also contain a sufficient number/proportion of cells expected to carry the abnormality of interest (e.g. cells from the malignant clone in the cancer setting).

Viable cells are required because the majority of cells in most specimen types are in interphase. At this point in the cell cycle, the DNA of the cell exists in a decondensed, elongated form. If one were to examine an interphase cell nucleus under the microscope, individual chromosomes cannot be distinguished or counted. To perform a karyotype, one must allow the cells to go through the cell cycle, and catch them in metaphase when the chromatin is condensed into separate identifiable chromosomes. For peripheral blood specimens, this is accomplished by adding a mitogen such as phytohaemagglutinin (PHA) to stimulate T cell proliferation. For malignancies, stimulation may or may not be necessary, since the cells may already be actively dividing, but mitogens may be used to expand specific cell populations (e.g. PHA for T cell stimulation, or the phorbol ester TPA for B cell stimulation). The cell culture is then incubated in a temperature, humidity, and CO_2 (pH) controlled environment for a period of time, usually around 48–72 h.

Another specimen type suitable for postnatal cytogenetics is skin biopsy, where the fibroblasts from the dermis are cultured. This requires special liaison with the receiving laboratory to ensure that the laboratory is prepared to receive this specimen type, that suitable transport medium is used (e.g. cell culture medium), and the specimen is delivered to the laboratory as soon as possible after collection. For prenatal diagnoses, cells from amniocentesis (amniocytes) or chorionic villus sampling (CVS) can be used.

At the end of culture, there should be sufficient numbers of actively dividing cells. A compound such as colcemid is added at this point to disrupt the cell spindle and block cells from exiting metaphase. This enriches the proportion of cells containing condensed (metaphase) chromosomes which can be analysed.

The cells are then swollen with a hypotonic solution, then fixed and washed in a combination of acid and alcohol. The fixed cell suspension is dropped onto a microscope slide and dried. The slide is aged at 60–65 °C overnight, then incubated with a trypsin solution briefly to preferentially digest the histone proteins at parts of the chromosome with an open conformation. The slides are then washed and stained with Giemsa stain, which produces 'G banding' of chromosomes. This stain binds to the histone proteins still attached to the chromosomal DNA, creating characteristic dark and light band patterns which allow different chromosomes to be recognised.

Alternative staining processes are also sometimes used, to create other banding patterns. For example, 'R banding' (reverse banding) which creates the opposite pattern to G banding, 'C banding' which stains centromeric regions of chromosomes, distamycin/DAPI which preferentially stains heterochromatin, and silver stains for nucleolar organising regions (NOR) on acrocentric chromosomes (13, 14, 15, 21, and 22).

After staining, slides are examined by light microscopy, usually at a final magnification of 1000×. Traditionally, microscopic examination is performed manually, and the whole slide is scanned by the human operator to identify cells that are in metaphase. Each chromosome is examined band by band and compared to its homologous partner to determine if there is any structural abnormality (such as translocation, inversion, and deletion) or numerical abnormality (i.e. aneuploidy). This process is repeated for multiple cells per specimen (e.g. 15–20 cells, but more if an abnormality is detected). This ensures that if an abnormality is only present in a subset of cells (i.e. not only a mosaic or somatic mutation, such as in cancer but also a possibility in constitutional disorders), it can still be detected with sufficient sensitivity. As one can imagine, this process is highly labour intensive and time consuming.

More recently, it has become possible to semi-automate part of this process, with robotic image capture instruments that can take whole batches of slides and automatically recognise and capture metaphases at high resolution. Software with image recognition algorithms can perform rudimentary analyses of the chromosomes, but the images and computer-generated karyograms still require trained cytogeneticists to analyse to verify correct chromosomal assignment, and when abnormalities are present, to 'call' the abnormality.

Molecular Cytogenetics

In addition to classical cytogenetics, the modern clinical cytogenetics laboratory usually performs a number of molecular cytogenetic techniques. These include FISH (fluorescence in situ hybridisation) and microarray analyses.

Fluorescence In Situ Hybridisation

Karyotyping provides a whole-genome survey of the chromosomal make-up of cells. FISH can generally be thought of as a targeted interrogation of a specific genomic location. It allows one to determine, for example, how many copies of a gene are present in a cell, or if a particular rearrangement is present [2–4].

For this technique, a piece of DNA which corresponds to the sequence of a genomic region of interest is used as a 'FISH probe'. This FISH probe is labelled with fluorescent dye. More than one genomic target (usually up to three in a routine clinical laboratory) can be interrogated concurrently by using different colour probes. A fixed cell suspension from the patient is spotted onto a microscope slide, and the FISH probes are added. The slide is incubated at 95 °C to denature the DNA of the cells, then the incubation temperature is reduced to 60 °C to allow the FISH probe to anneal overnight to its complementary region in the genome. Excess probes and non-specifically bound probes are washed off with a series of stringent washes. The signals from the FISH probes are detected by fluorescence microscopy, using lasers of different wavelengths and different filter sets for different colour FISH probes.

FISH can be used to interrogate the copy number of a genomic region of interest. If the copy number for an autosomal region is normal in a cell, then two fluorescent signals will be observed for the locus-specific FISH probe. If one copy is deleted, there will only be one signal. Conversely if the region is duplicated (or amplified), there will be three signals (or more). Usually, for copy number detection, in addition to the 'test' probe of interest, a 'control probe' labelled with a different colour which targets a control genomic region will be used (which is expected to have a normal copy number), to help interpret the signal pattern.

Although FISH can be performed on dividing metaphase cells (metaphase FISH), it can also be used on resting cells. This is called 'interphase FISH'. The FISH probes will hybridise to decondensed DNA just as well as to DNA in condensed metaphase chromosomes. This means that specimens not suitable for karyotyping such as formalin fixed paraffin embedded tissue (FFPE) and uncultured cells can be interrogated by interphase FISH.

When FFPE samples are used for FISH, a tissue section is mounted on a microscope slide and deparaffinised in xylene. Then, through a series of steps of xylene/alcohol solvent exchanges, the tissue is brought back to aqueous phase. A suitable FISH probe is then applied to a defined area of the section, and the rest of the procedure is similar to FISH on other types of specimens. One important point to note for FFPE FISH is that a suitably trained person (e.g. a histopathologist) should examine the slide and mark out the area of interest, e.g., by comparison to a H&E guide slide. This is an often neglected but important step. In the case of a tumour specimen, for example, there may only be a small area in the section containing the tumour of interest, surrounded by large areas of normal tissue. FISH probes are usually applied in a very small volume of buffer to a small defined area of the slide. It would be completely pointless to apply the FISH probes to the normal parts of the slide, which would only produce a false negative result.

In addition to determining copy number, FISH can be used to determine if a particular chromosomal locus is involved in a structural rearrangement, or whether a specific fusion event is present in a cell. To achieve this, 'break apart' or 'fusion' probes are used, which are really just clever ways of placing and tagging locus-specific FISH probes [5, 6]. A 'break-apart' probe can be used to determine if a specific chromosomal location of interest has been 'broken apart' by a rearrangement (e.g. a translocation). This strategy involves placing a red-labelled probe and a green-labelled probe in close proximity on either side of the potential breakpoint. If the breakpoint is intact, the red and green signals will co-localise, and merge as a yellow signal under fluorescent microscopy. If the breakpoint is involved in a translocation and thus 'broken apart', the green and red probes will be physically separated and appear as separate green and red signals. Therefore, for a breakpoint located on an autosome, there will be two yellow signals in a normal cell (one on each homologous chromosome), while in a cell with a translocation, there will be 1 green signal, 1 red signal, and 1 yellow signal (on the homologous unbroken chromosome). Break-apart probes can be used with either interphase cells or metaphase cells. Note that a break-apart probe only tells you that a chromosomal site is involved in a rearrangement. It does not tell you what the other partner is in the rearrangement. For this, one needs a fusion probe, which detects the presence of a specific translocation between two defined chromosomal locations.

In a fusion probe design, two FISH probes are used, one against each potential partner of the fusion. Each probe is labelled with a different colour (e.g. green/red). For example, to detect a reciprocal translocation between 9q34 and 22q11.2, a FISH probe straddling the breakpoint at 9q34 will be labelled red, and a probe for the 22q11.2 breakpoint will be labelled green. In a normal cell which does not have the t(9;22) reciprocal translocation, there will be two green signals and two red signals, corresponding to two normal chromosome 9 s and two normal chromosome 22 s. In a cell with a t(9;22), there will be one normal chromosome 9 (with a single red signal), one normal chromosome 22 (with a single green signal), one derivative chromosome 9 [der(9q)] joined to 22q11.2 (where half the green signal from one fusion partner will be juxtaposed with half the red signal from the other partner creating a merged yellow signal), and one derivative chromosome 22 [der(22q)] joined to 9q34 (also with a yellow signal). This is known as a 'dual fusion' design. 'Single fusion' designs are also sometimes used and are similar in principle. The difference is in the location of the FISH probes relative to the breakpoint. In single fusion FISH, the probe is placed to one side of the breakpoint rather than straddling it as in dual fusion FISH. This means that in the presence of a reciprocal translocation, only one yellow fusion signal will be produced (e.g. on the der(22q)). The reciprocal event (der(9q)) is not visualised.

There are many other variations to the basic FISH technique. One variant is 'whole chromosome painting' (WCP) [7]. In WCP, rather than using one locus-specific probe, a mixture of hundreds to thousands of probes which target different parts of one specific chromosome (e.g. chromosome 3) are labelled the same colour and used together to 'paint' the chromosome. This technique is typically used in metaphase cells to characterise rearrangements which are suspected to involve a particular chromosome.

A further extension of the WCP strategy is multicolour FISH (M-FISH), where each of the 24 different chromosomes (22 autosomes + X and Y) are painted with a different 'colour', using different ratios and combinations of multiple fluorophores to label the FISH probes. The image is acquired using special filter sets, and computer image analysis is used to pseudocolourise the different chromosomes to allow human interpretation of the image. This technique requires special hardware and software setup, and is available only in specialist centres. M-FISH is especially useful for interpretation of complex karyotypes and/or extra structurally abnormal chromosomes (ESACs) whose material of origin is uncertain and may be derived from multiple chromosomes.

All the FISH techniques discussed so far require fluorescence microscopy for detection of probe hybridisation signals. There are modifications to FISH which allow non-fluorescence-based signal detection. One example is chromogenic in situ hybridisation (CISH) [8], where the locus-specific DNA probes are labelled with a protein tag (e.g. digoxigenin) instead of fluorescent tags. This allows colour signal production using standard direct or indirect immunoreactions (e.g. using anti-digoxigenin antibodies coupled directly or via a secondary antibody to alkaline phosphatase or peroxidase, followed by incubation with chromogenic substrates). CISH is particularly useful for FFPE samples, since the ISH signals and tissue morphology can be concurrently examined under bright field microscopy.

Microarrays

In recent years, microarray testing is becoming an increasingly common method used in the clinical cytogenetics laboratory as an alternative to karyotyping. Microarrays have a number of advantages over conventional cytogenetics, but it should be noted from the outset that microarrays cannot detect all types of chromosomal abnormalities detectable by karyotyping. Specifically, while microarrays can detect copy number losses and gains (i.e. deletions and duplications), they cannot detect copy number neutral 'balanced' structural abnormalities (e.g. balanced translocations, inversions, etc.). Microarrays also have a lower sensitivity for mosaic changes. To understand why, let us take a look at what a microarray is and how the technology works.

A microarray consists of thousands of DNA molecules (probes) covalently immobilised in an ordered pattern on the surface of a solid substrate similar to a microscope slide. Each 'coordinate' on the microarray contains a DNA probe of known sequence, with different spots having probes with different sequences corresponding to different locations spread throughout the genome. Most cytogenetic microarrays commercially available today are 'oligo arrays', where the individual probes are oligonucleotides of ~50–60 bases in length.

Microarray designs are often designated by the number of different probe sequences, or features, on the array. For example, a '180 k array' has approximately 180,000 unique probe features. The sequences of the DNA probes are designed to map to genomic loci at more or less regularly spaced intervals throughout the human genome. Usually, arrays for cytogenetic use are designed to have a 'backbone' coverage across the genome at a certain density (e.g. one probe every 25 kb for one particular 180 k array design), but with denser coverage (e.g. one probe approximately every 5 kb) in targeted genomic regions which are gene-rich and/or known to be associated with disease. The higher the number of features on an array, the denser the genome coverage, and the higher the technical resolution of the array. Some arrays in clinical use today have more than a million 'features', and are technically capable of detecting genomic deletions and duplications only a few kilobases in size. To put this in context, traditional karyotyping cannot detect deletions/duplications smaller than ~3–5 megabases, and will sometimes miss changes 5–10 megabases in size (depending on the microscopic appearance of the cytogenetic band, the quality of the chromosome sample/preparation, and the skills of the operator), which are a few hundred times bigger than those detectable by microarray. In fact, in most clinical cytogenetics laboratories today, the reporting resolution of microarrays are less limited by the raw technical capabilities of the array than practical considerations, such as the difficulties with validating a microarray method to a satisfactory standard for very small changes, the lack of availability of a suitable alternative method to confirm a small abnormality, or the difficulties with interpreting the clinical significance of very small changes. As a result, most clinical laboratories set a reporting resolution for microarray studies somewhere between 100 kb and 400 kb.

There are two main types of microarray platforms currently used in clinical cytogenetic laboratories: array comparative genomic hybridisation (aCGH) and single nucleotide polymorphism array testing (SNP arrays). The two methods have similarities in principle but also have some methodological differences which result in different workflows, analytical outputs, and capabilities. These methods will be explained in more detail below.

Array Comparative Genomic Hybridisation

For aCGH, the principle behind determination of copy number status is to compare the patient's sample against a gender-matched reference sample ('normal control') taken from phenotypically normal individuals. Conceptually, if the patient's sample contains less DNA hybridising to a probe compared to the reference sample, then the patient has a 'copy number loss' relative to the reference, at the genomic location corresponding to the probe [9].

In practice, the comparison cannot be undertaken directly since there are other factors which influence the inter-sample signal ratio, such as differences in amount and quality of input DNA between the patient and the reference, variability introduced at different points in the process such as efficiency of the sample DNA labelling reaction, differences in binding affinity between DNA at different genomic loci, stochastic differences in hybridisation kinetics, and non-specific background binding. A number of correction strategies need to be employed, e.g., fluorescent intensity normalisation, background correction, signal averaging across redundant probes, etc., before the signal from the patient or reference can be compared [10].

The basic workflow for aCGH is as follows. Genomic DNA is extracted from a suitable patient specimen (e.g. peripheral blood collected in EDTA for constitutional cytogenetics), and labelled with a fluorescent dye (e.g. green). Genomic DNA from a reference sample ('normal control') is labelled with a different fluorescent dye (e.g. red). Equal amounts of the two DNA are mixed together, then co-hybridised onto the microarray. Each DNA molecule will preferentially hybridise to the corresponding oligo probe on the microarray with the complementary nucleotide sequence. Unbound DNA is washed off and the microarray with the bound labelled DNA is scanned with a laser scanner in the two colour channels corresponding to patient and reference to determine how much DNA is bound to each probe on the array. The intensity of the fluorescent signal at each spot on the microarray will be proportional to the amount of captured sample or reference DNA. For each genomic locus, if the copy number is identical in the patient and reference, there will be a roughly 1:1 ratio between green and red fluorescent signals (ratio = 1.0). If there is a copy number loss in the patient, the green:red ratio will be approximately 0.5 (1:2). Conversely, if there is a copy number gain in the patient, the green:red ratio will be approximately 1.5 (3:2). The green:red ratio for each probe is plotted graphically by computer software across the genome. Chromosomal regions with consecutive probes which show deviation from a 1:1 ratio are interpreted as regions with copy number change in the patient. If the copy number losses and gains are mosaic, such as may be found in tumour specimens 'contaminated' by normal stromal tissue, the green:red ratios will be intermediate between 0.5 and 1.0 or between 1.0 and 1.5, respectively. As the level of mosaicism decreases, the ratio will get closer and closer to 1.0. Low level mosaicism (less than approximately 10–20 % abnormal cells) is difficult to detect by array CGH.

Single Nucleotide Polymorphism Arrays

In contrast to array CGH, only a single DNA sample is hybridised at a time to each SNP array chip [11]. Reference samples are run on a completely separate array chip to the patient sample. In an SNP array, the probes are designed to map to areas in the human genome which commonly show sequence variation in the general population. The type of sequence variation targeted by SNP arrays are, as the name suggests, single nucleotide polymorphisms (SNPs).

The SNP sites targeted by SNP array probes are typically biallelic (i.e. has two common forms, or alleles, in the population). For example, for a hypothetical SNP site 'N' in the genome flanked by the sequence ...CCNACG..., one allele of N might be a T (...CC**T**ACG...), the other allele might be a G (...CC**G**ACG...). Some individuals are homozygous for T at the SNP (genotype = T/T), some are homozygous for G (genotype = G/G), while some individuals are heterozygous (genotype = T/G).

One way that SNP arrays detect genotype at an SNP site is to use short probes approximately 25 nucleotides in length. Compared to the longer probes used on CGH arrays, these shorter probes show lower hybridisation affinity to sequence-mismatched DNA, to allow discrimination between alleles. With this method, there are usually four probes designed for each SNP site to detect the two alleles in both the forward and reverse direction, for example, two to detect the 'T' allele (one forward and one reverse), and two to detect the 'G' allele. If a sample shows hybridisation predominantly to the 'T' probes, the individual will be genotyped as 'T/T'. Conversely, if the sample shows roughly equal hybridisation to the 'T' and 'G' probes, the individual will be genotyped as heterozygous 'T/G' at the SNP. When a sample is run on an SNP array, each of the thousands of SNP sites across the genome which is targeted by the array is genotyped.

In addition to genotype information, SNP arrays can also provide copy number information by inter-array comparisons, i.e., by comparing the signal from each SNP site in the patient's sample against the corresponding signal from reference samples. Conceptually this is similar to array CGH. However, in addition to the inter-array comparison, the genotype information from a sample also helps to confirm copy number changes. In a region with normal copy number (CN = 2), each SNP in the region have only 3 possible genotypes: AA, AB, or BB (if we arbitrarily call the alleles 'A' and 'B'). In a genomic region which shows a deletion (CN = 1), only a single allele is left, which means there will not be any heterozygous genotype calls within a deletion, resulting in a stretch of apparent homozygosity. For a duplicated region (CN = 3), there are three alleles present with the possible genotype combinations being AAA, AAB, ABB, and BBB. The 'heterozygous' SNPs within the duplicated region will show fluorescent intensity ratios of 2:1 rather than 1:1, thus providing support to the detection of a copy number gain. SNP arrays can also detect regions with copy neutral absence of heterozygosity, which are relevant for disease conditions which involve uniparental disomy or loss of heterozygosity.

Limitations of Microarrays

Although microarrays allow detection of deletions and duplications at a much higher resolution than standard karyotyping, there are limitations to their use. The most fundamental limitation is that arrays provide no information on the positional relationship between probes. This means that if two probes which are normally adjacent to each other on a normal genome are split up by a structural variation like a translocation or inversion, arrays will not detect the abnormality unless there is also a copy number imbalance as a result of the structural rearrangement. In contrast, karyotype and FISH can potentially provide positional information.

Also, arrays cannot provide any information on the parts of the genome that are not targeted by the array probes. Therefore, even though SNP arrays provide some genotype information at common SNP sites, they cannot detect most sequence variations in the intervening sequences in genes or triplet-repeat expansions such as in Fragile X syndrome.

Microarrays detect the averaged signal from the DNA extracted from many cells. Therefore, as explained earlier, if an abnormality is present only in a small proportion of cells in the sample (low level mosaicism, or low mutation burden in a cancer sample), it may not be detected by array. In contrast, standard karyotyping and FISH are in effect 'single cell analyses'. Provided that a sufficient number of cells are examined, FISH and karyotype will have a higher sensitivity for abnormalities at a low level of mosaicism, with the caveat that the disease clone meets the other requirements of karyotyping and FISH, such as the ability for the cells to divide in culture (for karyotyping), or that the target chromosomal aberration is known (for FISH).

The implications of these limitations in terms of choice of test methodology for particular disease conditions will be discussed further in the next chapter.

> The use of cDNA microarrays for gene expression profiling (GEP), i.e., to determine which genes are actively transcribed, or SNP arrays for genome wide association studies (GWAS), are quite different applications of microarray technology than those described in this chapter, and are not commonly performed by the clinical cytogenetic laboratory.

Molecular Genetics

Polymerase Chain Reaction + Direct DNA Sequencing (Sanger Sequencing)

This is one of the most commonly used techniques in the molecular genetic laboratory. This method can be used to directly determine the DNA sequence in a particular genomic region of interest in the patient's sample. The methodology most widely used today is based on modifications of the pioneering techniques of Fred Sanger [12, 13].

In this technique, two rounds of DNA amplification are performed. In the first round, the genomic region that needs to be sequenced is first amplified by standard polymerase chain reaction (PCR), using a pair of PCR primers which flank the region. In this PCR reaction, the patient's DNA is incubated with a mixture of the PCR primers, deoxynucleotides (dNTPs), and a thermostable DNA polymerase enzyme (e.g. Taq polymerase), and undergoes multiple rounds of heating and cooling ('thermocycling') where the genomic region of interest between the primers is exponentially amplified using the patient's DNA as template.

At the end of the PCR reaction, the PCR product (amplicon) is cleaned up to remove any excess reagents remaining (such as excess primers and dNTPs). The PCR product is then taken into a second round 'sequencing reaction', where only one of the PCR primers is used in each reaction (either the 'forward' or 'reverse' primer), and in addition to dNTPs, a mixture of fluorescently tagged dideoxy-nucleotides (ddNTPs) is added. A different colour is used for each of the ddNTPs, e.g., red for G, green for A, blue for T, and yellow for C. The DNA polymerase will extend the sequencing primer by incorporating the unlabelled dNTPs in the growing strand using the first round PCR product as template, but when a labelled ddNTP is eventually incorporated, further extension cannot take place and the chain is terminated. Therefore, at the end of the sequencing reaction, a mixture of different sized sequencing products will be present, each terminating in a fluorescently labelled end with a colour corresponding to the identity of the last base incorporated. To 'read' the DNA sequence, these sequencing products are run on a genetic analyser instrument, where the molecules are separated according to size using capillary electrophoresis. The colour of the fluorescent label on each molecule is read by a detector as the molecules move past a detection window, starting from the shortest molecule. The output of the instrument is a 'chromatogram' consisting of a series of peaks corresponding to the order of the colours of the nucleotide bases. The chromatogram is interpreted by a computer program to determine the DNA sequence, which is compared to a 'normal' reference sequence to determine if a sequence variation is present.

Sanger sequencing can detect sequence variations including point mutations and small insertions/deletions (indels). However, deletions and insertions/duplications which span a range larger than the size of the PCR product cannot be detected, i.e., Sanger sequencing cannot detect the types of copy number changes detected by cytogenetics techniques, even if the copy number change affects the region covered by the PCR product.

Multiplex Ligation-Dependent Probe Amplification

In contrast to Sanger sequencing, MLPA [14, 15] detects copy number changes, but does not provide sequence information.

In this technique, for each genomic region of interest, two oligonucleotide probes are designed which bind to DNA sequences directly adjacent to each other within the region. These MLPA probes are mixed with the patient's DNA to allow hybridisation overnight at 60 °C. The next day, a DNA ligase is added to join together those pairs of probes which have hybridised to the patient's DNA. Unhybridised free-floating probes are not ligated. The ligation products are then used as templates for a second round PCR reaction, using fluorescently tagged PCR primers. For analysis, the PCR products are resolved by capillary electrophoresis on a genetic analyser. The fluorescently labelled PCR products are detected as peaks on the electropherogram, where the area under each peak corresponds to the amount of PCR products generated.

The amount of PCR product from each genomic locus is proportional to the amount of ligation product generated from the locus, which in turn is proportional to the number of copies of the genomic region in the sample used in the first part of the reaction. By comparing the result from the patient's sample against a set of known normal reference samples (after normalising for signals from control genes), the copy number status of the genomic regions can be calculated. Multiple genomic regions can be tested in a single MLPA reaction, by designing the hybridisation probes such that the length of the PCR product for each region is different, hence the 'multiplex' in the name of the technique.

Southern Blot

This is one of the classic techniques of molecular biology. Although it is quite a labour-intensive method, it is still used in many molecular genetics laboratories today.

The basic premise of a Southern blot [16, 17] is to determine the presence and/or size of a piece of DNA fragment from a known genomic location. The process is as follows. First, the genomic DNA from the patient and control samples is digested with a DNA restriction endonuclease (e.g. EcoRI, BamHI, etc.). These enzymes recognise short specific palindromic DNA sequence motifs which recur at somewhat irregular intervals throughout the genome. The restriction digestion converts the long genomic DNA into a mixture of shorter DNA fragments of varying sizes. These fragments are separated by gel electrophoresis. The gel containing the separated DNA fragments is then placed on a nylon or nitrocellulose membrane and sandwiched in an assembly of filter paper and paper towels wetted with transfer buffer solution. This 'blotting' process allows the DNA in the gel to transfer onto the membrane.

To detect the DNA fragment that we are specifically interested in, a DNA probe which corresponds to a part of the sequence of the genomic region of interest is radioactively labelled (e.g. with ^{32}P) and incubated with the membrane. This allows the probe to hybridise to its complementary DNA fragment on the membrane. The membrane is washed to remove excess probe, then exposed to an X-ray film or phosphorimaging screen to detect the radioactive probe. The pattern and location of band(s) on the film for the patient are compared with a set of molecular weight standards and with the patterns for negative and positive control samples. Depending on the indication for testing, one might be interested in an abnormal shortening of the target fragment (indicating a deletion), or lengthening of the fragment (indicating a duplication or expansion). Southern blots can detect fragments which are thousands of bases in length.

PCR + Fluorescent Fragment Sizing

This is another method to determine if there is a change in the size of a genomic region. The method is quite simple. A pair of PCR primers are designed to flank the genomic region of interest. One of the PCR primers is covalently labelled with a fluorescent dye. Standard PCR is performed using the patient's sample DNA as template. At the end of the PCR reaction, the PCR product, which will be fluorescently labelled, is mixed with a set of molecular size standards labelled with a different colour, and run out on a genetic analyser by capillary electrophoresis. The molecular size of the PCR product (in base pairs) is determined by comparison with the size standards, and can in turn be compared to the size of known negative and positive control samples.

Compared to Southern blotting, this method provides a much higher resolution in size estimate, with the ability to determine the size of a fragment down to the single base level. However, it is limited by the ability of PCR to amplify a region. Using standard PCR on DNA extracted by standard methods, this is usually limited to a few hundred base pairs or so, depending on sample quality. PCR reaction will fail to produce a product if the primers are spaced further than the amplifiable size. Therefore, for disorders where it is possible for a genomic region to grossly expand in size (e.g. triplet-repeat disorders where the number of repeats can go into the thousands), it is generally necessary to use a combination of fragment sizing and Southern blot to cover the possible size ranges of the expanded region.

Real-Time PCR (Quantitative PCR)

This is a method used to quantify the number of copies of a DNA target of interest in a specimen. It is commonly used for gene expression analysis. To quantify the number of transcripts for a gene in a specimen, the total RNA in the specimen is first converted to cDNA by reverse transcription. The number of copies of cDNA corresponding to the gene of interest is then measured by real-time PCR using primer pairs which target the transcript.

There are some differences between standard PCR and real-time PCR. In standard PCR, analysis of the PCR product is undertaken at the end point of the reaction. In contrast, in real-time PCR, the amplification reaction is monitored as it happens (in real time), with the amount of PCR product measured continuously at each cycle. The cycle number at which the amount of accumulated product exceeds a predefined threshold is called the threshold cycle value (Ct value). The higher the number of target templates (transcripts) in the input DNA, the lower the Ct value will be in a reaction. The Ct value of a patient specimen can be compared against the Ct values of a set of standards with known concentrations, to determine the concentration of targets in the patient specimen.

Emerging/Translational Techniques

Massively Parallel Sequencing

MPS, also sometimes called next generation sequencing, is a technique which has seen very rapid development in research laboratories, and is only just beginning to be used in clinical laboratories [18]. It is a very powerful technique with huge potential, but also produces many challenges for implementation in a clinical diagnostic environment.

There are a number of vendors offering MPS platforms. Each is based on a slightly different technology. It is beyond the scope of this chapter to describe each of these in detail, so we will only provide a basic overview of the common elements which define MPS, and look at the types of genetic variations it can detect.

In basic terms, MPS determines the sequences and quantity of a representative sample of all the DNA molecules in a specimen. A 'library' is first constructed from the sample DNA, for example, by ligating adaptors to each fragment. The whole library of DNA molecules is sequenced simultaneously (in parallel) on the MPS instrument, using a sequencing-by-synthesis approach or sequencing-by-ligation approach (depending on vendor/platform), both of which involve monitoring the sequencing reaction as it occurs in real time. For example, with sequencing by synthesis, a camera in the instrument takes a snapshot image after each base is added to the growing strand to identify which base is added to each of the millions of DNA molecules. The raw output of the MPS sequencing run is the set of snapshot images, which is then computationally analysed to determine the nucleotide sequence ('base calls') of each DNA molecule. This in turn produces a computer file with a list of millions of 'reads', each read being the sequence of a part of a DNA molecule in the original specimen. These reads are then mapped to the reference genomic sequence. Deviations from the reference genome sequence are noted as sequence variations (such as point mutations and small indels) in the specimen. Using additional techniques, such as algorithms which note the genomic distance and location of mate-pair and/or paired-end reads, or which measure the read depth of a genomic region in the patient's sample compared to a reference control sample, it is also possible to detect copy number changes or chromosomal structural rearrangements. MPS can be used for whole-genome sequencing (WGS), or to sequence a subset of the genome, for example, all coding exons (WES) or selected genes (targeted gene panel), by targeting the selected genomic regions using capture methods or amplicon-based methods.

In other words, MPS can potentially detect all types of genetic aberrations, spanning the spectrum from small changes traditionally detected by molecular genetic techniques to large changes traditionally detected by cytogenetic techniques. However, although the developments in MPS are rapid, there are many challenges in clinical implementation of the technology. The different platforms have different limitations in terms of read length, error rates, or the ability to handle homopolymeric sequences. Due to the large number of sequences produced and the large size of files created by each sequencing run, the need for bioinformatic data analysis is crucial. The lack of standardised analysis pipelines and issues of data storage are non-trivial problems. Although the cost per base sequenced by MPS is plummeting, the cost per sequence run is still in the order of thousands of dollars at the time of writing. There are also issues at the post-analytical stage of testing such as interpretability of results and handling of incidental findings. While these issues remain, there are already further developments in the technology such as long-read MPS or the so-called third generation sequencing. MPS clearly has tremendous potential in revolutionising genetic testing. It will be very interesting to see how the field evolves.

References

1. Gardner RJM, Sutherland GR, Shaffer LG (2011) Chromosome abnormalities and genetic counselling. Oxford monographs on medical genetics, vol 61, 4th edn. Oxford University Press, New York
2. Trask BJ (2002) Human cytogenetics: 46 chromosomes, 46 years and counting. Nat Rev Genet 3(10):769–778. doi:10.1038/nrg905
3. Tonnies H (2002) Modern molecular cytogenetic techniques in genetic diagnostics. Trends Mol Med 8(6):246–250
4. Price CM (1993) Fluorescence in situ hybridization. Blood Rev 7(2):127–134
5. Volpi EV, Bridger JM (2008) FISH glossary: an overview of the fluorescence in situ hybridization technique. Biotechniques 45(4):385–386, 388, 390 passim. doi:10.2144/000112811
6. Ventura RA, Martin-Subero JI, Jones M, McParland J, Gesk S, Mason DY, Siebert R (2006) FISH analysis for the detection of lymphoma-associated chromosomal abnormalities in routine paraffin-embedded tissue. J Mol Diagn 8(2):141–151. doi:10.2353/jmoldx.2006.050083
7. Ried T, Schrock E, Ning Y, Wienberg J (1998) Chromosome painting: a useful art. Hum Mol Genet 7(10):1619–1626
8. Tanner M, Gancberg D, Di Leo A, Larsimont D, Rouas G, Piccart MJ, Isola J (2000) Chromogenic in situ hybridization: a practical alternative for fluorescence in situ hybridization to detect HER-2/neu oncogene amplification in archival breast cancer samples. Am J Pathol 157(5):1467–1472. doi:10.1016/S0002-9440(10)64785-2

9. Holcomb IN, Trask BJ (2011) Comparative genomic hybridization to detect variation in the copy number of large DNA segments. Cold Spring Harb Protoc 2011(11):1323–1333. doi:10.1101/pdb.top066589
10. Chari R, Lockwood WW, Lam WL (2006) Computational methods for the analysis of array comparative genomic hybridization. Cancer Inform 2:48–58
11. Cooper GM, Mefford HC (2011) Detection of copy number variation using SNP genotyping. Methods Mol Biol 767:243–252. doi:10.1007/978-1-61779-201-4_18
12. Sanger F, Nicklen S, Coulson AR (1977) DNA sequencing with chain-terminating inhibitors. Proc Natl Acad Sci U S A 74(12):5463–5467
13. Smith LM, Sanders JZ, Kaiser RJ, Hughes P, Dodd C, Connell CR, Heiner C, Kent SB, Hood LE (1986) Fluorescence detection in automated DNA sequence analysis. Nature 321(6071):674–679. doi:10.1038/321674a0
14. Shen Y, Wu BL (2009) Designing a simple multiplex ligation-dependent probe amplification (MLPA) assay for rapid detection of copy number variants in the genome. J Genet Genomics 36(4):257–265. doi:10.1016/S1673-8527(08)60113-7
15. Schouten JP, McElgunn CJ, Waaijer R, Zwijnenburg D, Diepvens F, Pals G (2002) Relative quantification of 40 nucleic acid sequences by multiplex ligation-dependent probe amplification. Nucleic Acids Res 30(12), e57
16. Southern EM (1975) Detection of specific sequences among DNA fragments separated by gel electrophoresis. J Mol Biol 98(3):503–517
17. Southern E (2006) Southern blotting. Nat Protoc 1(2):518–525. doi:10.1038/nprot.2006.73
18. Rizzo JM, Buck MJ (2012) Key principles and clinical applications of "next-generation" DNA sequencing. Cancer Prev Res 5(7):887–900. doi:10.1158/1940-6207.CAPR-11-0432

Cytogenetics: Applications

Chiyan Lau

Introduction

In the previous chapter, we discussed the laboratory techniques used to detect various types of genetic abnormalities, from single nucleotide changes to changes that affect entire chromosomes. We discussed the principles behind some of the more commonly used cytogenetic and molecular genetic techniques. In this chapter, we will look at how these techniques are used clinically in various diseases/conditions including cancer and constitutional disorders, and we will discuss some of the considerations that may go into deciding what would be appropriate genetic tests to perform in these clinical scenarios.

Cancer Cytogenetics

One important area of application for cytogenetic techniques is in cancer management, to detect somatic genetic changes in the neoplastic cells. This is particularly relevant for haematological malignancies, but there are increasing numbers of solid tumours where cytogenetics has a role. In cancer, cytogenetic investigations can be used to help with diagnosis, inform prognosis, or help prioritise treatment options. The following are some examples. We shall also discuss situations where molecular genetic techniques may be more appropriate.

C. Lau (✉)
SEALS Genetics, NSW Health Pathology, Level 4, Campus Centre,
Prince of Wales Hospital, Randwick, NSW 2031, Australia
e-mail: chiyan.lau@health.nsw.gov.au

S.R. Lakhani, S.B. Fox (eds.), *Molecular Pathology in Cancer Research*,
DOI 10.1007/978-1-4939-6643-1_4

Diagnosis and Monitoring

Chronic Myelogenous Leukaemia

Chronic myelogenous leukaemia (CML) is a myeloproliferative neoplasm, a clonal proliferation of haematopoietic stem cells of one of the myeloid lineages. One of the defining features of CML is the presence of the Philadelphia chromosome (Ph), which is present in 90–95 % of CML cases [1]. The Ph chromosome is the result of a reciprocal translocation between the long arms of chromosomes 9 and 22, t(9;22)(q34;q11.2). This structural rearrangement results in the creation of a fusion gene formed from the 5′ part of BCR on chromosome 22 and the 3′ part of ABL1 on chromosome 9. ABL1 is a tyrosine kinase, and the active promoter element of BCR in myeloid cells overactivate the transcription of the fusion BCR-ABL1 product, leading to constitutive activation of tyrosine kinase signalling, resulting in deregulated cell proliferation [2].

The Ph chromosome looks like a shortened chromosome 22, and was historically detected by standard karyotyping on bone marrow aspirate [3]. It is now more commonly detected by FISH using fusion probes, where the ABL1 breakpoint region on chromosome 9 is targeted with a FISH probe labelled with one colour (e.g. red), and the BCR breakpoint region on chromosome 22 is targeted with a probe labelled with a different colour (e.g. green). In a normal cell with no Ph chromosome, there will be two red and two green signals. If a t(9;22) translocation is present in the cell, there will be two yellow fusion signals, together with one red and one green signal. The use of FISH has an advantage over karyotyping in that ~5–10 % of CML cases do not have a typical Ph chromosome. Some of these are atypical translocations involving a third or even fourth chromosome, while others are cryptic translocations which cannot be detected by karyotyping [4]. In these atypical Ph-negative cases, BCR-ABL1 fusion is still present, and therefore can be detected by FISH. Another advantage of FISH is that it can be performed on interphase cells, thus eliminating the need for cell culture.

At the time of diagnosis, the number of CML cells in the patient's blood is high, therefore BCR-ABL1 fusions are quite easy to detect by FISH. However, once the patient has undergone treatment, e.g., with the kinase inhibitor imatinib (Glivec), the number of cells from the CML clone would decrease and eventually fall below the limit of detection of karyotyping and FISH if treatment is successful ('Complete Cytogenetic Response') [5]. In order to continue to monitor the patient for signs of relapse at this stage of the disease, a more sensitive method is required, such as real-time PCR [6]. This molecular method quantifies the number of copies of BCR-ABL1 transcripts in the patient sample, and allows early detection of relapse before the CML clone has expanded above cytogenetically detectable levels. Loss of disease control may be due to ABL1 kinase site mutations leading to acquired resistance to therapy, and can be detected by ABL1 sequencing [7]. Depending on the mutation, changes in treatment (e.g. to newer generation kinase inhibitors such as nilotinib or dasatinib) may be possible to maintain control over the CML [8, 9]. Therefore, while cytogenetic methods are useful at the time of CML diagnosis, molecular methods are the investigations of choice during minimal residual disease monitoring [10].

Prognosis and Management

Multiple Myeloma

Multiple myeloma (MM) is a malignant proliferation of plasma cells. It is a heterogeneous disorder, with several subtypes differing in prognosis and the underlying genetics. There are a number of recurrent chromosomal abnormalities in MM which are predictive of prognosis, therefore standard cytogenetics and FISH analyses can be useful in MM.

Multiple myeloma can be broadly divided by genetic changes into two main groups: hyperdiploid and non-hyperdiploid. The hyperdiploid group (h-MM) shows numerous trisomies (resulting in chromosome number >46) and generally has a better prognosis. This group has a low prevalence for translocations involving the immunoglobulin heavy chain gene locus (IGH) at 14q32. In contrast, the non-hyperdiploid group (nh-MM), which includes hypodiploid, pseudodiploid, and near-tetraploid cases, tends to have a poorer prognosis. This group is enriched for 14q32 IGH translocations, although not all IGH translocations confer a poor prognosis. Specifically, the t(11;14) IGH/CCND1 translocation appears to have a neutral or even favourable prognosis, while the t(4;14) IGH/FGFR3-MMSET translocation and t(14;16) IGH/MAF translocation are associated with poorer survival. Other cytogenetic changes which are markers of poor prognosis in MM include deletion of the TP53 gene at 17p13 which codes for the p53 tumour suppressor protein, 1q21 gains, 1p21 deletions, and cytogenetically detected monosomy 13 and 13q deletions [11, 12].

Interphase FISH can be used to detect these chromosomal changes. One strategy is to use a 14q32 break-apart probe to determine if the IGH locus is involved in a translocation, and if so, specific fusion probes for t(11;14), t(4;14), and t(14;16) can be used to determine the fusion partner. Karyotyping can also provide useful information on other chromosomal abnormalities which may be present, but can be problematic because plasma cells from the MM clone often show poor growth in culture. This results in an apparently normal karyotype result since the metaphases are dominated by normal cells. Sole use of karyotyping for MM cytogenetics is therefore not recommended.

For FISH testing in MM, to maximise sensitivity, it is recommended that either purified plasma cells are used, or that FISH analysis/scoring is confined to plasma cells by performing cytoplasmic immunoglobulin-enhanced FISH (cIg-FISH) [13] or CD138 immunostaining [14]. The reason for this is that the bone marrow aspirate specimen for cytogenetics often has a low concentration of plasma cells from the MM clone, and this is further lowered by the effects of haemodilution. Standard interphase FISH without cell enrichment would have a low sensitivity of abnormality detection on such a sample. Enrichment of the sample for plasma cells can be achieved by using anti-CD138-conjugated beads. Alternatively, the use of cIg-FISH allows identification of plasma cells during FISH analysis by immunofluorescently staining cytoplasmic kappa or lambda light chains. This allows FISH scoring to be restricted to plasma cells. These strategies maximise the chance that FISH analysis is obtained on cells relevant to the disease process.

Microarrays also play a role in current research in MM, for example, to identify new cytogenetic changes which may have prognostic significance [15]. However, standard arrays are unable to characterise translocations which play an important prognostic role in MM, and at the present time, arrays are not widely used for clinical testing in MM.

Chronic Lymphocytic Leukaemia

Chronic lymphocytic leukaemia (CLL) is a neoplastic disorder of mature B lymphocytes, where the neoplastic cells are commonly present in peripheral blood, bone marrow, spleen, and lymph nodes [16]. A number of recurrent cytogenetic abnormalities are commonly found in CLL cases, including deletions at 13q14.3, trisomy 12, and deletions at 11q22-23 (including the ATM gene) and 17p13 (including the TP53 gene) [17]. Among these, the deletions at 11q and 17p are associated with adverse prognosis, while isolated deletion at 13q or the absence of cytogenetic changes is associated with a more favourable disease outcome. These cytogenetic changes, when used in conjunction with other clinical and laboratory information, may also help to guide choice of therapy [18]. Commonly, a FISH panel consisting of probes which map to 13q14.3, ATM, TP53, and CEP12 (centromere of chromosome 12) are used for detection of the common cytogenetic changes in interphase peripheral blood cells. Conventional karyotyping in CLL is more difficult and less sensitive than FISH owing to the poor growth of CLL cells in culture, and karyotyping has a lower resolution for small deletions. Microarray-based testing (especially SNP arrays) may become a viable alternative to FISH [19], since it also does not require dividing cells, has a high enough resolution for small deletions and is a survey of the entire genome (and can therefore detect additional chromosomal abnormalities other than those targeted by specific FISH probes), although currently it has not yet been widely adopted in clinical CLL testing outside of research settings.

In addition to cytogenetic changes, somatic hypermutation status at the IGHV locus (immunoglobulin heavy chain V region) also has prognostic significance, with CLLs which show hypermutation having a better prognosis than unmutated cases [20, 21]. Also, there is emerging evidence that pathogenic mutations in the TP53, BIRC3, SF3B1, and NOTCH1 genes also influence prognosis [22, 23]. These changes cannot be detected by cytogenetic techniques, but require molecular sequencing methods for detection.

Acute Leukaemias

In patients affected by acute leukaemias, including acute myeloid leukaemia (AML) and acute lymphoblastic leukaemia (ALL), disease classification and prognosis depend substantially on the genetic abnormalities in the malignant clone [24]. There are a number of recurrent cytogenetic rearrangements which define particular subtypes of AML. These include translocations and inversions such as t(8;21)(q22;q22)

[RUNX1-RUNX1T1], inv(16)(p13.1q22)/t(16;16)(p13.1;q22) [CBFB-MYH11], and t(15;17)(q22;q12) [PML-RARA], which are associated with better chances of long-term survival [25, 26]. The PML-RARA rearrangement is specifically associated with a subtype of AML known as acute promyelocytic leukaemia (APL), which responds to treatment with all-trans retinoic acid (ATRA) [27]. Other rearrangements such as inv(3)(q21q26.2)/t(3;3)(q21;q26.2) RPN1-EVI1 and t(6;9)(p23;q34) DEK-NUP214, as well as monosomies of chromosome 5 and 7, or complex karyotype (defined as 3 or more clonal cytogenetic abnormalities) are associated with poorer outcomes [25, 26]. Likewise, in ALL of precursor B cell origin (B-ALL), some cytogenetic changes have prognostic significance and are also used to define disease subgroups and determine treatment options. For example, rearrangements involving the MLL gene at 11q23 (especially t(4;11) MLL-AF4 translocations) are associated with poor prognosis [28]. The presence of the Philadelphia chromosome t(9;22) also confers a poor prognosis in adult ALL, although the availability of imatinib treatment for this subgroup has improved patient survival [29].

Conventional karyotyping of bone marrow aspirate specimens is the standard method for detecting these cytogenetic abnormalities in acute leukaemias. However, in malignancy karyotyping, often the chromosome quality is poor, therefore FISH testing (e.g. using fusion probes for specific rearrangements) is sometimes used to confirm karyotype findings.

In addition to cytogenetic changes, mutations in genes such as FLT3, NPM1, and CEBPA also have prognostic significance in acute leukaemias, but these require molecular methods for detection [26, 30].

The prognostic information from cytogenetic (and molecular) studies allow the clinician to adjust the aggressiveness of treatment and balance the risks and benefits of offering allogeneic stem cell transplant, which is ultimately the treatment option that offers a chance of cure in poor risk patients but is associated with significant morbidity and mortality.

Non Small Cell Lung Cancer

Non-small cell lung cancer (NSCLC) is an aggregate category of lung cancers which includes adenocarcinoma, squamous cell carcinoma, large cell carcinoma as well as a number of rarer histologic subtypes. Most cases are discovered at an advanced stage of disease and prognosis is poor. Traditionally, in metastatic NSCLC the main pharmacological treatment option was combination cytotoxic chemotherapy including platinum-based agents. Recently a number of genetic abnormalities have been found in NSCLC (especially adenocarcinoma) which allow targeted treatment of those subgroups of patients whose NSCLC carry specific mutations [31].

One of the specific mutations in NSCLC is a chromosomal structural rearrangement involving the anaplastic lymphoma kinase (ALK) gene, which codes for a receptor tyrosine kinase. This rearrangement is found in approximately 2–7 % of NSCLCs, although it appears to be enriched in never- or light-smokers

[32, 33]. The most common form is a paracentric inversion on the short arm of chromosome 2 which creates a fusion product between the 5′ end of the EML4 gene at 2p21 and the 3′ end of the ALK gene at 2p23.2 [34]. This leads to activation of kinase signalling in the affected cell. The clinical importance of this ALK rearrangement is the availability of a small molecule ALK inhibitor crizotinib which has been shown to improve outcomes in patients who carry the ALK rearrangement [33].

The most common way to detect ALK rearrangements is by the use of a dual colour break-apart FISH probe located at the 3′ end of ALK [32]. Karyotyping is inappropriate because commonly the only specimen type available in NSCLC are formalin fixed paraffin embedded (FFPE) tumour biopsies, which do not contain viable cells capable of dividing. The use of break-apart FISH also has the advantage of being able to detect atypical ALK rearrangements where the fusion partner is not EML4. Theoretically, a molecular strategy (e.g. real-time PCR) could also be designed to detect the specific EML4-ALK fusion event, but due to potential variability in breakpoints particularly in EML4 [35], this would be impractical especially in a clinical laboratory setting.

Another type of tumour-specific mutation in NSCLC is activating mutations of the epidermal growth factor receptor (EGFR) gene. The majority of reported mutations are small in-frame deletions in exon 19, and point mutations in exon 21 of the gene, including a missense mutation which replaces a leucine at amino acid position 858 by arginine (p.Leu858Arg) [36]. These mutations lead to constitutive activation of kinase activity in EGFR. Patients with activating mutations in EGFR show improved response to anti-EGFR therapy, such as the tyrosine kinase inhibitors (TKIs), gefitinib, and erlotinib [37]. Detection of these activating mutations require molecular methods, since the DNA changes are too small to detect by cytogenetic methods including FISH. Direct PCR/Sanger sequencing of the EGFR gene is one way to identify the mutations. This method has the advantage of identifying the exact mutation in the gene, and is also able to detect rare mutations. The disadvantage is that if the mutation load in the specimen is low (e.g. less than ~20 %), then sequencing may not be able to detect the change, although there is some evidence that the analytical sensitivity of Sanger sequencing may be higher for at least some mutations [38]. Also, tumour DNA extracted from FFPE specimens tend to be lower quality and more fragmented, and may be difficult to sequence. Therefore a more commonly used method in the clinical laboratory is a targeted mutation panel using strategies such as real-time PCR, with specific PCR primers and probes which target a panel of common activating EGFR mutations (and/or mutations which confer resistance to TKI inhibitors) [39].

In addition to sequence variations, some NSCLCs have amplifications in the copy number of EGFR genes. For these tumours, FISH probes against EGFR may be used to detect the gene amplification, which will show up as multiple signals per cell under fluorescence microscopy [40]. However, molecular methods are required to determine if the amplified EGFR contains activating sequence variations.

Breast Cancer

Breast cells express receptors on the cell surface which respond to extracellular growth signals. These include oestrogen receptors (ER), progesterone receptors (PR), and HER2 receptors. Human epidermal growth factor receptor 2 (HER2) is a receptor tyrosine kinase which belongs to the same protein family as EGFR, and is encoded by the ERBB2 gene on chromosome 17. In normal cells, there are two copies of the ERBB2 gene, one on each chromosome 17. In some breast cancers (~20 %), there is amplification in the copy number of the ERBB2 gene, leading to overexpression of HER2 receptors on the cell surface. These HER2-positive breast cancers have an aggressive disease course, but respond favourably to monoclonal antibodies directed against HER2, such as trastuzumab (Herceptin) and pertuzumab (Perjeta®) [41]. Testing for ERBB2 copy number has clinical utility since patients with HER2-negative breast cancer do not benefit from anti-HER2 treatment, and trastuzumab is also known to have cardiac toxicity [42]. Therefore currently trastuzumab therapy is only recommended in patients with HER2-positive cancers.

ERBB2 amplification can be detected by FISH on an FFPE tumour specimen. HER2-positive cells show multiple signals for the ERBB2 FISH probe, while normal cells only show two signals. Alternatively, some centres use a related non-fluorescence in situ hybridisation (ISH) method for detection of ERBB2 copy number, such as CISH (chromogenic ISH). These methods allow detection of signal in bright field microscopy rather than requiring fluorescence microscopy. Yet another alternative is the detection of HER2 protein overexpression rather than gene amplification. This method utilises immunohistochemistry (IHC) staining for the HER2 protein. CISH and IHC allow co-examination of tissue morphology and HER2 status, and are less expensive compared to FISH analyses, although some tumours show discordance between protein expression and gene amplification results [43, 44].

In some breast cancer patients, there is a strong family history (e.g. with multiple closely related relatives affected, who may also have developed breast cancer at a younger age than average), which suggests a familial rather than sporadic form of breast cancer. Germline mutations in some genes have been associated with familial breast cancer. The most recognised of these are BRCA1 and BRCA2 which are associated with autosomal dominant forms of breast (and ovarian) cancer predisposition. In contrast to detection of ERBB2 amplification, which is performed on tumour material, mutation screening of BRCA1/BRCA2 genes for familial cancer predisposition requires germline DNA (e.g. from peripheral blood specimens), because the aim here is to determine if there is a *heritable* mutation in these genes. The method used is most commonly a combination of direct Sanger sequencing and MLPA, since the types of mutations reported in BRCA1 and BRCA2 include sequence variations as well as whole exon deletions/duplications. FISH and karyotyping are generally not applicable, since the deletions/duplications are usually below the resolution of these techniques. Microarrays, on the other hand, can technically detect the exonic deletions/duplications, but is seldom used in the clinical laboratory for this indication because of factors such as cost.

Some centres are moving to a massively parallel sequencing (MPS) approach for BRCA1 and BRCA2 mutation screening [45]. Theoretically, MPS can detect both sequence variations and whole exon deletions/duplications with the one technique. However, at the time of writing, there are still technical issues with clinical implementation of MPS, especially with detection of copy number changes and bioinformatic analyses, so this is still at a research and development stage in most clinical centres.

Germline Disorders

Another major area of application for cytogenetic techniques is the diagnosis of constitutional genetic disorders, either in the postnatal or prenatal period of life. The following are some examples of how cytogenetics are used in these clinical settings.

Postnatal Testing

Intellectual Disability/Developmental Delay

Intellectual disability (ID) and developmental delay (DD) are common presentations in the paediatric population with a wide range of severity. These encompass disorders in one or more neurodevelopmental domains, including motor skills (gross and fine), psychosocial, language, and cognitive development. Both environmental factors and genetic factors may contribute to ID/DD.

One of the most common genetic causes of ID/DD is Down syndrome (DS), or trisomy 21 [46]. DS patients commonly present with very typical and recognisable facial features (dysmorphism), intellectual disability, and a range of other complications which may include heart and other organ defects, immune deficiency, etc. The majority of Down syndrome patients have three separate copies of chromosome 21 in every cell of the body. However, in some DS patients, the extra copy of chromosome 21 is fused to another acrocentric chromosome (chromosome 13, 14, 15, 21 or 22) at the centromere (as a Robertsonian translocation), rather than being free in the cell. The translocation may have arisen de novo, but may also be inherited from one of the parents, in which case the recurrence risk of aneuploidy in a subsequent pregnancy would be increased. Therefore the detection of translocation is important for genetic counselling. Also, a small proportion of DS patients have mosaic trisomy 21, where some of the patient's cells have two chromosome 21 s but other cells have three copies. The presentation of mosaic DS is variable and may be milder than typical DS patients.

The most informative and appropriate test for Down syndrome is conventional karyotyping, because of the known possibility of mosaicism and translocation. Microarrays and FISH would also be able to detect the extra chromosome 21.

However, microarrays have a lower sensitivity for mosaic Down syndrome, and neither array nor FISH can detect a Robertsonian translocation. On the other hand, a karyotype can, in the one test, detect and confirm the presence of the extra chromosome 21, and determine if the extra copy is free or translocated to another chromosome. If mosaicism is present, karyotype has a higher sensitivity for detecting the abnormality than microarray.

Apart from Down syndrome, many other chromosomal deletions and duplications also cause ID/DD. The clinical presentations in these cases may often be non-specific. In the past, karyotyping was used as the standard screening test for non-specific ID/DD, with an abnormality detection rate of ~3–5 %. More recently, microarray testing has become the preferred test, due to its ability to detect submicroscopic copy number changes (CNCs), i.e., microdeletions/microduplications. Many of these CNCs also show significant association with autism spectrum disorder (ASD) and/or multiple congenital anomalies (MCA). Therefore, in the setting of ID/DD, ASD, or MCA, microarrays are now recommended as a first line investigation, with an abnormality detection rate of up to ~15 % [47].

However, with the increased resolution of microarrays come new issues. Some of the many CNCs detected by microarrays turn out to be relatively common in the general population and are now believed to be benign 'normal' variation. However, some of the detected CNCs appear to be rare, and have not been reported either in the normal population or in affected patients. These CNCs are called variants of uncertain clinical significance (VUCS). Other CNCs have been reported at a low frequency in normal individuals, but appear to be 'enriched' in individuals with certain phenotypes such as ASD and schizophrenia. Some of these are now thought to be 'susceptibility variants' with variable penetrance or expressivity for the associated phenotypes. In addition, microarray testing sometimes uncovers 'incidental findings' such as deletion of genes associated with familial cancer syndromes (e.g. BRCA1, BRCA2, APC, etc.). These findings do not explain the patient's presenting complaint of ID/DD, but may have important clinical implications for other members in the extended family or for the patient later in life. These findings raise issues of consent and disclosure, and present challenges for clinical management of the patient and family. They highlight the importance of adequate counselling and informed consent prior to embarking on genetic testing [48].

Syndromic Presentations

Sometimes ID/DD patients present with additional phenotypic features and/or facial dysmorphism which suggest a specific syndromic diagnosis. In some cases where the syndrome is associated with a specific chromosomal deletion/duplication, it is possible to target the genomic region with a locus-specific FISH probe. For example, in a child with conotruncal heart defects and cleft lip/palate, it is possible to perform FISH using a probe which localises to the 22q11.2 DiGeorge/velocardiofacial syndrome (DGS/VCFS) critical region [49]. Observation of only one signal for this probe would indicate heterozygous deletion of the locus and confirm the clinical diagnosis of VCFS. However, depending on the clinical circumstances, microarray testing may in fact be a more efficient investigation, especially if there is genetic heterogeneity (i.e. several genomic loci associated with the syndrome).

Single Gene Disorders

Apart from deletions/duplications, other genetic defects such as sequence variations or triplet repeat expansion may also lead to ID/DD. These are not detectable by cytogenetic methods, and will require the application of molecular techniques. One important example is Fragile X syndrome.

The molecular mechanism underlying Fragile X is an expansion in the CGG trinucleotide repeat at the 5′ upstream region of the FMR1 gene. The number of CGG repeats is variable in the population, but is normally <45. In affected individuals, the number of CGG repeats expand to >200 ('full mutation' range), which leads to methylation of the promoter region of the FMR1 gene and silencing of expression of the FMR1 protein product (FMRP). Intermediate repeat sizes (56–200 repeats) are known as pre-mutations, and can be found in asymptomatic males or carrier females. Pre-mutation expansions do not cause ID/DD, but may have other late-onset health implications such as premature ovarian failure or tremor-ataxia syndrome [50].

The expansion in the trinucleotide repeat creates a folate-sensitive fragile site (FRAXA) on the affected X chromosome, and historically a special cytogenetic method (karyotype after culture of cells in a modified folate-deficient media) was used to detect the fragile site [51, 52]. However, this method is rarely performed today due to costs, low sensitivity, and slow turn-around time. It should be noted that routine karyotyping using standard culture techniques cannot detect the fragile site. FISH or microarray testing also cannot detect the triplet repeat expansions. Instead, molecular methods are now the method of choice for diagnosis of Fragile X. A common approach is to use two complementary molecular methods to detect the entire range of possible triplet repeat sizes. For smaller repeats, PCR followed by fragment analysis is used. PCR primers are designed to amplify the genomic region containing the CGG repeats. The size of the PCR product is determined by fragment analysis, which is used to calculate the number of repeats. This method provides a highly precise estimate of the repeat size (usually to ±1 to 3 repeats) at the low end of the repeat size range (up to a maximum of ~100 repeats), but for individuals with full mutations (>200 repeats), the repeat tract is too long to amplify by PCR. Therefore, if no amplification product is detected by PCR/fragment analysis in an affected male, Southern blot would be performed, using a probe which binds to the restriction fragment containing the CGG repeat. The size of the restriction fragment provides only a rough estimate of the repeat size, but is able to detect expansions in excess of a thousand repeats [53, 54].

It is worth pointing out that in extremely rare instances, FMR1 gene deletions and sequence variations have also been reported to cause Fragile X syndrome [55, 56]. In these very rare cases, FISH, microarray, and MLPA could be used to detect deletions, and PCR/Sanger sequencing could be used to detect sequence variations. But for the vast majority of Fragile X patients (>99 %), fragment analysis and Southern blot are the mainstay of diagnosis.

Infertility/Recurrent Pregnancy Loss

Another application of cytogenetic testing is for couples who suffer from infertility or recurrent pregnancy loss. There are many possible underlying causes for this clinical presentation, including anatomical, endocrine, and genetic factors either in the male or female partner. One important genetic factor which may contribute to this presentation is balanced chromosomal rearrangements, such as balanced reciprocal translocations, which may be present in the male or female partner.

In a balanced translocation, there is no net gain or loss of chromosomal material in the carrier. Therefore, in most cases there are no phenotypic consequences in the carrier unless the breakpoints interrupt an important gene. However, when the gonadal cells in a carrier undergo meiosis, there is a high likelihood that some of the gametes produced will be unbalanced, depending on the way the chromosomes segregate. If the resulting unbalanced gametes were used in fertilisation, the zygote formed would contain a chromosomal imbalance (typically a partial trisomy and a concomitant partial monosomy of the chromosomes involved in the translocation). The level of imbalance in these zygotes is such that many are not compatible with full-term gestation, resulting in recurrent miscarriage.

Microarray analysis cannot detect balanced rearrangements in carriers since there is no net copy number gain or loss. FISH testing is also impractical as a screening test since there is no way to predict which chromosomes are potentially involved. Karyotype analysis is therefore the most appropriate method in this clinical setting.

In addition to providing a precise diagnosis for the couple, cytogenetic investigation is also of value for planning subsequent use of artificial reproductive techniques including pre-implantation genetic diagnosis (PGD). With PGD, knowing the specific rearrangement in the carrier parent allows specific FISH probes to be designed to screen in vitro fertilised embryos. It then allows selective implantation of only those embryos which contain a balanced chromosomal complement.

Prenatal Testing

In the prenatal setting, cytogenetic testing is often requested as a result of concerns over the risk of aneuploidy, due to advanced maternal age or 'high risk' results of biochemical maternal serum screening. Other common indications include the finding of abnormalities on foetal ultrasound.

The most common prenatal diagnostic test is a conventional karyotype conducted on an amniotic fluid (AF) specimen or chorionic villus sample (CVS). These are invasive tests and carry a finite risk of miscarriage (estimated to be 1/100 to 1/200, depending on the centre) [57, 58]. These types of samples contain cells of fetal origin that will divide in culture, thus allowing the use of standard cytogenetic techniques to directly visualise the chromosomes in metaphase and confirm a trisomy or any other chromosomal abnormality if present. The turn-around time for a result is typically in the order of 7–10 days.

Many expectant mothers who have a 'high risk' for aneuploidy are understandably anxious and often wish to have a faster answer. Another method which may be used in this setting is aneuploidy FISH testing. This test is performed as interphase FISH using probes which target the centromeres or specific loci on chromosomes 13, 18, and 21 (and optionally chromosome X and Y) on uncultured fetal cells [59]. This test provides a faster turn-around time, typically within 24–48 h. The cells are scored for the number of signals for each probe. Two signals for each of the chromosome 13, 18, and 21 probes is the 'normal' pattern. The presence of three signals for the chromosome 21 probe would strongly suggest Trisomy 21. However, this test cannot be considered a definitive test. The absence of a third signal does not exclude the potential for duplication of a part of the chromosome which is not targeted by the FISH probe. Abnormalities of chromosomes other than 13, 18, 21, X, and Y, or unusual rearrangements, cannot be detected. Also, if trisomy is detected, it cannot distinguish between translocation trisomy and a free extra chromosome.

For these reasons, aneuploidy FISH testing should only be considered an extra screening step to fast-track an abnormal result. Whether the FISH result is positive or negative, it should always be followed up with karyotype for a definitive diagnosis.

More recently, some centres have started to offer microarrays for prenatal testing [60]. The technical principles of prenatal array are similar to the use of microarrays postnatally. As discussed already, microarray may have less sensitivity for mosaic results compared to karyotyping or FISH, especially if the level of abnormal cells is low (e.g. below 10–20 %), but can detect submicroscopic copy number changes. This increased resolution may be perceived as an advantage, but it can also lead to challenges to interpretation of results. For example, if a copy number change corresponds to a well-known microdeletion or microduplication syndrome, and is consistent with malformations seen on fetal ultrasound, then the interpretation may be straightforward, in which case the microarray testing has provided a diagnosis where karyotyping could not. However, if the copy number change is a VUCS or susceptibility variant or incidental finding, then interpretation of the finding and counselling of the parents will be challenging.

References

1. Vardiman JW, Melo JV, Baccarani M, Thiele J (2008) Chronic myelogenous leukaemia, BCR-ABL1 positive. In: Swerdlow SH, Campo E, Harris NL et al (eds) WHO classification of tumours of haematopoietic and lymphoid tissues. WHO classification of tumours, 4th edn, vol 2. IARC, Lyon, pp 32–39
2. Deininger MW, Goldman JM, Melo JV (2000) The molecular biology of chronic myeloid leukemia. Blood 96(10):3343–3356
3. Rowley JD (1973) Letter: a new consistent chromosomal abnormality in chronic myelogenous leukaemia identified by quinacrine fluorescence and Giemsa staining. Nature 243(5405):290–293
4. Melo JV (1996) The diversity of BCR-ABL fusion proteins and their relationship to leukemia phenotype. Blood 88(7):2375–2384

5. Fausel C (2007) Targeted chronic myeloid leukemia therapy: seeking a cure. J Manag Care Pharm 13(8 Suppl A):8–12
6. Branford S, Hughes TP, Rudzki Z (1999) Monitoring chronic myeloid leukaemia therapy by real-time quantitative PCR in blood is a reliable alternative to bone marrow cytogenetics. Br J Haematol 107(3):587–599
7. Hughes T, Deininger M, Hochhaus A, Branford S, Radich J, Kaeda J, Baccarani M, Cortes J, Cross NC, Druker BJ, Gabert J, Grimwade D, Hehlmann R, Kamel-Reid S, Lipton JH, Longtine J, Martinelli G, Saglio G, Soverini S, Stock W, Goldman JM (2006) Monitoring CML patients responding to treatment with tyrosine kinase inhibitors: review and recommendations for harmonizing current methodology for detecting BCR-ABL transcripts and kinase domain mutations and for expressing results. Blood 108(1):28–37. doi:10.1182/blood-2006-01-0092
8. Talpaz M, Shah NP, Kantarjian H, Donato N, Nicoll J, Paquette R, Cortes J, O'Brien S, Nicaise C, Bleickardt E, Blackwood-Chirchir MA, Iyer V, Chen TT, Huang F, Decillis AP, Sawyers CL (2006) Dasatinib in imatinib-resistant Philadelphia chromosome-positive leukemias. N Engl J Med 354(24):2531–2541. doi:10.1056/NEJMoa055229
9. Kantarjian H, Giles F, Wunderle L, Bhalla K, O'Brien S, Wassmann B, Tanaka C, Manley P, Rae P, Mietlowski W, Bochinski K, Hochhaus A, Griffin JD, Hoelzer D, Albitar M, Dugan M, Cortes J, Alland L, Ottmann OG (2006) Nilotinib in imatinib-resistant CML and Philadelphia chromosome-positive ALL. N Engl J Med 354(24):2542–2551. doi:10.1056/NEJMoa055104
10. Hughes T (2006) ABL kinase inhibitor therapy for CML: baseline assessments and response monitoring. Hematology Am Soc Hematol Educ Program: 211–218.doi:10.1182/asheducation-2006.1.211
11. Fonseca R, Bergsagel PL, Drach J, Shaughnessy J, Gutierrez N, Stewart AK, Morgan G, Van Ness B, Chesi M, Minvielle S, Neri A, Barlogie B, Kuehl WM, Liebisch P, Davies F, Chen-Kiang S, Durie BG, Carrasco R, Sezer O, Reiman T, Pilarski L, Avet-Loiseau H; International Myeloma Working Group (2009) International Myeloma Working Group molecular classification of multiple myeloma: spotlight review. Leukemia 23(12):2210–2221. doi:10.1038/leu.2009.174
12. Munshi NC, Anderson KC, Bergsagel PL, Shaughnessy J, Palumbo A, Durie B, Fonseca R, Stewart AK, Harousseau JL, Dimopoulos M, Jagannath S, Hajek R, Sezer O, Kyle R, Sonneveld P, Cavo M, Rajkumar SV, San Miguel J, Crowley J, Avet-Loiseau H; International Myeloma Workshop Consensus Panel 2 (2011) Consensus recommendations for risk stratification in multiple myeloma: report of the International Myeloma Workshop Consensus Panel 2. Blood 117(18):4696–4700. doi:10.1182/blood-2010-10-300970
13. VanWier S, Fonseca R (2005) Detection of chromosome 13 deletions by fluorescent in situ hybridization. Methods Mol Med 113:59–69. doi:10.1385/1-59259-916-8:59
14. Cook JR, Hartke M, Pettay J, Tubbs RR (2006) Fluorescence in situ hybridization analysis of immunoglobulin heavy chain translocations in plasma cell myeloma using intact paraffin sections and simultaneous CD138 immunofluorescence. J Mol Diagn 8(4):459–465. doi:10.2353/jmoldx.2006.050149
15. Avet-Loiseau H, Li C, Magrangeas F, Gouraud W, Charbonnel C, Harousseau JL, Attal M, Marit G, Mathiot C, Facon T, Moreau P, Anderson KC, Campion L, Munshi NC, Minvielle S (2009) Prognostic significance of copy-number alterations in multiple myeloma. J Clin Oncol 27(27):4585–4590. doi:10.1200/JCO.2008.20.6136
16. Muller-Hermelink HK, Montserrat E, Catovsky D, Campo E, Harris NL, Stein H (2008) Chronic lymphocytic leukaemia/small lymphocytic lymphoma. In: Swerdlow SH, Campo E, Harris NL et al (eds) WHO classification of tumours of haematopoietic and lymphoid tissues. WHO classification of tumours, 4th edn, vol 2. IARC, Lyon, pp 180–182
17. Dohner H, Stilgenbauer S, Benner A, Leupolt E, Krober A, Bullinger L, Dohner K, Bentz M, Lichter P (2000) Genomic aberrations and survival in chronic lymphocytic leukemia. N Engl J Med 343(26):1910–1916. doi:10.1056/NEJM200012283432602
18. Hallek M, Cheson BD, Catovsky D, Caligaris-Cappio F, Dighiero G, Dohner H, Hillmen P, Keating MJ, Montserrat E, Rai KR, Kipps TJ; International Workshop on Chronic Lymphocytic Leukemia (2008) Guidelines for the diagnosis and treatment of chronic lymphocytic leukemia:

a report from the International Workshop on Chronic Lymphocytic Leukemia updating the National Cancer Institute-Working Group 1996 guidelines. Blood 111(12):5446–5456. doi:10.1182/blood-2007-06-093906
19. Pfeifer D, Pantic M, Skatulla I, Rawluk J, Kreutz C, Martens UM, Fisch P, Timmer J, Veelken H (2007) Genome-wide analysis of DNA copy number changes and LOH in CLL using high-density SNP arrays. Blood 109(3):1202–1210. doi:10.1182/blood-2006-07-034256
20. Damle RN, Wasil T, Fais F, Ghiotto F, Valetto A, Allen SL, Buchbinder A, Budman D, Dittmar K, Kolitz J, Lichtman SM, Schulman P, Vinciguerra VP, Rai KR, Ferrarini M, Chiorazzi N (1999) Ig V gene mutation status and CD38 expression as novel prognostic indicators in chronic lymphocytic leukemia. Blood 94(6):1840–1847
21. Hamblin TJ, Davis Z, Gardiner A, Oscier DG, Stevenson FK (1999) Unmutated Ig V(H) genes are associated with a more aggressive form of chronic lymphocytic leukemia. Blood 94(6):1848–1854
22. Rossi D, Rasi S, Spina V, Bruscaggin A, Monti S, Ciardullo C, Deambrogi C, Khiabanian H, Serra R, Bertoni F, Forconi F, Laurenti L, Marasca R, Dal-Bo M, Rossi FM, Bulian P, Nomdedeu J, Del Poeta G, Gattei V, Pasqualucci L, Rabadan R, Foa R, Dalla-Favera R, Gaidano G (2013) Integrated mutational and cytogenetic analysis identifies new prognostic subgroups in chronic lymphocytic leukemia. Blood 121(8):1403–1412. doi:10.1182/blood-2012-09-458265
23. Wang L, Lawrence MS, Wan Y, Stojanov P, Sougnez C, Stevenson K, Werner L, Sivachenko A, DeLuca DS, Zhang L, Zhang W, Vartanov AR, Fernandes SM, Goldstein NR, Folco EG, Cibulskis K, Tesar B, Sievers QL, Shefler E, Gabriel S, Hacohen N, Reed R, Meyerson M, Golub TR, Lander ES, Neuberg D, Brown JR, Getz G, Wu CJ (2011) SF3B1 and other novel cancer genes in chronic lymphocytic leukemia. N Engl J Med 365(26):2497–2506. doi:10.1056/NEJMoa1109016
24. Mrozek K, Heerema NA, Bloomfield CD (2004) Cytogenetics in acute leukemia. Blood Rev 18(2):115–136. doi:10.1016/S0268-960X(03)00040-7
25. Grimwade D, Walker H, Oliver F, Wheatley K, Harrison C, Harrison G, Rees J, Hann I, Stevens R, Burnett A, Goldstone A (1998) The importance of diagnostic cytogenetics on outcome in AML: analysis of 1,612 patients entered into the MRC AML 10 trial. The Medical Research Council Adult and Children's Leukaemia Working Parties. Blood 92(7):2322–2333
26. National Comprehensive Cancer Network (2013) Acute myeloid leukemia. NCCN clinical practice guidelines in oncology
27. Tallman MS, Andersen JW, Schiffer CA, Appelbaum FR, Feusner JH, Ogden A, Shepherd L, Willman C, Bloomfield CD, Rowe JM, Wiernik PH (1997) All-trans-retinoic acid in acute promyelocyticleukemia.NEnglJMed337(15):1021–1028.doi:10.1056/NEJM199710093371501
28. Mrozek K, Harper DP, Aplan PD (2009) Cytogenetics and molecular genetics of acute lymphoblastic leukemia. Hematol Oncol Clin North Am 23(5):991–1010. doi:10.1016/j.hoc.2009.07.001, v
29. Schultz KR, Bowman WP, Aledo A, Slayton WB, Sather H, Devidas M, Wang C, Davies SM, Gaynon PS, Trigg M, Rutledge R, Burden L, Jorstad D, Carroll A, Heerema NA, Winick N, Borowitz MJ, Hunger SP, Carroll WL, Camitta B (2009) Improved early event-free survival with imatinib in Philadelphia chromosome-positive acute lymphoblastic leukemia: a children's oncology group study. J Clin Oncol 27(31):5175–5181. doi:10.1200/JCO.2008.21.2514
30. How J, Sykes J, Minden MD, Gupta V, Yee KW, Schimmer AD, Schuh AC, Kamel-Reid S, Brandwein JM (2013) The prognostic impact of FLT3-ITD and NPM1 mutations in patients with relapsed acute myeloid leukemia and intermediate-risk cytogenetics. Blood Cancer J 3, e116. doi:10.1038/bcj.2013.14
31. National Comprehensive Cancer Network (2013) Non-small cell lung cancer. NCCN clinical practice guidelines in oncology
32. Perner S, Wagner PL, Demichelis F, Mehra R, Lafargue CJ, Moss BJ, Arbogast S, Soltermann A, Weder W, Giordano TJ, Beer DG, Rickman DS, Chinnaiyan AM, Moch H, Rubin MA (2008) EML4-ALK fusion lung cancer: a rare acquired event. Neoplasia 10(3):298–302
33. Kwak EL, Bang YJ, Camidge DR, Shaw AT, Solomon B, Maki RG, Ou SH, Dezube BJ, Janne PA, Costa DB, Varella-Garcia M, Kim WH, Lynch TJ, Fidias P, Stubbs H, Engelman JA,

Sequist LV, Tan W, Gandhi L, Mino-Kenudson M, Wei GC, Shreeve SM, Ratain MJ, Settleman J, Christensen JG, Haber DA, Wilner K, Salgia R, Shapiro GI, Clark JW, Iafrate AJ (2010) Anaplastic lymphoma kinase inhibition in non-small-cell lung cancer. N Engl J Med 363(18):1693–1703. doi:10.1056/NEJMoa1006448
34. Shaw AT, Solomon B (2011) Targeting anaplastic lymphoma kinase in lung cancer. Clin Cancer Res 17(8):2081–2086. doi:10.1158/1078-0432.CCR-10-1591
35. Lindeman NI, Cagle PT, Beasley MB, Chitale DA, Dacic S, Giaccone G, Jenkins RB, Kwiatkowski DJ, Saldivar J-S, Squire J, Thunnissen E, Ladanyi M (2013) Molecular testing guideline for selection of lung cancer patients for EGFR and ALK tyrosine kinase inhibitors: guideline from the College of American Pathologists, International Association for the Study of Lung Cancer, and Association for Molecular Pathology. J Mol Diagn 15(4):415–453
36. Sharma SV, Bell DW, Settleman J, Haber DA (2007) Epidermal growth factor receptor mutations in lung cancer. Nat Rev Cancer 7(3):169–181. doi:10.1038/nrc2088
37. Paz-Ares L, Soulieres D, Melezinek I, Moecks J, Keil L, Mok T, Rosell R, Klughammer B (2010) Clinical outcomes in non-small-cell lung cancer patients with EGFR mutations: pooled analysis. J Cell Mol Med 14(1–2):51–69. doi:10.1111/j.1582-4934.2009.00991.x
38. Young EC, Owens MM, Adebiyi I, Bedenham T, Butler R, Callaway J, Cranston T, Crosby C, Cree IA, Dutton L, Faulkes C, Faulkner C, Howard E, Knight J, Huang Y, Lavender L, Lazarou LP, Liu H, Mair D, Milano A, Sandell S, Skinner A, Wallace A, Williams M, Spivey V, Goodall J, Frampton J, Ellard S, Clinical Molecular Genetics Society (CMGS) Scientific Subcommittee (2013) A comparison of methods for EGFR mutation testing in non-small cell lung cancer. Diagn Mol Pathol 22(4):190–195. doi:10.1097/PDM.0b013e318294936c
39. Wong AT, To RM, Wong CL, Chan WK, Ma ES (2013) Evaluation of 2 real-time PCR assays for in vitro diagnostic use in the rapid and multiplex detection of EGFR gene mutations in NSCLC. Diagn Mol Pathol 22(3):138–143. doi:10.1097/PDM.0b013e31827fedcc
40. Wang F, Fu S, Shao Q, Zhou YB, Zhang X, Zhang X, Xue C, Lin JG, Huang LX, Zhang L, Zhang WM, Shao JY (2013) High EGFR copy number predicts benefits from tyrosine kinase inhibitor treatment for non-small cell lung cancer patients with wild-type EGFR. J Transl Med 11(1):90. doi:10.1186/1479-5876-11-90
41. Hudis CA (2007) Trastuzumab — mechanism of action and use in clinical practice. N Engl J Med 357(1):39–51. doi:10.1056/NEJMra043186
42. Telli ML, Hunt SA, Carlson RW, Guardino AE (2007) Trastuzumab-related cardiotoxicity: calling into question the concept of reversibility. J Clin Oncol 25(23):3525–3533. doi:10.1200/JCO.2007.11.0106
43. Di Palma S, Collins N, Faulkes C, Ping B, Ferns G, Haagsma B, Layer G, Kissin MW, Cook MG (2007) Chromogenic in situ hybridisation (CISH) should be an accepted method in the routine diagnostic evaluation of HER2 status in breast cancer. J Clin Pathol 60(9):1067–1068. doi:10.1136/jcp.2006.043356
44. Penault-Llorca F, Bilous M, Dowsett M, Hanna W, Osamura RY, Ruschoff J, van de Vijver M (2009) Emerging technologies for assessing HER2 amplification. Am J Clin Pathol 132(4):539–548. doi:10.1309/AJCPV2I0HGPMGBSQ
45. Feliubadalo L, Lopez-Doriga A, Castellsague E, del Valle J, Menendez M, Tornero E, Montes E, Cuesta R, Gomez C, Campos O, Pineda M, Gonzalez S, Moreno V, Brunet J, Blanco I, Serra E, Capella G, Lazaro C (2013) Next-generation sequencing meets genetic diagnostics: development of a comprehensive workflow for the analysis of BRCA1 and BRCA2 genes. Eur J Hum Genet 21(8):864–870. doi:10.1038/ejhg.2012.270
46. Gardner RJM, Sutherland GR, Shaffer LG (2011) Chromosome abnormalities and genetic counselling. Oxford monographs on medical genetics, 4th edn, vol 61. Oxford University Press, New York
47. Miller DT, Adam MP, Aradhya S, Biesecker LG, Brothman AR, Carter NP, Church DM, Crolla JA, Eichler EE, Epstein CJ, Faucett WA, Feuk L, Friedman JM, Hamosh A, Jackson L, Kaminsky EB, Kok K, Krantz ID, Kuhn RM, Lee C, Ostell JM, Rosenberg C, Scherer SW, Spinner NB, Stavropoulos DJ, Tepperberg JH, Thorland EC, Vermeesch JR, Waggoner DJ, Watson MS, Martin CL, Ledbetter DH (2010) Consensus statement: chromosomal microarray is a first-tier

clinical diagnostic test for individuals with developmental disabilities or congenital anomalies. Am J Hum Genet 86(5):749–764. doi:10.1016/j.ajhg.2010.04.006
48. Schaaf CP, Wiszniewska J, Beaudet AL (2011) Copy number and SNP arrays in clinical diagnostics. Annu Rev Genomics Hum Genet 12:25–51. doi:10.1146/annurev-genom-092010-110715
49. Yu S, Graf WD, Shprintzen RJ (2012) Genomic disorders on chromosome 22. Curr Opin Pediatr 24(6):665–671. doi:10.1097/MOP.0b013e328358acd0
50. Kronquist KE, Sherman SL, Spector EB (2008) Clinical significance of tri-nucleotide repeats in Fragile X testing: a clarification of American College of Medical Genetics guidelines. Genet Med 10(11):845–847. doi:10.1097/GIM.0b013e31818b0c8a
51. Sutherland GR (1977) Fragile sites on human chromosomes: demonstration of their dependence on the type of tissue culture medium. Science 197(4300):265–266
52. Harvey J, Judge C, Wiener S (1977) Familial X-linked mental retardation with an X chromosome abnormality. J Med Genet 14(1):46–50
53. Human Genetics Society of Australasia (2012) Best practice Fragile X testing and analysis guidelines for Australasian laboratories
54. Spector EB, Kronquist KE (2006) Technical standards and guidelines for Fragile X testing. ACMG Standards and Guidelines for Clinical Genetics Laboratories
55. Hammond LS, Macias MM, Tarleton JC, Shashidhar Pai G (1997) Fragile X syndrome and deletions in FMR1: new case and review of the literature. Am J Med Genet 72(4):430–434
56. Wang YC, Lin ML, Lin SJ, Li YC, Li SY (1997) Novel point mutation within intron 10 of FMR-1 gene causing fragile X syndrome. Hum Mutat 10(5):393–399. doi:10.1002/(SICI)1098-1004(1997)10:5<393::AID-HUMU10>3.0.CO;2-V
57. Tabor A, Alfirevic Z (2010) Update on procedure-related risks for prenatal diagnosis techniques. Fetal Diagn Ther 27(1):1–7. doi:10.1159/000271995
58. Scott F, Peters H, Boogert T, Robertson R, Anderson J, McLennan A, Kesby G, Edelman D (2002) The loss rates for invasive prenatal testing in a specialised obstetric ultrasound practice. Aust N Z J Obstet Gynaecol 42(1):55–58
59. Klinger K, Landes G, Shook D, Harvey R, Lopez L, Locke P, Lerner T, Osathanondh R, Leverone B, Houseal T et al (1992) Rapid detection of chromosome aneuploidies in uncultured amniocytes by using fluorescence in situ hybridization (FISH). Am J Hum Genet 51(1):55–65
60. Wapner RJ, Martin CL, Levy B, Ballif BC, Eng CM, Zachary JM, Savage M, Platt LD, Saltzman D, Grobman WA, Klugman S, Scholl T, Simpson JL, McCall K, Aggarwal VS, Bunke B, Nahum O, Patel A, Lamb AN, Thom EA, Beaudet AL, Ledbetter DH, Shaffer LG, Jackson L (2012) Chromosomal microarray versus karyotyping for prenatal diagnosis. N Engl J Med 367(23):2175–2184. doi:10.1056/NEJMoa1203382

Genomic Analysis

Sally M. Hunter, Amy E. McCart Reed, Ian G. Campbell, and Kylie L. Gorringe

Introduction

Cancer is a genetic disease, from the predisposing alleles carried in the constitutive genome to the random somatic events selected for during tumorigenesis. In the last 15 years, the analysis of cancer genomes has dramatically improved in scope and level of detail. Low resolution and low-throughput methods such as G-banded karyotyping and comparative genomic hybridization have been superseded, first by array-based and more recently by sequencing-based technologies that enable affordable genome-wide single nucleotide resolution analysis of hundreds and even thousands of tumors. In a research setting this has led to novel insights regarding the initiation and evolution of cancer, and the genetic events detected are increasingly having clinically relevant implications.

This chapter introduces the main classes of genetic events that are commonly seen in cancer genomes and discusses the contemporary methodologies with which they are detected. Applying these methods has led to a number of discover-

S.M. Hunter
Cancer Genetics Laboratory, Peter MacCallum Cancer Centre, East Melbourne, VIC, Australia

A.E. McCart Reed
The University of Queensland, UQCCR, Herston, QLD, Australia

I.G. Campbell • K.L. Gorringe (✉)
Cancer Genetics Laboratory, Peter MacCallum Cancer Centre, East Melbourne, VIC, Australia

Sir Peter MacCallum Department of Oncology, University of Melbourne, Parkville, VIC, Australia

Department of Pathology, University of Melbourne, Parkville, VIC, Australia
e-mail: ian.campbell@petermac.org; kylie.gorringe@petermac.org

S.R. Lakhani, S.B. Fox (eds.), *Molecular Pathology in Cancer Research*,
DOI 10.1007/978-1-4939-6643-1_5

ies with implications for molecular pathology, including using genetic events to evaluate cancer risk, refine diagnoses, provide prognostic information, and most critically, determining genetic events against which molecular therapeutics can be targeted.

Classes of Genetic Events in Cancer

Cancer has long been recognized as a genetic disease, since the earliest observations of deranged chromosomes [1] and familial clustering of cases [2–5]. Predisposition to cancer, initiating events, and progression are all influenced by genetics whether they be constitutive or somatic aberrations. There are many different types of genetic alteration that can occur, each of which arise through different mechanisms and each having varying consequences. The vast majority of somatic changes that occur in a tumor are thought to have little functional effect and are consequently described as "passengers," carried along by coincidence upon selection of a co-existing "driver" in the same cell [6, 7]. Discerning driver mutations from passenger mutations remains a major challenge in translating genomic data into the clinic.

Somatic Mutation

Acquired changes in the constitutional DNA sequence are common in most cancer types and include base-pair substitutions and small (<1 kb) insertion–deletions (indels). Mutations are caused by a failure of one or more of the DNA repair pathways to recognize or accurately repair DNA following a genetic insult, which can include inherent replication errors, deamination of methylated cytosine, and mutagenic exposures such as UV light. The rate of such mutations per cancer genome varies greatly depending on the cancer type and has been estimated at 0.57/Mb of coding sequence for acute lymphoblastic leukemia [7], 0.19/Mb for breast cancer [7], ~1.8/Mb for high grade ovarian cancer [8], ~18/Mb for mutagen-exposed cancers like melanoma and non-small cell lung cancer [7, 9], and as high as 400/Mb for cancers with a DNA repair defect such as loss of mismatch repair in colorectal cancer [10]. Different cancer types often have unique mutation signatures in terms of the type of mutation, e.g. UV-exposed cancers have high rates of C:G>T:A transitions resulting in an enrichment of dipyrimidines.

The impact of a somatic mutation will vary widely depending on its location (coding/non-coding/splice site/regulatory) and type of change following translation (missense, nonsense, frameshift, etc.) (Fig. 1). Some mutations will have an immediate impact and are considered dominant while others may require loss of the remaining normal allele.

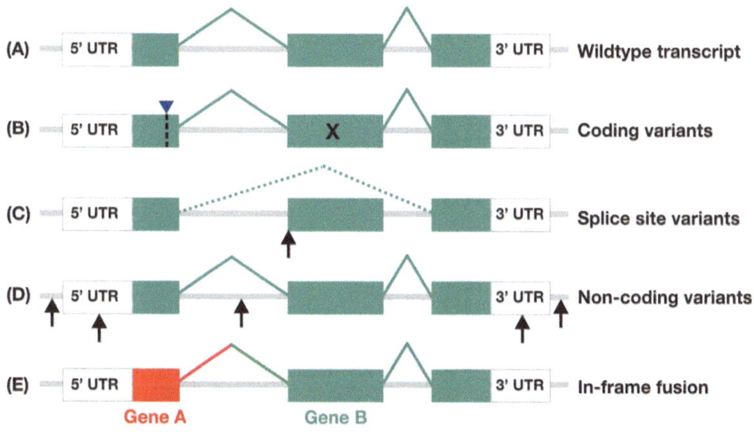

Fig. 1 Mutation types. (**a**) Wildtype transcript. *Shaded boxes* depict coding exons, *white boxes* depict the untranslated regions (UTRs) of the transcript, and intervening *grey lines* indicate intergenic and intronic regions (**b**) Coding variants: Frameshift (*triangle*) and nonsense (*X*) variants are often overtly deleterious due to protein truncation. Missense variants may be deleterious depending on the function of the specific amino acid changed and the effect on protein folding, or have no effect. Synonymous mutations do not change the amino acid identity, but may influence splice site function or binding of regulatory proteins, or have no effect. (**c**) Essential splice site variants (2 bp ± intron–exon boundary, *arrow*) can result in exon skipping or cryptic exons being transcribed. (**d**) Non-coding variants (intergenic, UTRs, intronic, *arrows*) may have an effect on transcription regulatory regions, transcript splicing, and mRNA stability, or often no effect on transcript function. (**e**) Translocations resulting in an in-frame fusion can produce functional protein

Copy Number

An abnormal chromosome number (aneuploidy) is a common feature of many carcinomas. The subsequent imbalance in gene copies is thought to lead to global changes in gene expression with wide-ranging effects on cell phenotype. Aneuploidy is caused by errors in chromosome segregation during mitosis or cytokinesis, leading to gain or loss of whole chromosomes, and not uncommonly duplication of the entire chromosome complement leading to tetraploidy.

Copy number aberrations can also occur at a sub-chromosomal level through various mechanisms, often involving compromised repair of double strand (ds) DNA breaks [11, 12], a breakage-fusion-bridge cycle subsequent to dsDNA breaks or telomere dysfunction [13], and less commonly chromothripsis [14]. Copy number changes include losses of material, either hemizygous or homozygous deletions, and gains of material, which can be low level changes, such as a duplication, or high-level amplification (from five to possibly hundreds of copies) (Fig. 2), as is often observed with *ERBB2* in breast cancer.

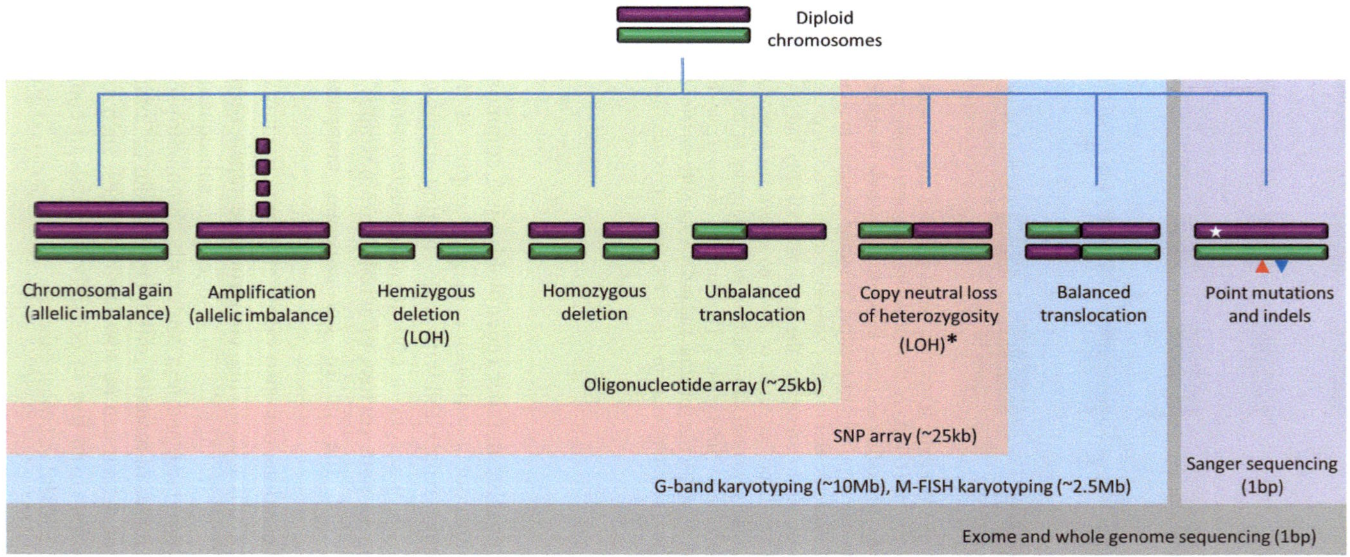

Fig. 2 Copy number aberrations and mutations. The *boxes* indicate the types of aberrations each technology is capable of detecting. The resolution of the technology indicates the smallest size of aberrations that can be robustly detected. (*Asterisk*) Detected as a copy number aberration by array methods, however, the site of the translocation fusion cannot be determined

Loss of Heterozygosity

Loss of heterozygosity (LOH) refers to the change in genotype from heterozygous to homozygous of polymorphic alleles that arises through chromosome loss, sub-chromosomal deletion or gene conversion via homologous recombination and DNA repair. LOH is often associated with copy number loss; however, gene conversion or duplication of chromosomes can lead to copy number neutral LOH (Fig. 2). LOH is distinct from allelic imbalance (AI), in which both alleles are still present but in a ratio different from 1:1 following copy number gain. The effect of LOH is to unmask recessive alleles, either inherited (e.g., *BRCA1*) or somatic (e.g., *TP53*), leading to loss of tumor suppressor gene function.

Structural Chromosome Changes

Any event involving inappropriate repair of a dsDNA break can lead to structural changes in chromosomes, including not only the aforementioned sub-chromosomal copy number changes, but also inversions and translocations. These latter events lead to the novel juxtaposition of genetic material, which can cause inappropriate gene regulation or novel protein products, such as the *BCR-ABL* translocation in chronic myelogenous leukemia [15, 16].

Germline Variation

The constitutive genetic variation carried by individuals is extensive and encompasses many of the same forms observed as somatic events. The most common class of germline aberrations most relevant to cancer are considered to be single nucleotide polymorphisms (SNPs) and small indels. It is likely that larger copy number variations and structural changes such as inversions are also important, but there are few conclusive incidences reported to date. SNPs and indels can vary widely in population frequency, from common (>10 % minor allele frequency) to rare (1–10 % frequency) to extremely rare (<1 % frequency). Most are inherited from parents, but some occur *de novo*, at a frequency estimated at 13×10^{-3}/Mb for SNPs and 0.78×10^{-3}/Mb for indels per generation [17, 18].

Methods of Genomic Analysis

High-resolution screening for genetic lesions on a genome-wide scale has only recently become feasible, i.e., on a kilobase-down to base-pair level. Prior to the invention of array and massively parallel sequencing-based methods (MPS; also referred to as next-generation sequencing, Sanger sequencing is considered first generation sequencing), analysis resolution was limited to tens of megabase-pairs using comparative genomic hybridization (CGH) or G-banded karyotyping. In addition, these methods were time-consuming and required a high degree of individual skill to interpret the chromosome spreads, limiting the number of samples that could be studied. Other, more targeted techniques could be readily applied to multiple samples such as fluorescence in situ hybridization (FISH), Sanger sequencing, and microsatellite genotyping by PCR, but these were not easily scaled up to a genome-wide analysis. These methodologies have now been supplemented by several whole genome techniques that have delivered myriad research findings and are increasingly been applied in clinical settings (Table 1).

Karyotyping

The advent of fluorescence-based chromosome painting in the 1990s enabled a more automated karyotyping procedure compared to traditional G-banding. Variously known as Spectral karyotyping (SKY), M-FISH, and 24-color FISH, this technique uses paints made from individual flow-sorted chromosomes each labeled with a different mix of fluorophores. This paint is applied to metaphase chromosome spreads usually generated from primary tumors after short term cell culture. SKY is able to resolve complex marker chromosomes and is the only method discussed here that can measure exact ploidy (Fig. 3a). It can also give some indication of tumor genetic heterogeneity, as each nucleus is individually analyzed. However, it is still a low-resolution method (~10 Mb) and relies heavily on good quality metaphase spreads. Thus its use is limited to fresh tumor material and to laboratories with a cell culture facility.

Array-Based

All array-based methods for genomic analysis operate on the same basic principle: hybridization of a labeled DNA sample to complementary probe sequences that are immobilized to a solid substrate in known locations. The strength of the signal is proportional to the amount of each target sequence in the sample, however, with most platforms, the dynamic range tends to become compressed at high copy number and generally cannot accurately distinguish, for example, 8 copies from 12 copies.

Array-based systems all require some kind of normalization strategy to calculate copy number, where the signal intensity of the tumor sample is converted to a ratio using normal diploid samples. Normalization can be performed against matched normal DNA, which is useful for discriminating constitutional copy number variants, or against the average of multiple normal DNA samples. All data for a sample

Table 1 Methods of genomic analysis

Platform	Detects	Advantages	Disadvantages
Spectral karyotyping (SKY), M-FISH	Ploidy; translocations	Single-cell analysis; only technology not requiring comparison to normal baseline	Requires cell culture; low resolution
Multiplex ligation probe amplification (MLPA, e.g., MRC-Holland)	Ploidy; LOH	PCR-based; fast; inexpensive; simple analysis	Low resolution; each fragment requires a unique primer pair
BAC aCGH	CN	Internal control from two-color analysis	Low resolution; irregular spacing
SNP arrays (e.g., Affymetrix, Ilumina)	CN; LOH; SNP genotyping	Detect copy number neutral LOH; genotyping	Irregular probe spacing
Oligonucleotide arrays (e.g., Agilent, NimbleGen)	CN	Relatively cheap	Cannot detect copy number neutral LOH events
Molecular inversion probe arrays (e.g., Affymetrix)	CN; LOH; SNP genotyping; limited mutations	Applicable to low-quality and limited DNA; good dynamic range	Lower-density arrays
Exome sequencing (e.g., Illumina, NimbleGen)	SNP; somatic mutation data; LOH; CN	Comprehensive analysis	Complex bioinformatics to extract CN; expensive; need high quality DNA; need matching normal DNA; irregular spacing; risk of incidental findings
Whole genome sequencing (low read depth)	CN; translocations	Good dynamic range	Expensive; low sensitivity for SNP and mutation data
Whole genome sequencing (high read depth)	SNP; somatic mutation data; LOH; CN; translocation	Comprehensive analysis; base-pair resolution of breakpoints	Very expensive; complex bioinformatics; risk of incidental findings

M-FISH multi-color fluorescence in-situ hybridization, *BAC* bacterial artificial chromosome, *aCGH* array comparative genomic hybridization, *CN* copy number, *LOH* loss of heterozygosity, *SNP* single nucleotide polymorphism

are median- or mean-centered, which means that it is not possible to distinguish between perfectly tetraploid and diploid samples, and exact ploidy cannot be determined unless genotype information is also available (see below).

The level of genomic resolution of array-based platforms is inherently limited by the number and type of probe sequences selected. Initially, detection of copy number was done using cDNA arrays produced for expression analysis but these were quickly superseded by superior platforms including bacterial artificial chromosome (BAC) arrays, oligonucleotide arrays, and SNP arrays.

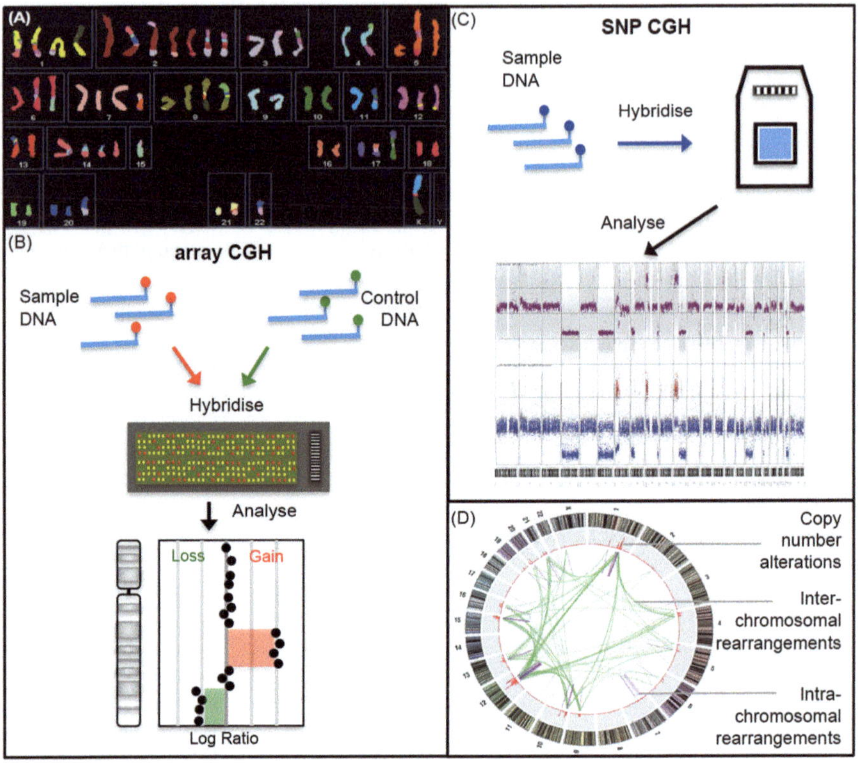

Fig. 3 Genomic analysis data. (**a**) Spectral karyotyping of the breast cancer cell line VP229, demonstrating aneuploidy and extensive translocations and structural rearrangements. (**b**) Array CGH method overview, with data analysis of one representative chromosome (log ratio data plotted), indicating where regions of chromosome have been gained (*red*) or lost (*green*). (**c**) SNP CGH method overview, with Partek® plot of Affymetrix® SNP6™ data from an ovarian tumor, showing chromosomes linearly mapped from 1 through X showing total copy number (*upper*) and allele-specific copy number (*lower*). (**d**) *Circos plot* demonstrating typical genomic copy number aberrations and structural rearrangements in the 94778 cell line derived from a retroperitoneal relapse of a well-differentiated liposarcoma [19] 94778 cells were provided by Florence Pedeutour (Laboratory of Solid Tumors Genetics, Nice University Hospital, Nice, France). The data was analyzed and figure generated by Anthony Papenfuss (Bioinformatics Division, The Walter and Eliza Hall Institute of Medical Research, Parkville, VIC, Australia, courtesy of D. Thomas and D. Garsed. *CGH* comparative genomic hybridization, *SNP* single nucleotide polymorphism

BACs are large plasmid vectors with inserts of tens to hundreds of kilobases but because of the way they are constructed, they tend to be irregularly spaced across the genome, although whole genome tiling arrays with better uniformity have been produced. The resolution of BAC arrays is also limited by the large insert size. However, the signal-to-noise ratio and dynamic range are generally good.

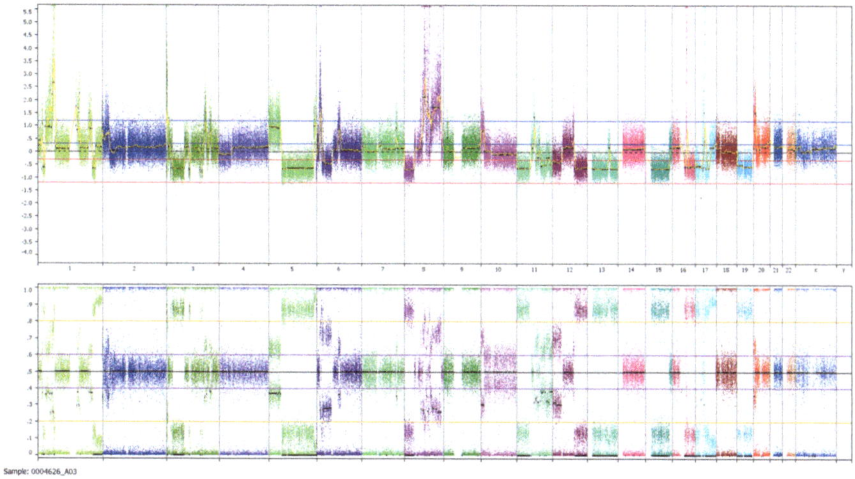

Fig. 4 Copy number analysis of FFPE DNA. DNA extracted from ductal carcinoma in situ assayed using OncoScan™ (Affymetrix®). The *upper panel* illustrates total copy number, while the *bottom panel* illustrates allele ratios. Detectable copy number aberrations include whole chromosome and chromosome arm gains and losses, focal deletions, and high-level amplifications

As a consequence of the completion of the sequence of the entire human genome in the early 2000s, synthetic oligonucleotide-based arrays have become possible. These arrays utilize relatively short (25–60 bp) synthesized oligonucleotides, overcoming issues with appropriately spacing probes throughout the genome, and increasing the resolution of breakpoint detection. Array comparative genomic hybridization (aCGH; Fig. 3b) typically refers to the use of non-polymorphic oligonucleotide arrays, which can detect changes in total copy number but not changes in allelic ratios and therefore cannot identify copy number neutral LOH (Fig. 2). Single nucleotide polymorphism (SNP) arrays, a type of oligonucleotide array, allow the detection of both total copy number and LOH by designing probes spanning known common polymorphisms in the human genome (Fig. 3c).

Oligonucleotide-based arrays perform well on good quality DNA from fresh or frozen tissues and lymphocyte DNA; however, they are less useful for degraded DNA such as that obtained from formalin-fixed, paraffin-embedded (FFPE) tissue.

DNA from FFPE sources is predominantly highly fragmented and alternative approaches have been successfully developed for its analysis (Fig. 4). Molecular inversion probe (MIP) technology has been incorporated into the Affymetrix® OncoScan™ assay. This approach involves circularizable "padlock" probes with two terminal sequences that bind to homologous sequences either side of an SNP, followed by highly specific closing of the padlock through incorporation of a nucleotide complementary to the SNP. This approach circumvents the issue of having to directly digest, ligate, and amplify the fragmented DNA. An alternative approach, taken by Illumina®, is to "restore" the FFPE DNA by ligating the fragmented DNA together to generate fragments large enough for whole genome amplification prior to labeling and hybridization. This DNA can then be hybridized to a standard bead array.

Sequencing-Based

Sanger Sequencing

In the mid-1970s, Fredrick Sanger first described "Sanger Sequencing," and since this time huge advances in the technology and its applications have been made, with the technology notably underpinning the Human Genome Project. The current iteration of the technology is based on sequencing by synthesis, with the products resolved using capillary electrophoresis and laser optics. Selective incorporation of fluorescently labeled, chain terminating dideoxynucleotides occurs during modified PCR amplification, and following purification of the products, they are resolved based on size using a capillary sequencer. Lasers excite the fluorescent dyes, and sequences of up to 700–900 bases can be determined, which are exported as a chromatogram (Fig. 5a).

Massively Parallel Sequencing

The advent of massively parallel sequencing (MPS), or the so-called next-generation sequencing, represents an enormous step forward in the genomics field. While continuing to "sequence by synthesis," these technologies vastly increase throughput by simultaneously sequencing multiple DNA strands (in "parallel"). Unlike Sanger sequencing, MPS generates millions of random short reads (35–700 bp) that must then be mapped to a reference genome (Fig. 5b). The most commonly employed technologies in cancer research are ion semi-conductor sequencing and optics-based dye sequencing. Semi-conductor sequencing (Ion Torrent Systems Inc.) measures the hydrogen release following incorporation of a nucleotide as determined by the template strand of DNA. Conversely, dye sequencing (Illumina®) relies on the immobilization of template DNA clusters onto a solid surface, upon which fluorescently labeled nucleotides competitively bind; a laser excites the label, and images of incorporated bases are recorded. Prior to sequencing, DNA samples are prepared into libraries, representing all of the desired DNA target sequences, be they the entire genome (whole genome sequencing, WGS), only the exons (whole exome sequencing, WES), or a targeted panel (Fig. 6). A targeted panel allows for specific enrichment of certain sequences; the enrichment can be performed using either a nucleic acid bait ("capture") or through PCR-based amplification. It is at the library preparation stage that samples can be barcoded and pooled to increase throughput.

Using MPS, it is possible to simultaneously detect single nucleotide mutations and SNPs, as well as small indels, large copy number aberrations, LOH, and structural rearrangements, depending on the type of sequencing performed (Fig. 2). Variants are identified ("called") by programs that assess the sequence evidence for the particular variant (read depth, base quality, etc.) and are exported into a mutation annotation format (MAF) file. This calling procedure is typically performed against a matched normal for the detection of somatic variants as the number of non-reference

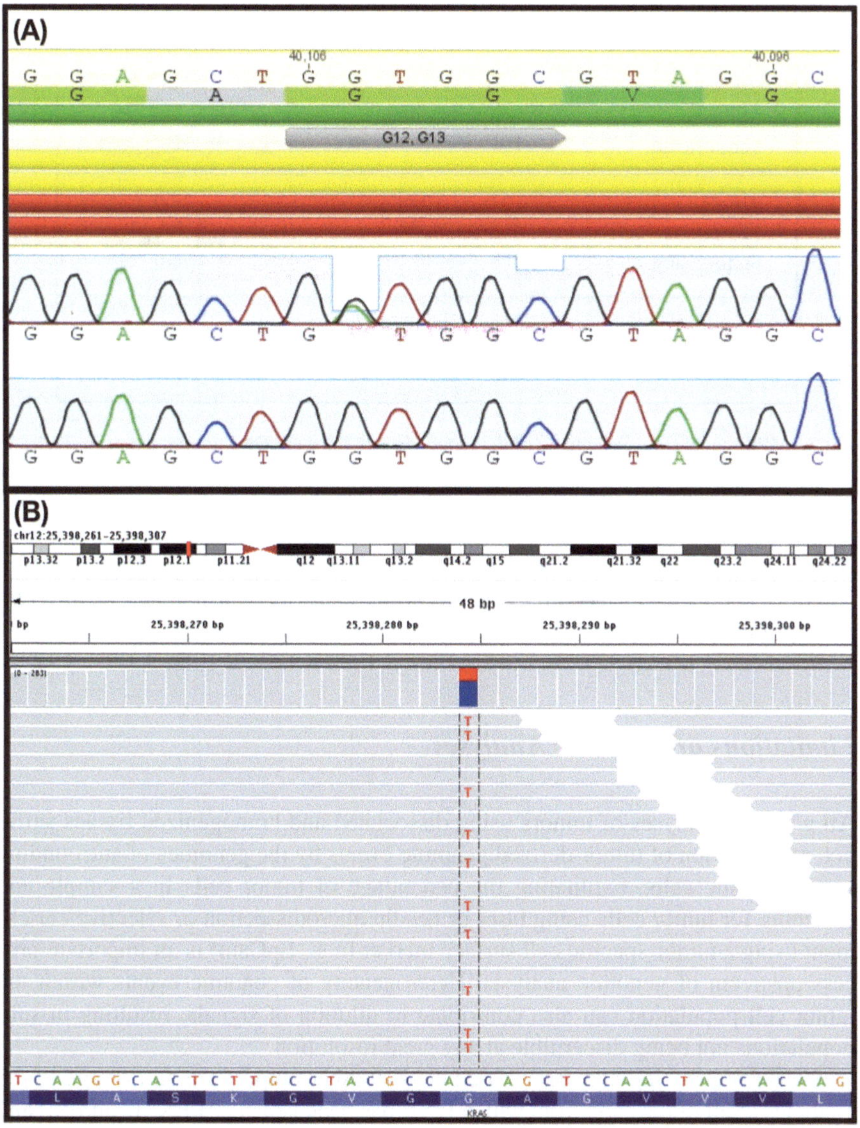

Fig. 5 Sequencing data. (**a**) Geneious (Biomatters) browser view of Sanger sequencing traces of wildtype *KRAS* (*lower*) and *KRAS* c.34G>A, p.G12D mutant (*upper*). (**b**) IGV browser view of MPS reads mapped to *KRAS* demonstrating wildtype and mutant (T) reads (*KRAS* c.34G>A, p.G12D mutant)

germline variants in any given individual is substantial. Copy number and LOH events are assessed by comparing read depths and allele frequencies from the tumor to those from the matched normal sample. Structural rearrangements (translocations) and large indels are identified by assessing paired reads that did not map

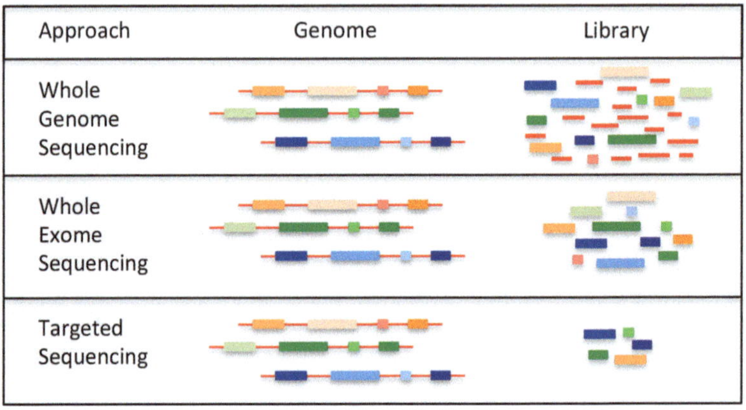

Fig. 6 MPS sequencing. Schematic of the input genome (intronic DNA in *red*, exons in shades of *blue*, *green* and *orange*) and the sequences represented in libraries generated for the three major types of sequencing; whole genome, whole exome, and targeted

within the expected distance from each other (determined by the average fragment length of the DNA library prepared for sequencing) or that mapped to different chromosomes (Fig. 3c).

Limitations of Genomic Analyses

All genomic analyses of tumors using dissociated and homogenized tissues suffer potential dilution of tumor-derived genomic events by the genomes of surrounding non-neoplastic cells. Estimating the percentage of tumor cells in a sample and enriching for tumor cells using laser or needle microdissection or selective enrichment using a tumor-specific cell surface marker (e.g., EpCam) is an important process upstream of genomic analysis. Heterogeneity of genomic events within the tumor cell population can also contribute to dilution of signals, resulting in subpopulations not being discernible at any great resolution.

Sanger sequencing and array-based copy number outputs are an average of the genomes of all of the cells from which DNA has been isolated (Figs. 3b and 5a) and therefore have limited sensitivity to detect events occurring only in a subpopulation of tumor cells. MPS, particularly at very high read depth, offers greater scope to resolve events occurring in a small subpopulation of cells and gives a digital count of variant reads (Fig. 5b). Paired-end sequencing also allows the mapping of translocation events, which cannot be resolved using array technology where chromosomes are linearly mapped (Fig. 3b, c). MPS is not without its own issues; a problematic area of MPS is the accurate mapping of reads to a reference genome and the calling of variants. Reads become difficult to correctly map to areas of the genome that are highly homologous to other regions, have repetitive sequence, or when there is an indel in the read or region relative to the reference.

Whole genome and exome sequencing provide the possibility of simultaneously detecting all variants in the genome or all coding variants. Along with real variants, PCR and sequencing artifacts are also detected, resulting in a huge number of potential variants to analyze for validity, recurrence, and functional impact. Differentiating between mutations that are driving tumorigenesis ("drivers") from those that are not anticipated to have any involvement in the development of a tumor ("passengers") is difficult and is made more onerous because passenger mutations are predicted to far outnumber driver mutations. The bioinformatic analysis burden of MPS should not be underestimated, although efforts to improve software design and usability are ongoing (see Chapter "Bioinformatics Analysis of Sequence Data").

Applications of Genomic Analysis in Cancer

Genome-Wide Association Studies

One of the most common uses of SNP arrays has been to the application of genome-wide association studies (GWAS), where linkage of SNPs to an increased (or decreased) risk of disease is assessed across the genome. Due to cost constraints, these studies are most often performed in a staged process, where a few hundred or thousand individuals with features suggestive of a genetic predisposition to cancer such as family history or early age of onset are first compared to age- and ethnicity-matched controls. Significant hits from this analysis are then validated in tens of thousands of cases and controls. This strategy has been applied to all common cancer types, with multiple predisposing SNPs identified in breast, colorectal, lung, and ovarian cancers. The risks associated with individual SNPs are usually low (1.2–1.5-fold above the general population), however, the polygenic risk when multiple low-risk SNPs are inherited together can reach much greater significance. In addition, because the resolution of the studies is limited by the array density, the SNPs with the highest risk association may not be the causative variant, but only closely linked. Thus, fine mapping is required for more precise information on the gene affected and the possible mechanism of the increase in risk. Nonetheless, some risk alleles, which have been validated in multiple independent cohorts, are now being utilized for testing in familial cancer clinics [20].

More recently, as the cost of MPS continues to fall, exome and genome sequencing of large cohorts are being undertaken. For example, large databases have been established with a focus on thoroughly characterizing common cancers [International Cancer Genome Consortium (ICGC), The Cancer Genome Atlas (TCGA)] and providing a population baseline for common and rare variants [Exome Variant Server (EVS), 1000 Genomes, dbSNP, Exome Aggregation Consortium (ExAC)]. These studies will have the advantage of being able to detect rare variants and causative alleles; however, it may be some years before they are powerful enough to identify rare, low to moderate risk alleles.

Mapping of Oncogenes and Tumor Suppressor Genes

A major goal of genomic analyses in the research setting is the discovery of the full repertoire of genes with a role in tumorigenesis. These genes can be tumor promoting when their activity is deregulated (oncogenes) or when they are inactivated (tumor suppressors [TSG]). Some genes can act as either oncogene or TSG depending on the cellular context and the pathways driving tumorigenesis; for example, *NOTCH1* is targeted by activating mutations in hematopoietic malignancies [21] and inactivating mutations in solid tumors such as head and neck squamous cell carcinoma [22].

Oncogenes commonly act in a dominant fashion, with the genetic aberration ranging from copy number increase (e.g. *ERBB2*), recurrent activating point mutation (e.g. *KRAS*, *BRAF*) (Fig. 7a, b), translocation (e.g. *BCR-ABL*), or other structural chromosomal changes leading to loss of transcriptional (e.g. *MYC*) or post-translational control (e.g. *EGFR*). These types of recurrent activating events typically make oncogenes easier to design clinical tests for (compared to TSGs) because there is a limited number of functionally relevant mutational events.

Methods for discovering new oncogenes include mapping regions of copy number gain, exome sequencing for somatic mutations, and karyotyping or genome sequencing for structural chromosome changes. Regardless of methodology, a common challenge is distinguishing the driving genetic events from benign passenger events.

Identifying genes affected by copy number alterations has been most effectively achieved using array technologies. The increases in copy number are mapped in multiple samples, and those regions of the genome that most often display increases in copy number are short-listed as potential sites of oncogenes. However, this is complicated by the degree of copy number change—should any increase be investigated even if only a single copy, or should only high-level amplifications be considered? Both methods have been applied, and bioinformatic techniques that balance both possibilities have been developed (e.g. GISTIC [23, 24]). The list of genes in minimal regions of copy number change can still be long, and expression and functional analyses are then required to identify putative drivers. For example, integrated copy number and expression analysis identified novel growth promoting genes in ovarian carcinomas [25] and candidate oncogenes driving ovarian cancer were functionally investigated using RNA interference [26].

Full genome sequencing is the most comprehensive and sensitive method to identify structural chromosome changes, although to date fusion genes have also been detected using the much cheaper approach of RNAseq to short-cut to those translocations with an expressed gene moiety. For example, RNAseq analysis identified the MHC II transactivator CIITA as a recurrent fusion partner in lymphoid cancers [27].

Tumor suppressor genes are characterized by loss-of-function genetic events (Fig. 7c, d). Apart from a few examples where dominant negative mutations can be selected for (e.g. *TP53*), it is usually expected that both copies of a tumor suppressor gene must be inactivated, either through bi-allelic point mutation, homozygous deletion, methylation, or a combination of mutation and LOH. Thus, mapping of

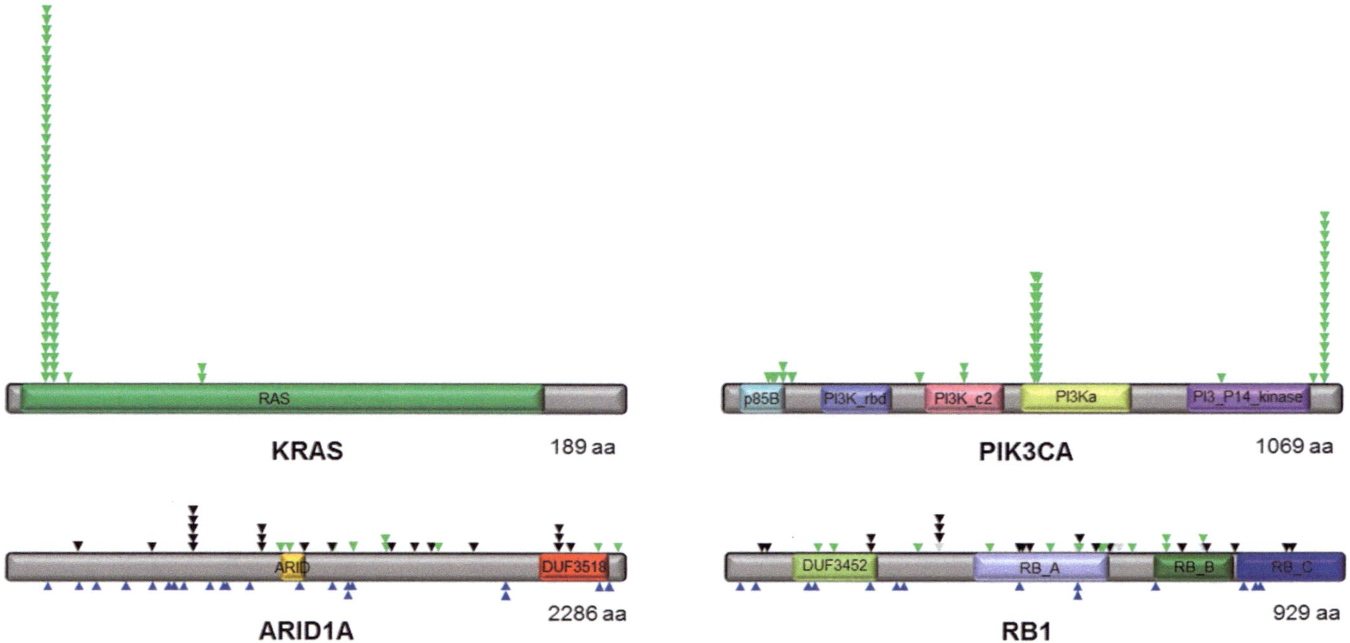

Fig. 7 Oncogene and tumor suppressor gene mutation patterns. Distribution of mutations in the oncogenes (**a**) *KRAS* and (**b**) *PIK3CA* and TSGs (**c**) *ARID1A* and (**d**) *RB1*. *Green arrowheads* indicate missense mutations, *black arrowheads* indicate nonsense mutations, *grey arrowheads* synonymous mutations, while *blue arrowheads* indicate truncating indels. Mutation patterns are based on 50 randomly selected mutations from the COSMIC database (COSMIC http://cancer.sanger.ac.uk/cosmic)

copy number loss and LOH has been applied to try and identify new TSGs. While early successes included genes where the initial mutation event was inherited (e.g. *RB1* [28]), in the genomic age there have been very few genes identified through this method [29].

Exome sequencing studies have been applied to multiple cancer types to identify both oncogenes and tumor suppressor genes. Initially, only a small number of samples were investigated, with candidates followed up in larger cohorts (e.g., CANgenes [30]). More recently, as sequencing has become relatively cheap, cohorts of hundreds of samples have been analyzed. Interestingly, apart from a few histologically defined tumor types (e.g., granulosa cell tumors) there have been very few genes identified that are mutated at high frequency. It seems that for solid tumors each tissue type has 1–5 commonly mutated genes (>10 % frequency), and a long tail of genes each with a mutation frequency of just a few percent. Thus, the issue of identifying drivers versus passengers is again a problem. One strategy used to enrich for potential driver mutations is the employment of algorithms to predict the deleteriousness of an SNV given the nature of the amino acid change, the position within the protein sequence and the level of conservation of the protein sequence compared to other species [31–33]. Another strategy is the use of statistical methods to assess the mutation rate for a given gene relative to the background mutation rate and gene size [34, 35]. Increasingly, gene discovery studies are applying algorithms to identify common pathways that are affected which can assist in identifying the likely driver genes. For example, pathway analysis identified axonal guidance pathway aberrations in pancreatic cancer, revealing novel tumorigenic roles for these proteins [36].

Association of Genetic Events with Clinical Features

Genetic events are intrinsic to the development of malignant characteristics, thus, it is logical to assume that differences in clinical behavior may be attributed to specific genetic aberrations. Many studies have investigated the association of clinical with genetic features on a genome-wide scale. Associated features may then assist in prediction and risk management, diagnosis, prognosis, or treatment.

Germline Predisposition to Cancer

Many of the well-known cancer predisposition genes, such as *APC* (familial adenomatous polyposis), *BRCA1* and *BRCA2* (hereditary breast and ovarian cancer), and *MLH1* (hereditary non-polyposis colon cancer), were identified through linkage analysis and candidate gene approaches [37–40]. This was possible because of their relative commonness and high penetrance in these hereditary conditions. Identifying additional candidates is now primarily undertaken through large-scale exome and genome sequencing of multiple members of high-risk families. However, the task of identifying these genes remains difficult since pathogenic mutations are often vanishingly rare, as encountered with *RAD51C* mutations in *BRCA1/2* mutation-negative breast/ovarian cancer families [41–43], and definitive classification as a cancer predisposition requires very large case and control cohorts to achieve sufficient power.

Genome-wide association studies (GWAS) have identified many more common genomic variants with much lower individual effect on cancer risk. Although the functionally relevant genes may not be identified, the SNPs from these studies can prove useful for risk prediction. Common risk alleles may act in concert to produce a multiplicative polygenic risk or act as risk modifiers [20].

In order to effectively incorporate new risk alleles into the clinic, current practices for genetic testing are undergoing a shift towards gene panels, where all known cancer susceptibility genes and SNPs can be sequenced simultaneously using MPS. This approach substantially decreases the time and cost per gene tested.

Molecular Subtyping and Diagnostics

Most subtyping studies have used expression microarrays to determine classes of tumors with distinct characteristics; however, it is becoming clear that these expression subtypes are often correlated with specific underlying genetic profiles. For example, a number of prognostic tests have been developed for breast cancer subtyping (e.g., OncotypeDX, MammaPrint, and PAM50) and these reflect both histological and genetic differences [44–46]. Targeted sequencing panels are increasingly being used to inform clinical decision-making by matching patients with appropriate conventional therapies or to direct patients to relevant clinical trials. Many biotechnology companies offer companion diagnostic cancer gene panels, enriched for the so-called druggable mutations and those associated with prognostication, including, for example, Illumina® (TruSight® Cancer/Tumor; TruSeq® Cancer Amplicon [47]); Foundation One™ [48]; and Ion Torrent (IonAmpliSeq™ Comprehensive Cancer/Cancer HotSpot [49]).

Prognostic Markers

Treatment of cancer tends to be aggressive, with side effects that can have a severe impact on the quality of life both in the short and long term, including radical surgery leading to scarring and loss/reduction of organ function, radiotherapy-induced burns and increased risk of subsequent malignancy, systemic cytotoxics leading to hair loss, nausea, etc. The consequences of disease progression or recurrence are sufficiently severe that these outcomes are accepted as a necessary evil. However, not all patients are at the same risk of progression, even after controlling for known prognostic factors such as stage, grade, and histological subtype. Genomic analysis has been applied to attempt to identify robust prognostic markers that may indicate that an aggressive treatment regime may not be necessary. For example, the presence of microsatellite instability in colorectal cancer has been shown to be a good prognostic indicator, identifying a proportion of colorectal tumors that do not respond to 5-fluorouracil (5-FU) systemic treatment, the mainstay of colorectal cancer systemic therapy [50, 51]. At the clinical level these data mean that individuals with stage II mismatch-deficient colorectal cancer are unlikely to be treated with systemic 5-FU treatment compared with mismatch repair proficient tumors as the clinical benefits do not outweigh the complications associated with this treatment.

Pharmacogenomics

Response to Conventional Therapies

In a similar manner to identifying prognostic markers of general tumor aggressiveness, studies have also tried to find markers that indicate a likely response to chemotherapies. Such a marker could be constitutional, for example, polymorphisms in cell transporter channels that affect the rate of drug efflux are strong determinants of chemotherapy toxicity and tolerable dosage [52–54]. Alternatively, deleterious germline mutations in *BRCA1* and *BRCA2* that are cancer predisposing paradoxically tend to improve the patient's response to treatment due to a heightened susceptibility to the DNA damage caused by chemotherapy. Alternatively, response to therapies could be tumor-intrinsic, for example, *CCNE1* gene amplification was determined to be an intrinsic resistance mechanism to platinum-taxol-based chemotherapy in high grade serous ovarian cancer [55].

Targeted Molecular Therapeutics

Recently, targeted molecular therapies have emerged with the potential to transform cancer treatment by personalizing drug regimens to the genetic "Achilles heels" of each tumor. Genome-wide analyses are key to identifying such targets in a research setting, and could be used clinically in the future, especially to identify the cause of therapy resistance. Obvious candidates for targeted therapies are over-active oncogenes as reducing activity is theoretically straightforward. Some prime examples of successful targeted treatments are imatinib (Glivec), which acts as an inhibitor of several tyrosine kinases including the BCR-ABL fusion, trastuzumab (Herceptin), targeted against overexpressed HER2 (first used in HER2+ breast cancers), and PLX 4032 (Vemurafenib), which targets the constitutively active form of BRAF (BRAF V600E) frequently mutated in melanoma. Targeted therapies for gene products where function has been lost tend to rely on unique weaknesses arising as a side-effect of the loss of gene function. For example, deleterious mutations in homologous recombination repair genes *BRCA1/BRCA2/PALB2* that impede the efficient repair of DNA double-stranded breaks leave the cells susceptible to both conventional DNA damaging chemotherapies (e.g. cisplatin) and more molecularly targeted poly ADP ribose polymerase (PARP) inhibitors, which affect alternative DNA repair pathways and impede subsequent DNA replications [56].

Despite the breakthroughs in targeted molecular therapies, these almost always induce drug resistance and are often not directly transferable to other tumor types characterized by the same mutation or pathway alteration. For example, attempts to treat BRAF V600E positive colon cancers with the same BRAF inhibitors that had been successful in melanoma resulted in poor clinical response rates due to feedback activation of EGFR in response to BRAF inhibition [57]. However, in this case combination therapy with BRAF inhibitors and EGFR or PI3K inhibitors looks more promising [58].

Cancer cells can become resistant through a range of mechanisms, finding alternative ways to activate pathways or undergoing secondary mutations that reverse susceptibility. For example, initially successful treatment of colon cancer patients with EGFR inhibitors has been found to select for cancer cells with activating *KRAS* mutations, leading to bypassing of the receptor tyrosine kinase signal and resistance to EGFR inhibition [59]. Secondary mutations in *BRCA1* and *BRCA2* have been detected in chemotherapy resistant ovarian cancers, that result in restoration of protein function through reestablishment of the reading frame, mutation of deleterious nonsense codons to missense codons, and gene conversion where the mutant allele is lost [60, 61]. Detection of these resistance mechanisms is crucial for patient prognosis and identifying effective treatments for patients to progress.

Future and Near-Term Clinical Applications

New technologies are expediting the identification of cancer driver genes and potential new therapeutic targets, leading genomics to take center stage in diagnosis, prognosis, treatment planning, and the search for new treatment options. The possibility of affordable whole genome sequencing is likely to result in many current clinical tests becoming ancillary and potentially redundant.

With the advent of accessible genome-wide molecular analysis, the molecular subtyping of all cancer types using next-generation DNA and RNA sequencing, and copy number and expression arrays is currently being realized. This offers the possibility of mutation, copy number, and expression profiles superseding histological classification, particularly concerning selection of the most effective treatment options and prediction of recurrence risk.

Whole genome and whole exome sequencing of germline and tumor DNA are likely to become standard practice, both for the identification of predisposing genetic variation and to identify molecular targets for treatment and potential resistance mechanisms. Before these technologies become standard clinical techniques, however, there are the ethical and legal hurdles of incidental findings and patents concerning certain cancer predisposition genes. In the meantime, clinical tests are being converted to these modern technologies with the creation of high-throughput panels of cancer genes.

Advancements in genomics technologies that allow very limiting amounts of DNA to be sequenced are providing future potential for real-time monitoring of treatment response and development of resistance. Isolation of circulating tumor DNA from plasma offers a non-invasive "liquid biopsy" that gives an indication of tumor burden and provides a more representative sampling of the tumor cell population than traditional core biopsies [62]. Highly sensitive monitoring of patients at the molecular level as they progress through treatment and altering treatment based on resistance mutations as they arise could drastically alter the outcome for many cancer patients. These applications are currently under investigation and offer a paradigm shift in cancer screening and treatment in the near future [62].

Acknowledgements This work was supported by the Australian National Health and Medical Research Council (NHMRC), the Cancer Council Victoria (CCV), and the Emer Casey Research Foundation.

References

1. Boveri T (2008) Concerning the origin of malignant tumours by Theodor Boveri. Translated and annotated by Henry Harris. J Cell Sci 121(Suppl 1):1–84
2. Lynch HT, Shaw MW, Magnuson CW, Larsen AL, Krush AJ (1966) Hereditary factors in cancer. Study of two large midwestern kindreds. Arch Intern Med 117:206–212
3. Li FP, Fraumeni JF Jr (1969) Soft-tissue sarcomas, breast cancer, and other neoplasms. A familial syndrome? Ann Intern Med 71:747–752
4. Macklin MT (1959) Comparison of the number of breast-cancer deaths observed in relatives of breast-cancer patients, and the number expected on the basis of mortality rates. J Natl Cancer Inst 22:927–951
5. Macklin MT (1960) Inheritance of cancer of the stomach and large intestine in man. J Natl Cancer Inst 24:551–571
6. Bozic I, Antal T, Ohtsuki H, Carter H, Kim D, Chen S, Karchin R, Kinzler KW, Vogelstein B, Nowak MA (2010) Accumulation of driver and passenger mutations during tumor progression. Proc Natl Acad Sci U S A 107:18545–18550
7. Greenman C, Stephens P, Smith R, Dalgliesh GL, Hunter C, Bignell G, Davies H, Teague J, Butler A, Stevens C, Edkins S, O'Meara S, Vastrik I, Schmidt EE, Avis T, Barthorpe S, Bhamra G, Buck G, Choudhury B, Clements J, Cole J, Dicks E, Forbes S, Gray K, Halliday K, Harrison R, Hills K, Hinton J, Jenkinson A, Jones D, Menzies A, Mironenko T, Perry J, Raine K, Richardson D, Shepherd R, Small A, Tofts C, Varian J, Webb T, West S, Widaa S, Yates A, Cahill DP, Louis DN, Goldstraw P, Nicholson AG, Brasseur F, Looijenga L, Weber BL, Chiew YE, DeFazio A, Greaves MF, Green AR, Campbell P, Birney E, Easton DF, Chenevix-Trench G, Tan MH, Khoo SK, Teh BT, Yuen ST, Leung SY, Wooster R, Futreal PA, Stratton MR (2007) Patterns of somatic mutation in human cancer genomes. Nature 446:153–8
8. Bell D, Berchuck A, Birrer MJ, Thomson E (2011) Integrated genomic analyses of ovarian carcinoma. Nature 474:609–15

9. Lee W, Jiang Z, Liu J, Haverty PM, Guan Y, Stinson J, Yue P, Zhang Y, Pant KP, Bhatt D, Ha C, Johnson S, Kennemer MI, Mohan S, Nazarenko I, Watanabe C, Sparks AB, Shames DS, Gentleman R, de Sauvage FJ, Stern H, Pandita A, Ballinger DG, Drmanac R, Modrusan Z, Seshagiri S, Zhang Z (2010) The mutation spectrum revealed by paired genome sequences from a lung cancer patient. Nature 465:473–477
10. Cancer Genome Atlas Network (2012) Comprehensive molecular characterization of human colon and rectal cancer. Nature 487:330–7
11. Myllykangas S, Himberg J, Bohling T, Nagy B, Hollmen J, Knuutila S (2006) DNA copy number amplification profiling of human neoplasms. Oncogene 25:7324–7332
12. Weaver Z, Montagna C, Xu X, Howard T, Gadina M, Brodie SG, Deng CX, Ried T (2002) Mammary tumors in mice conditionally mutant for Brca1 exhibit gross genomic instability and centrosome amplification yet display a recurring distribution of genomic imbalances that is similar to human breast cancer. Oncogene 21:5097–5107
13. Lo AW, Sabatier L, Fouladi B, Pottier G, Ricoul M, Murnane JP (2002) DNA amplification by breakage/fusion/bridge cycles initiated by spontaneous telomere loss in a human cancer cell line. Neoplasia 4:531–538
14. Stephens PJ, Greenman CD, Fu B, Yang F, Bignell GR, Mudie LJ, Pleasance ED, Lau KW, Beare D, Stebbings LA, McLaren S, Lin ML, McBride DJ, Varela I, Nik-Zainal S, Leroy C, Jia M, Menzies A, Butler AP, Teague JW, Quail MA, Burton J, Swerdlow H, Carter NP, Morsberger LA, Iacobuzio-Donahue C, Follows GA, Green AR, Flanagan AM, Stratton MR, Futreal PA, Campbell PJ (2011) Massive genomic rearrangement acquired in a single catastrophic event during cancer development. Cell 144:27–40
15. Heisterkamp N, Stam K, Groffen J, de Klein A, Grosveld G (1985) Structural organization of the bcr gene and its role in the Ph′ translocation. Nature 315:758–761
16. Shtivelman E, Lifshitz B, Gale RP, Canaani E (1985) Fused transcript of abl and bcr genes in chronic myelogenous leukaemia. Nature 315:550–554
17. Lynch M (2010) Rate, molecular spectrum, and consequences of human mutation. Proc Natl Acad Sci U S A 107:961–968
18. Kong A, Frigge ML, Masson G, Besenbacher S, Sulem P, Magnusson G, Gudjonsson SA, Sigurdsson A, Jonasdottir A, Wong WS, Sigurdsson G, Walters GB, Steinberg S, Helgason H, Thorleifsson G, Gudbjartsson DF, Helgason A, Magnusson OT, Thorsteinsdottir U, Stefansson K (2012) Rate of de novo mutations and the importance of father's age to disease risk. Nature 488:471–475
19. Pedeutour F, Forus A, Coindre JM, Berner JM, Nicolo G, Michiels JF, Terrier P, Ranchere-Vince D, Collin F, Myklebost O, Turc-Carel C (1999) Structure of the supernumerary ring and giant rod chromosomes in adipose tissue tumors. Genes Chromosomes Cancer 24:30–41
20. Sawyer S, Mitchell G, McKinley J, Chenevix-Trench G, Beesley J, Chen XQ, Bowtell D, Trainer AH, Harris M, Lindeman GJ, James PA (2012) A role for common genomic variants in the assessment of familial breast cancer. J Clin Oncol 30:4330–4336
21. Kridel R, Meissner B, Rogic S, Boyle M, Telenius A, Woolcock B, Gunawardana J, Jenkins C, Cochrane C, Ben-Neriah S, Tan K, Morin RD, Opat S, Sehn LH, Connors JM, Marra MA, Weng AP, Steidl C, Gascoyne RD (2012) Whole transcriptome sequencing reveals recurrent NOTCH1 mutations in mantle cell lymphoma. Blood 119:1963–1971
22. Agrawal N, Frederick MJ, Pickering CR, Bettegowda C, Chang K, Li RJ, Fakhry C, Xie TX, Zhang J, Wang J, Zhang N, El-Naggar AK, Jasser SA, Weinstein JN, Trevino L, Drummond JA, Muzny DM, Wu Y, Wood LD, Hruban RH, Westra WH, Koch WM, Califano JA, Gibbs RA, Sidransky D, Vogelstein B, Velculescu VE, Papadopoulos N, Wheeler DA, Kinzler KW, Myers JN (2011) Exome sequencing of head and neck squamous cell carcinoma reveals inactivating mutations in NOTCH1. Science 333:1154–1157
23. Beroukhim R, Getz G, Nghiemphu L, Barretina J, Hsueh T, Linhart D, Vivanco I, Lee JC, Huang JH, Alexander S, Du J, Kau T, Thomas RK, Shah K, Soto H, Perner S, Prensner J, Debiasi RM, Demichelis F, Hatton C, Rubin MA, Garraway LA, Nelson SF, Liau L, Mischel PS, Cloughesy TF, Meyerson M, Golub TA, Lander ES, Mellinghoff IK, Sellers WR (2007)

Assessing the significance of chromosomal aberrations in cancer: methodology and application to glioma. Proc Natl Acad Sci U S A 104:20007–20012
24. Mermel CH, Schumacher SE, Hill B, Meyerson ML, Beroukhim R, Getz G (2011) GISTIC2.0 facilitates sensitive and confident localization of the targets of focal somatic copy-number alteration in human cancers. Genome Biol 12:R41
25. Ramakrishna M, Williams LH, Boyle SE, Bearfoot JL, Sridhar A, Speed TP, Gorringe KL, Campbell IG (2010) Identification of candidate growth promoting genes in ovarian cancer through integrated copy number and expression analysis. PLoS One 5, e9983
26. Davis SJ, Sheppard KE, Pearson RB, Campbell IG, Gorringe KL, Simpson KJ (2013) Functional analysis of genes in regions commonly amplified in high-grade serous and endometrioid ovarian cancer. Clin Cancer Res 19:1411–1421
27. Steidl C, Shah SP, Woolcock BW, Rui L, Kawahara M, Farinha P, Johnson NA, Zhao Y, Telenius A, Neriah SB, McPherson A, Meissner B, Okoye UC, Diepstra A, van den Berg A, Sun M, Leung G, Jones SJ, Connors JM, Huntsman DG, Savage KJ, Rimsza LM, Horsman DE, Staudt LM, Steidl U, Marra MA, Gascoyne RD (2011) MHC class II transactivator CIITA is a recurrent gene fusion partner in lymphoid cancers. Nature 471:377–381
28. Cavenee WK, Hansen MF, Nordenskjold M, Kock E, Maumenee I, Squire JA, Phillips RA, Gallie BL (1985) Genetic origin of mutations predisposing to retinoblastoma. Science 228:501–503
29. Gorringe KL, Ramakrishna M, Williams LH, Sridhar A, Boyle SE, Bearfoot JL, Li J, Anglesio MS, Campbell IG (2009) Are there any more ovarian tumor suppressor genes? A new perspective using ultra high-resolution copy number and loss of heterozygosity analysis. Genes Chromosomes Cancer 48:931–942
30. Sjoblom T, Jones S, Wood LD, Parsons DW, Lin J, Barber TD, Mandelker D, Leary RJ, Ptak J, Silliman N, Szabo S, Buckhaults P, Farrell C, Meeh P, Markowitz SD, Willis J, Dawson D, Willson JK, Gazdar AF, Hartigan J, Wu L, Liu C, Parmigiani G, Park BH, Bachman KE, Papadopoulos N, Vogelstein B, Kinzler KW, Velculescu VE (2006) The consensus coding sequences of human breast and colorectal cancers. Science 314:268–274
31. Adzhubei I, Jordan DM, Sunyaev SR (2013) Predicting functional effect of human missense mutations using PolyPhen-2. Curr Protoc Hum Genet, Chapter 7: Unit 7.20
32. Kumar P, Henikoff S, Ng PC (2009) Predicting the effects of coding non-synonymous variants on protein function using the SIFT algorithm. Nat Protoc 4:1073–1081
33. Gonzalez-Perez A, Lopez-Bigas N (2011) Improving the assessment of the outcome of nonsynonymous SNVs with a consensus deleteriousness score, Condel. Am J Hum Genet 88:440–449
34. Dees ND, Zhang Q, Kandoth C, Wendl MC, Schierding W, Koboldt DC, Mooney TB, Callaway MB, Dooling D, Mardis ER, Wilson RK, Ding L (2012) MuSiC: identifying mutational significance in cancer genomes. Genome Res 22:1589–1598
35. Parmigiani G, Boca S, Lin J, Kinzler KW, Velculescu V, Vogelstein B (2009) Design and analysis issues in genome-wide somatic mutation studies of cancer. Genomics 93:17–21
36. Biankin AV, Waddell N, Kassahn KS, Gingras MC, Muthuswamy LB, Johns AL, Miller DK, Wilson PJ, Patch AM, Wu J, Chang DK, Cowley MJ, Gardiner BB, Song S, Harliwong I, Idrisoglu S, Nourse C, Nourbakhsh E, Manning S, Wani S, Gongora M, Pajic M, Scarlett CJ, Gill AJ, Pinho AV, Rooman I, Anderson M, Holmes O, Leonard C, Taylor D, Wood S, Xu Q, Nones K, Fink JL, Christ A, Bruxner T, Cloonan N, Kolle G, Newell F, Pinese M, Mead RS, Humphris JL, Kaplan W, Jones MD, Colvin EK, Nagrial AM, Humphrey ES, Chou A, Chin VT, Chantrill LA, Mawson A, Samra JS, Kench JG, Lovell JA, Daly RJ, Merrett ND, Toon C, Epari K, Nguyen NQ, Barbour A, Zeps N, Kakkar N, Zhao F, Wu YQ, Wang M, Muzny DM, Fisher WE, Brunicardi FC, Hodges SE, Reid JG, Drummond J, Chang K, Han Y, Lewis LR, Dinh H, Buhay CJ, Beck T, Timms L, Sam M, Begley K, Brown A, Pai D, Panchal A, Buchner N, De Borja R, Denroche RE, Yung CK, Serra S, Onetto N, Mukhopadhyay D, Tsao MS, Shaw PA, Petersen GM, Gallinger S, Hruban RH, Maitra A, Iacobuzio-Donahue CA, Schulick RD, Wolfgang CL, Morgan RA, Lawlor RT, Capelli P, Corbo V, Scardoni M, Tortora G, Tempero MA, Mann KM, Jenkins NA, Perez-Mancera PA, Adams DJ, Largaespada DA, Wessels LF, Rust AG, Stein LD, Tuveson DA, Copeland NG, Musgrove EA, Scarpa A,

Eshleman JR, Hudson TJ, Sutherland RL, Wheeler DA, Pearson JV, McPherson JD, Gibbs RA, Grimmond SM (2012) Pancreatic cancer genomes reveal aberrations in axon guidance pathway genes. Nature 491:399–405
37. Bronner CE, Baker SM, Morrison PT, Warren G, Smith LG, Lescoe MK, Kane M, Earabino C, Lipford J, Lindblom A et al (1994) Mutation in the DNA mismatch repair gene homologue hMLH1 is associated with hereditary non-polyposis colon cancer. Nature 368:258–261
38. Okamoto M, Sato C, Kohno Y, Mori T, Iwama T, Tonomura A, Miki Y, Utsunomiya J, Nakamura Y, White R et al (1990) Molecular nature of chromosome 5q loss in colorectal tumors and desmoids from patients with familial adenomatous polyposis. Hum Genet 85:595–599
39. Miki Y, Swensen J, Shattuck-Eidens D, Futreal PA, Harshman K, Tavtigian S, Liu Q, Cochran C, Bennett LM, Ding W et al (1994) A strong candidate for the breast and ovarian cancer susceptibility gene BRCA1. Science 266:66–71
40. Wooster R, Bignell G, Lancaster J, Swift S, Seal S, Mangion J, Collins N, Gregory S, Gumbs C, Micklem G (1995) Identification of the breast cancer susceptibility gene BRCA2. Nature 378:789–792
41. Meindl A, Hellebrand H, Wiek C, Erven V, Wappenschmidt B, Niederacher D, Freund M, Lichtner P, Hartmann L, Schaal H, Ramser J, Honisch E, Kubisch C, Wichmann HE, Kast K, Deissler H, Engel C, Muller-Myhsok B, Neveling K, Kiechle M, Mathew CG, Schindler D, Schmutzler RK, Hanenberg H (2010) Germline mutations in breast and ovarian cancer pedigrees establish RAD51C as a human cancer susceptibility gene. Nat Genet 42:410–414
42. Thompson ER, Boyle SE, Johnson J, Ryland GL, Sawyer S, Choong DY, kConFab, Chenevix-Trench G, Trainer AH, Lindeman GJ, Mitchell G, James PA, Campbell IG (2012) Analysis of RAD51C germline mutations in high-risk breast and ovarian cancer families and ovarian cancer patients. Hum Mutat 33:95–99
43. Osorio A, Endt D, Fernandez F, Eirich K, de la Hoya M, Schmutzler R, Caldes T, Meindl A, Schindler D, Benitez J (2012) Predominance of pathogenic missense variants in the RAD51C gene occurring in breast and ovarian cancer families. Hum Mol Genet 21:2889–2898
44. Pollack JR, Sorlie T, Perou CM, Rees CA, Jeffrey SS, Lonning PE, Tibshirani R, Botstein D, Borresen-Dale AL, Brown PO (2002) Microarray analysis reveals a major direct role of DNA copy number alteration in the transcriptional program of human breast tumors. Proc Natl Acad Sci U S A 99:12963–12968
45. Sorlie T, Tibshirani R, Parker J, Hastie T, Marron JS, Nobel A, Deng S, Johnsen H, Pesich R, Geisler S, Demeter J, Perou CM, Lonning PE, Brown PO, Borresen-Dale AL, Botstein D (2003) Repeated observation of breast tumor subtypes in independent gene expression data sets. Proc Natl Acad Sci U S A 100:8418–8423
46. Curtis C, Shah SP, Chin SF, Turashvili G, Rueda OM, Dunning MJ, Speed D, Lynch AG, Samarajiwa S, Yuan Y, Graf S, Ha G, Haffari G, Bashashati A, Russell R, McKinney S, Langerod A, Green A, Provenzano E, Wishart G, Pinder S, Watson P, Markowetz F, Murphy L, Ellis I, Purushotham A, Borresen-Dale AL, Brenton JD, Tavare S, Caldas C, Aparicio S (2012) The genomic and transcriptomic architecture of 2,000 breast tumours reveals novel subgroups. Nature 486:346–352
47. Jones NL, Xiu J, Reddy SK, Burke WM, Tergas AI, Wright JD, Hou JY (2015) Identification of potential therapeutic targets by molecular profiling of 628 cases of uterine serous carcinoma. Gynecol Oncol 138:620–626
48. Frampton GM, Fichtenholtz A, Otto GA, Wang K, Downing SR, He J, Schnall-Levin M, White J, Sanford EM, An P, Sun J, Juhn F, Brennan K, Iwanik K, Maillet A, Buell J, White E, Zhao M, Balasubramanian S, Terzic S, Richards T, Banning V, Garcia L, Mahoney K, Zwirko Z, Donahue A, Beltran H, Mosquera JM, Rubin MA, Dogan S, Hedvat CV, Berger MF, Pusztai L, Lechner M, Boshoff C, Jarosz M, Vietz C, Parker A, Miller VA, Ross JS, Curran J, Cronin MT, Stephens PJ, Lipson D, Yelensky R (2013) Development and validation of a clinical cancer genomic profiling test based on massively parallel DNA sequencing. Nat Biotechnol 31:1023–31

49. Bartels S, Schipper E, Kreipe HH, Lehmann U (2015) Comprehensive molecular profiling of archival bone marrow trephines using a commercially available leukemia panel and semiconductor-based targeted resequencing. PLoS One 10, e0133930
50. Sargent DJ, Marsoni S, Monges G, Thibodeau SN, Labianca R, Hamilton SR, French AJ, Kabat B, Foster NR, Torri V, Ribic C, Grothey A, Moore M, Zaniboni A, Seitz JF, Sinicrope F, Gallinger S (2010) Defective mismatch repair as a predictive marker for lack of efficacy of fluorouracil-based adjuvant therapy in colon cancer. J Clin Oncol 28:3219–3226
51. Ribic CM, Sargent DJ, Moore MJ, Thibodeau SN, French AJ, Goldberg RM, Hamilton SR, Laurent-Puig P, Gryfe R, Shepherd LE, Tu D, Redston M, Gallinger S (2003) Tumor microsatellite-instability status as a predictor of benefit from fluorouracil-based adjuvant chemotherapy for colon cancer. N Engl J Med 349:247–257
52. Deeken JF, Cormier T, Price DK, Sissung TM, Steinberg SM, Tran K, Liewehr DJ, Dahut WL, Miao X, Figg WD (2010) A pharmacogenetic study of docetaxel and thalidomide in patients with castration-resistant prostate cancer using the DMET genotyping platform. Pharmacogenomics J 10:191–199
53. Tian C, Ambrosone CB, Darcy KM, Krivak TC, Armstrong DK, Bookman MA, Davis W, Zhao H, Moysich K, Gallion H, DeLoia JA (2012) Common variants in ABCB1, ABCC2 and ABCG2 genes and clinical outcomes among women with advanced stage ovarian cancer treated with platinum and taxane-based chemotherapy: a Gynecologic Oncology Group study. Gynecol Oncol 124:575–581
54. Ni LN, Li JY, Miao KR, Qiao C, Zhang SJ, Qiu HR, Qian SX (2011) Multidrug resistance gene (MDR1) polymorphisms correlate with imatinib response in chronic myeloid leukemia. Med Oncol 28:265–269
55. Etemadmoghadam D, de Fazio A, Beroukhim R, Mermel C, George J, Getz G, Tothill R, Okamoto A, Raeder MB, Harnett P, Lade S, Akslen LA, Tinker AV, Locandro B, Alsop K, Chiew YE, Traficante N, Fereday S, Johnson D, Fox S, Sellers W, Urashima M, Salvesen HB, Meyerson M, Bowtell D (2009) Integrated genome-wide DNA copy number and expression analysis identifies distinct mechanisms of primary chemoresistance in ovarian carcinomas. Clin Cancer Res 15:1417–27
56. Helleday T (2011) The underlying mechanism for the PARP and BRCA synthetic lethality: clearing up the misunderstandings. Mol Oncol 5:387–393
57. Prahallad A, Sun C, Huang S, Di Nicolantonio F, Salazar R, Zecchin D, Beijersbergen RL, Bardelli A, Bernards R (2012) Unresponsiveness of colon cancer to BRAF(V600E) inhibition through feedback activation of EGFR. Nature 483:100–103
58. Mao M, Tian F, Mariadason JM, Tsao CC, Lemos R Jr, Dayyani F, Gopal YN, Jiang ZQ, Wistuba II, Tang XM, Bornman WG, Bollag G, Mills GB, Powis G, Desai J, Gallick GE, Davies MA, Kopetz S (2013) Resistance to BRAF inhibition in BRAF-mutant colon cancer can be overcome with PI3K inhibition or demethylating agents. Clin Cancer Res 19:657–67
59. Diaz LA Jr, Williams RT, Wu J, Kinde I, Hecht JR, Berlin J, Allen B, Bozic I, Reiter JG, Nowak MA, Kinzler KW, Oliner KS, Vogelstein B (2012) The molecular evolution of acquired resistance to targeted EGFR blockade in colorectal cancers. Nature 486:537–540
60. Sakai W, Swisher EM, Karlan BY, Agarwal MK, Higgins J, Friedman C, Villegas E, Jacquemont C, Farrugia DJ, Couch FJ, Urban N, Taniguchi T (2008) Secondary mutations as a mechanism of cisplatin resistance in BRCA2-mutated cancers. Nature 451:1116–1120
61. Swisher EM, Sakai W, Karlan BY, Wurz K, Urban N, Taniguchi T (2008) Secondary BRCA1 mutations in BRCA1-mutated ovarian carcinomas with platinum resistance. Cancer Res 68:2581–2586
62. Dawson SJ, Tsui DW, Murtaza M, Biggs H, Rueda OM, Chin SF, Dunning MJ, Gale D, Forshew T, Mahler-Araujo B, Rajan S, Humphray S, Becq J, Halsall D, Wallis M, Bentley D, Caldas C, Rosenfeld N (2013) Analysis of circulating tumor DNA to monitor metastatic breast cancer. N Engl J Med 368:1199–1209

Gene Expression Analysis: Current Methods

Zhi Ling Teo, Peter Savas, and Sherene Loi

Introduction

Cancer is a genetic disease characterised by multiple heterogeneous genetic and epigenetic changes. Recent studies have identified extensive heterogeneity between and within tumours [1–3]. The genes need to be studied as a functioning collective in order to tease apart and understand the myriad different levels of processes and interactions that are coordinated towards the common goal of assuring vital functioning of a cell. The study of the transcriptome of cancer cells, a fundamental link between genotype and phenotype, is essential to understanding the complexity of cancer evolution.

Traditional methods of gene expression measurements include Northern blot, quantitative reverse transcription PCR (qRTPCR), serial analysis of gene expression (SAGE) and DNA microarrays. Northern blot [4] analysis is a low throughput method that uses electrophoresis to separate RNA by size. The separated RNA is transferred onto a nylon membrane and immobilised to the membrane through covalent linkage by UV light or heat. A labelled short oligonucleotide sequence or probe that is complementary to a sequence in the target transcript is introduced onto the membrane and its hybridisation to the target transcript is detected via the use of X-ray. The Northern blot procedure is useful for determining RNA size and to detect alternative splice products. However, Northern blot uses RNA without conversion into complementary DNA (cDNA), therefore, the quality of quantification is compromised by even low levels of RNA degradation. Northern blot has also relatively low sensitivity, due to non-specific hybridisation, requires the use of radioactivity and requires greater amounts of RNA compared to qRTPCR [5].

Z.L. Teo • P. Savas • S. Loi (✉)
Translational Breast Cancer Genomics and Therapeutics Lab, Cancer Therapeutics Program, Division of Research, Peter MacCallum Cancer Centre, Melbourne, VIC 3000, Australia
e-mail: Sherene.loi@petermac.org

S.R. Lakhani, S.B. Fox (eds.), *Molecular Pathology in Cancer Research*,
DOI 10.1007/978-1-4939-6643-1_6

qRTPCR [6, 7] involves the reverse transcription of the target RNA into cDNA followed by PCR of the cDNA to amplify the signal for detection. It is a fluorescence-based real-time reaction method that allows for detection and relative quantitation of target RNAs. The qRTPCR method has improved to enable high throughput multiplexed reactions to quantify multiple genes in a single reaction [8, 9]. However, the throughput capability of the current technology of qRTPCR remains on the order of only hundreds of known transcripts in one assay, and is not adept for transcriptome-wide gene expression analysis.

SAGE [10] is a gene expression profiling method that involves creating cDNA which is biotin-labelled at the 3′-end of the cDNA. Short sequence tags of 14 or 21 bp that can uniquely identify specific transcripts are extracted using restriction enzymes (normally Nla III). Nla III cleaves the cDNA at the 5′-end of every CATG site. The cleaved sequence closest to the 3′-end of the cDNA is isolated using streptavidin beads. This isolated sequence is shortened again to contain only the CATG and the next ten nucleotides. This final sequence is the SAGE tag. These tags are ligated and cloned into a vector which are Sanger sequenced to identify the sequence tags. This method allows direct measurement of transcript abundance and comparison between multiple samples. SAGE does not require a priori knowledge of the transcript sequence and has been used for discovery of novel transcripts and alternative splice isoforms. Nonetheless, SAGE is a costly technique with a laborious cloning procedure.

The DNA microarray technology has superseded single-gene approaches, allowing the measurement of RNA expression levels of thousands of known or putative transcripts simultaneously [11] and was further developed to characterise the gene expression profile of a complete eukaryotic genome (*Saccharomyces cerevisiae*) [12]. Global gene expression analysis has been a significant revolution in biology. The advent of DNA microarray technology has enabled comprehensive characterisation and/or comparison of expression signatures of various cell types and disease phenotypes. Gene profiling with microarrays have provided evidence of molecular heterogeneity in cancer. Microarray technology was used to identify four subgroups in breast cancer, each with a distinct gene signature [13]. Subsequently, it has been shown that these molecular distinct subtypes of breast cancer are associated with prognosis and response to treatment [14]. In the past decade, numerous cancer gene expression signatures have been established but few have progressed to be available commercially and used to inform on prognosis or treatment choice in the clinic setting. The two more significant assays are the Mammaprint [15] and the Oncotype DX [16] assays which help identify early stage breast cancer patients who are at lower risk of recurrence and may not be subjected to unnecessary chemotherapy and ultimately reduce healthcare costs [17–19].

Despite the success of gene expression profiling using DNA microarrays in its contribution to the field of biology, the technique remains limited by its requirement of a priori knowledge of the genes of interest. To overcome this limitation, tiling arrays have been developed [20, 21]. Prior to the advent of massively parallel sequencing technology, tiling arrays was the method of choice to identify novel transcribed regions [22, 23]. At present, it is still used for transcriptome profiling or to identify novel genes in the human genome [24, 25]. Like the DNA microarrays, tiling arrays are also based on the hybridisation of labelled target transcripts to probes covalently attached to a solid surface. Probes on tiling arrays have been designed to map in an unbiased manner to contiguous sequences in the genome regardless of whether the regions have been previously annotated. RNA resequencing using tiling arrays has been made possible. Advancements to the microarray technology produced a high resolution (1 bp) high density array using a set of highly tiled and overlapping probes. Four probes (25 bp each) are used for the interrogation of one base; the central position of the probe carries one of the four possible bases (A, T, C, G), one of which represents the reference sequence. A single nucleotide variant (SNV) will reduce binding efficiency and hybridisation signal of the probe. The probe that best matches the sequence will have the highest fluorescence intensity [26]. Resequencing of both strands of a chromosome of length *N* requires *8 N* probes. The large number of probes required for the complete resequencing of a typical eukaryote genome makes the array-based resequencing seem impractical, costly and inefficient but has been shown to be feasible for resequencing small targeted sequences [27] and high throughput mutation detection [28–30]. Tiling array designing can lead to specific design and analysis problems. A drawback to using the hybridisation approach is cross-hybridisation between probes. Non-specific cDNA molecules which can result in significant background noise which is amplified with increased regions of coverage and shorter probe sequences [31]. The specificity of potential probes for tiling arrays can be verified by an alignment algorithm against a sequence database and repetitive sequences can be identified and masked before the probe design to minimise the potential for cross-hybridisation [31–33]. Nonetheless, intensive normalisation is required to counteract the cross-hybridisation and background noise [34]. Probes also need to hybridise at similar efficiencies at a given temperature and need to be designed to be non-palindromic to prevent self-hybridisation. While tiling arrays are a powerful tool for genomic analysis, there remain significant practical limitations in terms of the design of probes and numerous number of chips required to cover an entire genome [26].

Meanwhile, another breakthrough in the analysis of RNA has been brought about through the development of massively parallel sequencing technology. RNA sequencing provides a cost-effective and rapid approach to sequence the transcriptome [35]. It has obvious advantages over microarray technology. RNA sequencing directly sequences the reverse transcribed transcripts which allow resolution to a single base. Transcript sequences are mapped to a reference genome. The number of reads that are mapped is an indication of the level of gene expression. Direct access to the sequence removes the need for a priori knowledge of the transcript target(s) and allows detection of RNA editing events, novel alternate splicing. Moreover, RNA sequencing can be performed on species that do not have full genome sequence available. RNA sequencing has been used in conjunction with whole genome or whole exome sequencing technology to further our understanding of cancer, its evolution, or to advance our progress in the identification of genes with previously unknown implications of carcinogenesis or cancer progression [36, 37].

Transcriptome-wide gene expression analysis has allowed insights into global mechanisms, for instance, interactions between biological pathways in an orchestrated response to external stimuli, compensatory mechanisms in an event of the disruption of a specific pathway, or evolutionary process underlying the mechanisms of drug resistance. These tools have been shown to be applicable in both the clinical and research setting. The exponential advancement of our understanding of biological processes underlying various disease phenotypes will enrich our progress in the era of personalised medicine.

This chapter will focus on DNA microarray and RNA sequencing gene expression profiling techniques due to their ability to interrogate large numbers of transcripts simultaneously.

DNA Microarray

DNA Microarray Technology

A DNA microarray is a collection of single-stranded oligonucleotides or probes, complementary to target transcript sequences, which are covalently bonded on the 5′ end to chemically compatible matrices on a solid surface. The identity of each probe is defined by its location on the array. The transcripts to be analysed are reverse transcribed into cDNA. The cDNA are labelled with fluorescent dyes, radioisotopes, biotin, amine groups, or micro- and nanoparticles which are excited by lasers to allow detection of the hybridisation of the target transcript to the probe on the microarray [38, 39]. The fluorescence emitted on each probe is used as a measure of gene expression and its intensity correlates to the level of gene expression.

There are different gene expression microarray platforms available commercially. Affymetrix (Santa Clara, CA, USA) and Agilent Technologies (Santa Clara, CA, USA) are renowned providers of gene expression microarray solutions. The 25-mer oligonucleotide probes on Affymetrix GeneChip® microarrays are built up using light-directed oligonucleotide synthesis on glass surfaces [40, 41]. Attached on the glass surfaces are photolabile protecting groups. When illuminated through a photolithographic mask, the illuminated regions yield reactive hydroxyl groups. Deoxynucleosides which are protected at the 5′-hydroxyl end with a photolabile group are then introduced to the surface. Coupling occurs at sites that are exposed to light. Since photolithography is used, high-density microarrays, where each probe is in a spacing of only 5 μm, can be generated. The Affymetrix GeneChips® are designed for whole exome and whole transcriptome gene expression profiling. The GeneChips® are one-colour arrays and only allow one sample to be hybridised to the microarray and are read on GeneChip® Scanner 3000 7G.

Agilent Technologies provides human gene expression microarray solutions that interrogate the whole transcriptome, whole exome and also provides a web-based tool (Agilent eArray) to allow users to rapidly design custom microarrays. Agilent microarrays are manufactured using the Agilent 60-mer SurePrint technology [42]. The SurePrint technology is a non-contact in situ synthesis inkjet printing process.

It involves delivering monomers to defined positions by commercial inkjet printer heads onto hydrophobic glass wafers (25 mm × 75 mm) containing exposed hydroxyl groups for nucleotide coupling. At present, Agilent microarrays are available in a variety of formats allowing flexibility to suit specific experimental requirements. Agilent 60-mer oligonucleotide microarrays can be used for one- and two-colour experiments and standard 25 mm × 75 mm glass slides can be read using Agilent SureScan Microarray laser induced fluorescence scanner or the Agilent SureScan Dx Microarray Scanner that is marketed for in vitro diagnostic use in Europe.

For different platforms, standard operating procedures are provided by the manufacturers. These different protocols commonly require as low as 10 ng of RNA per sample which is used to generate cDNA or cRNA as targets for hybridisation. cDNA or cRNA can be labelled by covalently binding to biotin via terminal deoxynucleotidyl transferase, through incorporating fluorescently labelled dUTP or in vitro transcription using T7 polymerase that incorporates biotin-labelled nucleotides. For two-colour arrays, any two dyes with a good separable excitation and emission spectrum can be applied [43]. Cyanine-3 and Cyanine-5 are most commonly used fluorescent dyes in two-colour arrays. Biotin labels are detected using streptavidin phycoerythrin, the streptavidin binds to the biotin and the phycoerythrin exhibits bright fluorescence when excited. The labelled cDNA or cRNA are then introduced to the microarray and are allowed to hybridise to the covalently bonded probes on the microarray surface. Fluorescent scanners are used to excite the fluorescent dyes on the target sequence that is hybridised to the microarray. The fluorescence intensities at each position are measured by the scanner and transformed into a digital signal using a photomultiplier tube or a charge coupled device using a dye-specific emission filter [43]. The intensity levels at each probe position are the raw data which are the basis for further data analysis.

Experimental Design

The objective of an experimental design is to ensure that the analysis and interpretation of the data remains as simple and powerful as possible by working around limitations of cost and experimental material [44]. The main objectives of the experiment need to be well defined in order to select the most suitable design for the experiment.

Experimental Objectives

The objectives of a microarray experiment can be classified into three broad categories: class comparison, class discovery, and class prediction [45, 46]. Class comparison experiments look for genes that are differentially expressed among samples of different groups, for instance, experimental treatments, phenotypes, gender or age. Class discovery, as the name suggests, refers to the search for previously unknown taxonomy (i.e., previously unrecognised tumour subtypes) that can be identified by a gene expression signature. This is achieved through cluster analysis where microarray data is clustered and validated through gene annotations or by replicating results in other data sets. Class prediction refers to the identification and validation of a classifier that is predictive of a previously known class or phenotype (i.e. a gene signature that can be used to identify a specific tumour subtype or predict response to treatment). The aims of class prediction experiments are to (1) determine if there is a relationship between gene expression profile and a phenotype or clinical outcome and (2) develop a gene expression signature to predict target phenotype, prognosis or response to treatment.

Replication

There are several significant sources of random variation in microarray studies which can be classified into biological variation (e.g. the variation of gene expression between individuals within a population which could be influenced by environmental or genetic factors), technical variation (e.g. the variation of gene expression between specimens from the same individual which could be introduced during the extraction process, hybridisation or labelling) and measurement error associated with reading fluorescent signals [47].

Biological replicates are samples taken from different individuals, animals or primary cell cultures of different cell lines. These replicates are important to account for the variation between individuals. Each experimental condition should have multiple independent biological replicates in order for valid statistical testing and for extrapolation of conclusions that may be drawn from the experiment [47].

Technical replicates can refer to replication of the microarray hybridisation process, replication of probes on one microarray, dye-swap labelling (discussed further in the subsequent section), processing more than one specimen from the same individual, or repeating the processing of the same specimen more than once. Technical replicates are essential in all experiments to ensure that the experimental procedures, reagents and equipment are performing uniformly across samples and should show good agreement.

One- or Two-Colour Systems

There are two main approaches to the design of a DNA microarray gene expression profiling experiment: (1) one-colour (i.e. cDNA from one sample is hybridised to one microarray); (2) two-colour (i.e. cDNA from two samples are hybridised on a single microarray). In the one-colour system, the cDNA from one sample labelled with a single fluorophore (Cyanine-3, Cyanine-5, radioisotopes or phycoerythrin). The fluorescent intensity on each spot (i.e. one transcript) is compared to other spots within the sample, or reference normalising probes can be used to calibrate data across one microarray or across multiple microarrays to obtain a measurement of relative gene expression. To compare two sets of conditions (e.g. treatment versus control), two separate one-colour hybridisations are required. The two-colour system involves the cDNA from the reference sample to be labelled with a different fluorophore than that of the test sample (usually Cyanine-3, emitting at 570 nm, and Cyanine-5, emitting at 670 nm). The cDNAs from the two samples are then hybridised on a single microarray. The ratio of relative intensities of each fluorophore is then used to identify up- or down-regulated genes.

There are advantages and disadvantages associated with both one-colour and two-colour systems. A strength of the one-colour system is that a low-quality or aberrant sample is unable to affect the raw data of other samples as one microarray is exposed to only one sample. This is in contrast to the two-colour system where a single low-performing sample could significantly affect overall data precision even if the other sample is of high quality. In addition, the data from one-colour systems may be compared with microarrays from various experiments or between studies, as long as batch effects are accounted for, and is unaffected by time. However, a drawback of the one-colour system is that, compared to the two-colour system, twice as many microarrays are required to assess samples from different treatments within one experiment. The two-colour system compares two samples on one microarray which eliminates variability between runs and microarray, therefore, reducing the error that could be introduced by these variables as compared to the one-colour system. A disadvantage of the two-colour hybridisation system is that the use of two different dyes on one microarray could generate dye-specific biases, resulting from differences in the incorporation efficiency of the two fluorescent dyes used in the experiment, which can confound the true biological signal [44]. Dye-biases can be circumvented by dye-swap labelling or the use of split control hybridisations [48]. The dye-swap labelling method involves a technical replicate of a two-colour microarray where the same samples are hybridised by with reversed fluorescent labelling. Split control hybridisation is a more cost-effective alternative to the dye-swap labelling method where a control sample is split and each portion is labelled separately with either Cyanine-3 or Cyanine-5 before being combined and hybridised on the same microarray.

Although there are certain advantages and disadvantages associated with both approaches, studies comparing one- and two-colour microarrays have shown that both methods perform equally well in terms of data quality and accuracy of gene expression measurement [49, 50]. After using dye-reversed replicates to mitigate the effects of dye-specific biases associated with two-coloured microarrays, Patterson et al. [49] showed that the quality of data (reproducibility, sensitivity, specificity and accuracy) produced by the one- and two-colour systems was comparable and yielded highly concordant results regarding detection of differentially expressed genes within each platform. The two approaches were compared within the same platform (Aligent, CapitalBio and TeleChem were tested) and across multiple test sites (a total of five test sites). These results suggest that the decision between the one- and two-colour systems is not a primary factor regarding experimental microarray design. Therefore, the decision to use either the one- or two-colour approach will mainly be determined by cost and experimental design considerations [49].

Designs

Single-colour arrays involve one sample per microarray chip and therefore do not require designs for pairing and labelling of samples as is required for two-colour arrays. There are a few designs for two-colour arrays: the reference design, the balanced block design, the loop design and the reverse labelling design. The reference design and the loop design are the two main types of designs which will be the focus of this section.

The reference design is the most common design for two-colour microarrays. A common reference RNA sample is used as one of the two samples hybridised on one microarray. This means that only one experimental sample of interest is hybridised per microarray. The reference sample usually consists of a mixture of RNA from various tissues or cell lines to ensure that every probe on the array is hybridised by the reference sample. As the reference sample is used repeatedly across each microarray, there needs to be sufficient amounts of RNA of the reference sample available. The fluorescent intensity of one spot on the experimental sample of interest on the microarray is measured relative to the fluorescent intensity of the reference at the same location. It has been found that dye effects are confounded with treatment effects in the reference design [44] but this can be corrected by using dye swap on two arrays to compare each sample [47]. Despite the inefficiency of this design, it enables large number of samples to be assayed over a period of time as long as the same reference sample is being used for all microarrays [47].

The loop design [51] can be an efficient alternative to the reference design [44]. This design requires that two aliquots of an experimental sample of interest are each arrayed on different microarrays. The two microarrays that contain the same sample of interest can be used to control for variations that can exist between each microarray and between each run. However, the loop design is inferior to the reference design for use in cluster analysis [46].

DNA Microarray: Data Pre-processing

It has been shown repeatedly that there are several significant sources of systematic errors which can arise from RNA sample preparation [52], binding efficiency of target cDNA to probe (affected by GC content of the target sequence) [53] and spatial uniformity [54]. It is essential to identify and correct these sources of error from the data to ensure that the assessment of gene expression differences is precise and accurate.

Background Correction

There are several correction methods that can address non-specific hybridisation and/or spatial non-uniformity. The traditional correction method for non-specific hybridisation uses local background (ambient) intensity per spot, provided by the image analysis software, and it is removed from the overall measure of intensity of the same spot [55]. However, it has been shown to result in a significant number of false positives compared to other background correction methods [55]. GeneChip® microarrays (Affymetrix) incorporated probes that were of perfect match (PM) and mismatch (MM) to the target genes of interest onto each microarray [56]. The non-specific hybridisation signal, captured by the MM probes, is then used to infer the true specific hybridisation signal exhibited by the PM probes. However, it was shown that some of the MM intensities are greater than their PM pairs suggesting that they capture specific as well as non-specific signals and are therefore not optimal for this purpose [57, 58]. Comparing eight different background correction methods that produce markedly different bias and precision, Ritchie et al. [55] introduced and identified normexp + offset as the method that produced the lowest false discovery rate overall. The normexp + offset method was shown to be better in precision than most other methods despite the increase in bias (attenuation of hybridisation signals). The normexp + offset method is based on the Robust Multi-array Average (RMA) algorithm [57, 59] with changes made to make the method more suitable for use with two-colour microarrays and with the addition of a small positive offset, *k*, to more corrected intensities away from zero [55].

Spatial bias refers to the difference in intensity measures over the surface of a single array. Factors that may underlie spatial artefacts include fibres, scratches and uneven washing and temperature gradient. Normalisation methods do not specifically account for spatial biases even though these biases can affect analysis of results [60, 61]. Several methods addressing spatial biases in gene expression microarrays have been proposed. One method of spatial bias correction that has been widely accepted consists of applying loess-based intensity dependent bias correction method individually within each print tip group [62–64].

Normalisation

Dye intensity measures are affected by various factors such as image scanner settings and chemical characteristics of different dyes on the microarray [65, 66]. It is important that this dye intensity variability is corrected through normalisation methods to reduce system variability and to allow data from multiple microarrays to be directly compatible [67]. Locally weighted scatterplot smoothing (LOWESS) [67–69] and quantile normalisation algorithm [64] are widely used normalisation methods. LOWESS is a scatterplot-based normalisation that locally fits a line for each subset of probes with a similar intensity. It results in a non-linear transformation and does not require assumptions of the relationship between the different channels on the microarrays or between different microarrays [43]. The quantile normalisation algorithm is an extension of the LOWESS normalisation method. The quantile normalisation method aims to make the probe intensity distribution for each microarray in a set of microarrays the same [64]. This method has been shown to effectively reduce variability across microarrays [64].

Probes Summarisation

Some microarrays, such as Affymetrix GeneChips®, are designed to contain different probes that interrogate a single transcript (i.e. probe set) to reduce probe-specific effects. After background correction and normalisation, probes in each probe set are summarised to get a single-gene expression measure per probe set. There can be significant variability amongst the probes in each probe set [57]. RMA is a robust, one of the most widely used, procedure that can account for such outlier probes [57, 70]. It estimates an overall expression value for each probe set and probe-specific measurement error by fitting a linear model to the probe values. RMA was shown to outperform other methods of probe summarisation [57]. It can also be used to perform background corrections and normalisation in addition to probe summarisation. However, it is important to note that microarray spatial artefacts need to be corrected prior to using RMA, as despite its robustness, it was shown to be unable to accurately correct for such artefacts [71].

DNA Microarray: Quality Control

Quality control of microarray sample processing and microarray data is essential. Accurate sample processing is important and should be monitored by different quality checks of the sample, for example the Agilent Bioanalyzer or the Nanodrop Spectrophotometer, before hybridisation of the sample on the microarray.

The first microarray data quality check should be performed after scanning the microarray images. Quality reports of the microarray data should be provided after each scan which could be useful to detect corrupted chips or a flawed experiment. The samples deemed unreliable based on the quality control results can be removed from the analysis.

Bioconductor packages such as the Simpleaffy [72, 73] were developed to provide access to a series of quality control metrics such as:

- background level: where high background level can affect signal to noise ratio and is indicative of problems during sample processing,
- 3′/5′ ratio: the microarray contains probe sets that hybridise to the 5′ and 3′ ends of long transcripts, usually GAPDH and beta-actin. The 3′/5′ ratio of these genes is used to assess RNA quality and labelling as RNA degradation or problems during fluorescence labelling can lead to variability in the ratio, and
- percentage of genes called present: where percentage of present calls is used as an overall measure of quality and large variations of present calls between similar samples could be indicative of problems in the experiment.

For Agilent microarrays, GeneSpring GX [74] can be used to assess sample quality using various criteria such as:

- Principal Component Analysis (PCA) on samples: samples of the same experimental conditions are expected to be more similar to each other and group closer

together in a PCA plot than to samples of different experimental conditions. Deviation from this assumption could suggest poor quality samples or genuine biological variation of samples within the condition,
- Internal Controls: similar to Simpleaffy 3′/5′ ratio where ratios for beta-actin and GAPDH should be three or less otherwise indicating sample degradation, and
- Correlation Coefficients calculated for all pairwise comparisons of samples in the experiment: a pair of samples of the same experimental group should have a higher correlation coefficient than a pair of samples of different experimental groups.

Quality checks after data processing is also important to ensure that the normalisation was implemented as intended and to identify any outlier microarray after the normalisation process. PCA and simple correlation in GeneSpring and arrayQualityMetrics [75] from BioConductor [72] can be used to detect possible outlier microarrays [43].

DNA Microarray: Statistical Analysis

The appropriate statistical methods to use for microarray data analysis depend on the hypothesis of the experiment. There are three broad objectives that underlie microarray analyses that are discussed below. Before application of these statistical methods, the microarray data should be appropriately pre-processed to ensure that any systematic errors and confounding effects have been removed.

Class Comparison

Differential expression analysis is used to identify differences of gene expression between two experimental conditions. The t test is a simple statistical method for detecting gene differential expression in two conditions. $t = R/\mathrm{SE}$, where R is the mean log ratio of the expression levels of one gene and SE is the standard error. SE is computed by combining data across all genes [76]. However, it has low power in experiments that involve only a small number of samples in each experimental condition. In addition, the small sample size leads to inaccurate variance estimates for each gene which could lead to high t statistic despite very small fold changes in gene expression [76]. There have since been a few modifications to the t test to find a balance between power and bias in differential expression analysis. The significance analysis of microarrays (SAM) version of the t test, known as the S test [77] a constant positive term to the denominator of the standardised average, therefore, genes with small gene expression fold changes will not be selected as significant. The B statistic [78] is a Bayes log posterior odds ratio of differential expression versus non-differential gene expression. The B statistic uses data from all the genes which makes it more stable than the t statistic. Moderated t statistic [79] is similar to the t statistic except that the standard errors have been moderated to borrow information from all genes to allow inference about each individual gene so as to increase precision of variance estimates [80].

After statistical significance has been calculated for each gene, it is essential to correct for multiple testing. When the level of significance is set at 0.05, up to 5 % of the genes will be falsely scored as differentially expressed. In an experiment where thousands of genes are interrogated simultaneously, a substantial number of false positives may accumulate by chance. There are a few methods available to address this problem: the Family-wise error-rate (FWER) control methods and the False-discovery-rate (FDR) control method. FWER is the probability of accumulating one or more false positive errors over a number of statistical tests. Controlling FWER involves increasing the stringency applied to each test. The Bonferroni correction is the simplest FWER procedure where the significance level is divided by the number of tests. Other FWER control methods are the permutation-based one-step correction [81] and the Westfall and Young step-down adjustment methods [63] which are more powerful but more complex applications than the Bonferroni correction [76]. FWER criteria are stringent and may substantially result in low statistical power if the number of tests is too large. FDR is the expected proportion of false positives among all the genes initially identified as differentially expressed [82, 83]. An FDR control method proposed by Benjamini-Hochberg (BH method) is a simple Bonferroni-type procedure. It uses information from the microarray experiment to estimate the proportion of false positives that have been called. The BH method controls the expected proportion of falsely rejected hypotheses by adjusting the p value accordingly. The BH method of controlling FDR is a less stringent way to correct for multiple testing compared to methods that control for FWER. However, FDR also allows for a higher rate of false positive results and is less powerful than FWER control procedures.

Class Discovery

The basic methodology of class discovery is the cluster analysis algorithms which are mainly divided into hierarchical (Genesis [84]) or partitioning methods (*k*-means clustering [85]) or a hybrid of the two [43]. The goal of these algorithms is to find clusters of genes that are more similar to each other than to genes in other clusters.

Hierarchical clustering begins with each gene in its own cluster and each is represented in the tips of a dendrogram and the distance between two genes on the dendrogram (representing the similarities of expression patterns between the two connected genes or clusters, i.e. shorter distance equals more similar) is calculated [86]. The hierarchical clustering algorithms search the distance matrix for two or more genes that have the smallest distance. Applications of RNAseq to cancer biology between them and merge them into a cluster. The distance matrix is recalculated to include the new cluster containing more than one gene and the rest of the single-gene clusters. Hierarchical clustering algorithms do not compute a formal test statistic and therefore are unable to measure if the distances between the clusters are statistically different. External criteria are used to guide the number of clusters that can be made (e.g., if splitting the tree at one point leads to a cluster which is mostly made up of tumour samples and the other cluster which is mostly made up of normal samples, the split is considered appropriate). Nonetheless, this approach is subjective and may be prone to bias [86].

Partitioning algorithms normally require specification of a pre-defined number of classes to which the genes are partitioned into. The number of clustering parameters will also need to be defined. Partitioning clustering algorithm, *k*-means, begins with *k* initial clusters, each with a mean value. The aim is to assign each gene to the cluster whose mean has the least within-cluster sum of squares (i.e. the nearest mean). After which, the mean for each cluster is recalculated. This process is repeated until the assignment of genes to clusters no longer changes.

Class Prediction

The first step of class prediction is to identify a set of features (i.e. gene expression patterns) that has the highest discriminatory power of the class or phenotype of interest. Feature selection methods include the filter and wrapper approaches. The filter approach, in general, assesses the relevance of features by looking at the intrinsic properties of the data. The relevance is scored and the low-scoring features are removed [87]. The filter approach is computationally fast and simple but ignores the performance of the selected feature subset on the performance of the induction algorithm and is therefore, less accurate than the wrapper approach [88]. The wrapper feature selection approach involves using classifiers to evaluate subsets of features in the dataset [87]. The cross-validation, Bootstrap or split-sample approach is used to estimate the accuracy of the classifiers on an independent set of samples and the feature subset with the highest evaluation is chosen as the final set. The wrapper approach involves training a new classifier, using the induction algorithm, for each feature subset to identify the best feature subset which makes it very computationally intensive but also underlies its ability to identify feature sets with higher classification accuracies than filter approaches [88].

Once the features have been selected, a classifier is built to predict the class or phenotype of interest. This can be done using several different prediction methods such as the nearest neighbour approach [89], classification and regression trees [90] and the support vector machine [91].

A classifier can be evaluated through statistical measures such as variance or confidence intervals which can be obtained by split-sample or cross-validation procedures. In the split-sample procedure, the dataset containing samples of the class or phenotype of interest is randomly split into two; one is used to as a training set to build up the classifier whereas the other is used as a test set to measure the accuracy of the predictions made by the classifier. The cross-validation procedure is an iterative process. In each iteration process, a subset (normally one sample in the leave-one-out process [92]) of the dataset containing n samples of the class or phenotype of interest is left aside for testing, and the rest of the dataset ($n-1$) is used as a training set and a group of genes whose gene expression patterns are associated with the target class or phenotype is selected and is used to build the classifier. The classifier is used to test the left out test sample. A different subset of the dataset is used in each iteration process for n times until each sample has been left out once for testing. The series of classifiers built from each iteration process is then merged to form a fully developed classifier. One rule of assessing the results of the classifier is that the samples used for validation, using either split-sample or cross-validation procedures, must not be used for building the classifier [93].

Pathway Analysis

After identifying a subset of genes that are differentially expressed across experimental conditions, the next step is to put these findings into biological context (e.g. do these gene expression differences underlie any biological mechanism or are they predictive of the biological process that is being studied).

Gene ontology (GO) [94] is a database that provides standardised annotation of gene products that allows users of this database to query and retrieve genes and proteins based on their biological function. GO can be used to characterise the biological functions of the differentially expressed genes identified via microarray experiments. There are also other curated databases such as KEGG [95] and Biocarta [96] that provide stringently reviewed interactions between genes and gene products in different pathways. Pathway Explorer [97] and Ingenuity® Pathway Analysis (Ingenuity® Systems, http://www.ingenuity.com) are applications that enable users to analyse data from gene expression microarrays and allow mapping of the gene expression data to biological pathways. Chi-square or Fisher's exact test can be used to determine if the proportion of differentially expressed genes associated with particular pathways is larger than expected after correcting for multiple testing.

Gene Set Enrichment Analysis (GSEA) [98, 99] is a method that allows identification of sets of genes rather than single genes that are differentially expressed. The gene sets are defined based on previously known, published biological knowledge. Examples of gene sets include genes encoding products in a DNA repair pathway or genes that share the same GO category). The aim of GSEA is to determine if members of a gene set are significantly over- or under-expressed when comparing two experimental conditions and correlates the significant differential expression with phenotype. GSEA is applied to the whole data set from a microarray experiment. This is in contrast to the previously described methods which are applied only to genes identified to be significantly differentially expressed. The advantage of GSEA to single gene analysis is that GSEA can detect biological processes whose activity levels are significantly affected by expression changes across a whole network of genes despite the subtle expression changes of individual genes. Accurate definition of gene sets is crucial for a correct interpretation of the results.

Validation

Microarray technology is an efficient method for interrogating the transcriptome or a multitude of gene targets simultaneously. However, variability in microarray results exists and there is always a certain degree of uncertainty associated with each method used to detect differential gene expression despite employing the best statistical analysis tools available. Therefore, it is essential to use an orthogonal method to validate the accuracy of the technology (i.e. the target genes are in fact differentially expressed and the extent of differential expression) and also to extend the same experiment to different biological replicates from the same test population to validate the biological conclusions of the experiment (i.e. a test condition is associated with differential expression in the same sets of genes) [100]. The techniques traditionally used for validation of microarray data are qRT-PCR. The MicroArray Quality Control (MAQC) project has often used the TaqMan® Gene Expression Assays as a validation tool [49, 101].

RNASEQ

Introduction

Following the success of DNA microarrays, RNA sequencing as a technology platform has provided further insights into gene expression in cancer. It leverages rapid progress in the performance of short-read next-generation sequencing (NGS) technologies and the exponential decline in the cost of sequencing. It does, however, remain a technology under development. In particular, a gold standard software pipeline for quality control and interpretation of sequencing data is not well defined, but efforts to compare computational methods in a consistent fashion are underway [102, 103]. Improvements in RNA extraction and library preparation are also occurring constantly. Although it is not possible to describe a consensus protocol for RNAseq, an overview of methodological issues will be provided.

As has been mentioned, RNAseq does not require prior knowledge of mRNA sequence, which affords a unique advantage over DNA microarrays. This property is particularly relevant to the cancer transcriptome, which may contain unique mRNA transcripts generated through splice variants or genomic alterations. An example of this was seen in chemotherapy resistant colon cancer cell lines, where there was widespread splicing disruption detected with RNAseq [104]. RNAseq has also been used to detect novel gene fusions in tumours [105] that are difficult to detect with genome sequencing, and can represent particularly attractive drug targets [106, 107].

Table 1 Applications of RNAseq to cancer biology

Application	Examples	Reference
Global transcriptional activity	RNAseq of 79 cervical carcinoma samples	[111]
Alternate splicing	RNAseq of chemotherapy resistant and sensitive colon cancer cell lines. Widespread disruption of splicing seen in resistant cell lines	[104]
Novel transcript discovery	RNAseq of prostate cancer samples. Discovery of chimeric transcripts mediated by splicing events	[112]
Gene-fusion discovery	Discovery of gene-fusion events across multiple cancer types	[105]
Differential expression	RNAseq of different breast cancer subtypes identifies significant differential expression of transcripts	[113]

The most touted advantage of RNAseq for expression analysis per se, however, is the unlimited dynamic range when quantifying RNA transcripts. This property arises because RNAseq detects single RNA molecules, creating a linear relationship between the concentration of an RNA transcript in a sample and the number of sequence reads produced [108]. RNAseq has almost zero background signal for a given transcript, and has effectively no upper detection limit on the quantity of a transcript. This affords accurate quantification of transcripts over a very large range of expression levels. Studies in yeast suggest that greater than 9000 fold differences in expression are detectable with RNAseq [109]. In contrast microarrays have high background signal that masks detection of rare transcripts, and saturation occurs at high expression levels, limiting the dynamic range to 100 to 200 fold. This is reflected in the poor correlation of microarray data with RNAseq for low and highly expressed transcripts [110].

The aim of an RNAseq experiment is critical in refining the proposed methodology. RNAseq may be used to provide a global picture of the transcriptional activity of a tissue; examine alternative splicing of genes; discover novel transcripts, including gene-fusion events; compare transcriptional activity under different experimental conditions or between different samples, so-called differential expression. Examples of these applications of RNAseq to cancer biology are provided in Table 1.

Methods

Although the sequencing step in RNAseq is identical to that performed in next-generation sequencing of genomic DNA, RNAseq has unique requirements for sample preparation, RNA extraction and bioinformatics analyses.

Sample Preparation

RNAseq is a highly sensitive procedure. Sample input quality is a major determinant of subsequent data quality. Gene expression may change rapidly depending on external conditions, and therefore care must be taken to preserve a biologically valid transcriptome for the sample in question. This is particularly important if differential expression is to be studied between different samples.

For tumour samples harvested fresh from living tissues, it is optimal if tissue is immediately snap frozen in liquid nitrogen, or alternatively stored in commercially available RNA preservatives that operate at room temperature. Preservatives may be added to tissue before freezing, to minimise degradation during eventual thawing.

Formalin-fix paraffin-embedded (FFPE) samples may be used for RNAseq experiments, with several important caveats. Fixation itself degrades RNA in a complex fashion, which biases the transcript pool available for sequencing. Of more concern is that tissue handling for FFPE processing is usually carried out according to the needs of routine histological procedures, which are far removed from those of high fidelity RNAseq. There may be delays until fresh tissue is fixed due to the need for tissue transport and processing in the pathology department, and fixation procedures and subsequent tissue storage may be variable. Therefore, differential expression analysis between different FFPE samples may require large numbers of samples to control for non-biologic variability.

RNA Extraction

RNA extraction is carried out according to established procedures. Commercial kits are used for this purpose in published RNAseq protocols. The type of sample may determine the best method. For FFPE tumour samples, FFPE optimised RNA extraction kits may be more appropriate [114–116], although the performance of different kits has not been systematically compared for FFPE tissues. For tissues where RNA degradation is expected, this should be quantified prior to the next step.

RNA Purification

The primary RNA species of interest for RNAseq is messenger RNA (mRNA). mRNA generally comprises less than 5 % of the total RNA in a cell (Table 2). Sequencing of the gross total RNA pool will therefore produce only a small fraction of useful sequence reads. For this reason, it is desirable to enrich transcripts of interest, including mRNA and other non-coding regulatory RNAs. There are two methods currently in generalised use. The first is to subtract structural RNA (including ribosomal and transfer RNA) from the total RNA pool. Commercial kits available for this purpose remove rRNA using magnetic bead or microsphere separation [125]. Another method is to hybridise DNA to the rRNA and degrade the RNA-DNA duplexes using RNase H [126]. A less common alternative method, duplex specific nuclease (DSN) normalisation, involves a double stranded DNA specific DNase

Table 2 RNA species in mammalian cells

Species	Approximate relative abundance of total cellular RNA	Reference
Ribosomal RNA	80 %	[117]
Transfer RNA	15 %	[118]
mRNA	1–5 %	[118]
Low abundance non-coding RNAs		
Small nuclear RNA		[119]
Small nucleolar RNA		[120]
Long noncoding RNA		[121]
Antisense RNA		[122]
Telomerase RNA		[123]
Small interfering RNA		[124]

[127]. Abundant cDNAs re-anneal faster during cooling after heat denaturation, thus becoming susceptible to selective degradation by DSN, while less abundant cDNAs remain denatured and resistant to degradation. This step is carried out after reverse transcription to cDNA.

The second enrichment method exploits the fact that mRNA are poly-adenylated at their tails (poly(A)+RNA). These repeat adenine sequences can be used to purify mRNA by hybridising them to complementary thymine oligonucleotides (oligo(dT)) linked to a purification medium [128] such as a column or magnetic beads [129]. The method has sufficient specificity that it may be repeated multiple times to further enrich a sample.

Sample quality aside, the choice between rRNA depletion and poly(A)+enrichment rests on the need to detect non-coding RNA, which is only possible with rRNA depletion [130]. Alternatively, if coding mRNA are the RNA species of interest, then poly(A)+enrichment is preferable as it is a more cost-effective use of sequencing consumables. Poly(A)+enrichment is also more effective at excluding rRNA, reducing ribosomal reads to 2 % or less of total sequence reads, whereas rRNA depletion still permits 5–10 % ribosomal reads [126, 130]. Although rRNA sequences can be removed from the sequencing data post-hoc, this may reduce the power to discriminate novel transcripts [131].

Sample RNA quality is an important determinant of the optimal RNA enrichment method. Highly degraded RNA is not suitable for poly(A)+enrichment, as the 3′ end of transcripts will be selectively enriched, and discontiguous 5′ fragments from the same transcript depleted from the sample [132]. The degree of RNA fragmentation in a sample is therefore an important factor in selecting an enrichment method, and the most widely used metric of fragmentation is the RNA integrity number (RIN) [133]. The metric is calculated from various features of the RNA electropherogram of a sample, and is accessible in the analytic software accompanying the Agilent Bioanlyzer™. Samples with a RIN of 8–10 are ideal for sequencing library preparation [131]. The RIN of FFPE tumour samples is generally poor, ranging from 2–3, compared to 7–8 if the same tumour tissue is fresh frozen [115, 134]. There is no clear relationship between age of fixation and degree of RNA fragmentation, suggesting that most RNA fragmentation occurs at the time of fixation during tissue autolysis [135].

A further complication is that the RNA degradation induced by formalin fixation may not affect all RNA species equally. FFPE tissues display a reduction in exonic reads and an increase in intronic reads compared with fresh frozen tissues [126, 132]. In summary FFPE samples are highly fragmented and apparently depleted of coding transcripts, and thus poly(A)+enrichment is inappropriate for these archival samples with current technology. It should also be noted that for the same reasons discussed above, even with high quality input RNA, poly(A)+enrichment is sensitive to RNA fragmentation induced by experimental handling. It has been suggested that some older mRNA enrichment protocols employing centrifugation at 13,000 rpm may fragment transcripts and cause loss of 5′ end exons [130].

The performance of different rRNA depletion protocols was compared systematically in 2013 for low quality and degraded input RNA [132]. In this analysis, the RNase H method proved superior for low quality samples, followed by Ribo-Zero™ (Epicentre), a commercial capture hybridization method using magnetic beads. RNase H was also the cheapest method per sample, and had the lowest ribosomal reads which in this study were comparable to even poly(A)+depletion. RNase H is not available as a kit, however. Since publication of this comparison, several new rRNA depletion methods have been published, but their performance compared to other methods is unknown.

Generation of Complementary DNA (cDNA) Libraries

All commonly used next-generation sequencing platforms utilise DNA libraries. RNA must therefore be converted to cDNA to be sequenced. Although a method for directly sequencing RNA has been developed, it is not widely adopted [136]. Synthesis of cDNA has numerous variations in the published literature, and there are now RNAseq commercial kits available that incorporate this process. The essential elements are listed below

1. Priming of RNA for initiation of cDNA synthesis. This is commonly performed using random hexamer primers. For poly(A)+RNA, oligo(T) priming is less commonly used, and introduces a bias for the 3′ end of transcripts. Random primers also cause bias, as their coverage of transcript fragments is not uniform, which may affect downstream quantification of expression [137].
2. Fragmentation. This may be performed on extracted RNA or after cDNA synthesis, using chemical or mechanical fragmentation. Fragmentation of RNA has the theoretical advantage of disrupting RNA secondary structure that may prevent accurate reverse transcription thus reducing 5′ biases due to incomplete cDNA synthesis. Sequencing read coverage was more evenly distributed with RNA hydrolytic fragmentation than shearing of cDNA in one study [138].

All fragmentation methods produce their own biases, and should be tailored to the planned sequencing technology. Measurement of existing RNA degradation using the Bioanalyser or gel electrophoresis may show that FFPE or low quality samples do not require fragmentation. Removal of very small RNA fragments from degraded samples via bead purification can be performed here to improve library quality [132].

3. cDNA synthesis. Reverse transcriptase synthesis of cDNA following random priming removes information regarding the DNA strand of origin of the RNA transcript. This 'strandedness' of the RNA adds useful information to the analysis including: identification of regulatory antisense transcripts; more accurate assessment of gene expression levels; aiding detection of novel transcripts. Strand specific procedures rely on differentiating the primary RNA or cDNA strand. Preservation of strandedness has been achieved using selective adaptors to differentiate primary RNA or cDNA prior to subsequent cDNA synthesis or amplification, respectively. Chemical modification of the primary transcript has also been used. A number of methods were compared in Levin et al. [139], with dUTP second strand marking and Illumina RNA ligation performing the best. Any method of preserving strandedness will not be completely specific, and may introduce bias that should be accounted for when analysing sequence data [140]. The ongoing project of annotating non-coding RNA in the human genome will assist in these analyses. Commercial kits that preserve strandedness are available for major sequencing platforms. Stranded library preparation is not mandatory for differential gene expression analysis.
4. Library preparation. This is conducted using commercial kits optimised for the planned sequencing platform. It is decided here if paired-end or single-end sequencing will be performed, and the appropriate kit selected. Paired-end sequencing is preferred for novel transcripts and isoform discovery, but is not necessary for comparing expression between samples. Library preparation consists of attaching specific adapters to the cDNA fragment ends. This is discussed in more detail in the chapter on NGS. An important aspect of the process for RNAseq is restricting the fragment size. This may be done prior to or after adapter ligation. The optimal fragment size is generally 200 base pairs, although this may vary with the specific methodology and sequencing platform [113].

Sequencing

Next-generation sequencing of cDNA occurs according to methods used for sequencing genomic DNA, this would be discussed in more detail in the chapter on NGS. The majority of RNAseq experiments in cancer, including those performed by The Cancer Genome Atlas, have been conducted using the Illumina platform [141], and is the focus of this discussion. It should be noted that the maximum read length of any Illumina sequencer is 300 base pairs at present, although 100–150 base pair reads are usually employed. The length of these so-called short reads is generally much shorter than the length of RNA transcripts. Most of the work in interpreting Illumina sequencing data arises from assigning these short reads to the parent transcripts. The Pacific Biosciences Single Molecule, Real Time (SMRT) technology is able to sequence reads in excess of 1000 base pairs [142]. It thus has a theoretical advantage over Illumina technology in resolving comprehensive splicing architecture and fusion transcripts, but remains an immature platform.

The sequencing parameters for RNAseq depend on the aim of the experiment. The first consideration is multiplexing of samples. A sample may be barcoded and then mixed with other barcoded samples such that it can be sequenced in the same functional unit of the NGS platform. For Illumina sequencers, this refers to mixing samples on the same flow cell lane. This reduces the sequencing cost per sample. Barcoding is performed by adding a short nucleotide sequence to a sample during library preparation. Reads pertaining to that sample can then be identified after sequencing.

Read length and sequence depth are two important parameters in planning for a sequencing run. Read length refers to how much of each cDNA fragment is sequenced. Generally, shorter read lengths 35–50 bp are adequate for quantifying gene expression, and longer read lengths 100 bp or more are preferable for novel transcript discovery [131].

Sequencing depth refers to the number of reads arising from a particular RNA transcript. In practical terms, to generate a certain amount of read data a sequencing run uses fixed units of consumables that have an associated cost. The total number of reads therefore scales in a linear fashion with the cost of the experiment. Modern Illumina sequencers produce at least 100 million reads per flow cell lane. The number of samples multiplexed per lane will therefore determine the number of expected reads per sample and the sequencing cost per sample. Multiple factors will decide the number of desired reads per sample, which may range from ten million reads for simply establishing the presence or absence of abundant transcripts, to 20–40 million reads for quantification of abundantly expressed transcripts, to 80 million reads for quantification of rare transcripts. It has been estimated that to provide a comprehensive overview of all transcripts and isoforms in human samples, >200 million paired-end reads may be required [143]. An alternative approach to escalating sequencing depth for less abundant transcripts is to perform targeted enrichment of select RNA transcripts in a similar fashion to targeted exome enrichment [144, 145]. This approach provides very high read coverage for the selected transcripts and has been performed using commercially available exome capture kits applied to cDNA libraries after reverse transcription from RNA [146]. Capture efficiency does introduce another potential source of noise into quantification of gene expression.

Data Analysis

As discussed above, the majority of RNAseq data is in short read form produced on the Illumina platform. Analysis of short read RNAseq data will be discussed here.

Pre-processing and Alignment

Analysis of RNAseq data follows many of the usual steps in analysing any sequencing data. Sequencing data arrives as a list of sequence reads with an associated index of the quality of each base in the read. This list will often be a mixture of several samples. The first step is to de-multiplex the samples by segregating the reads based on the indexes which were attached during library preparation, and this is often performed by the sequencer. The next step is to trim any adapter sequences. There are many publicly available software solutions for adapter trimming, summarised in [147].

The next step in cancer RNAseq is to align the processed reads to the reference genome. In contrast to DNAseq, the sequence of RNA transcripts contains exon junctions, as intronic sequences have been spliced out during transcription. To map these exon junctions to the genome, the aligner must use additional information regarding the sequences of all known exon junctions, or be able to map exonic subsequences within a sequence read in a naïve fashion. There are many aligners available which may employ both of these methods, and several have been developed specifically for RNAseq data [148, 149]. Some aligners such as MapSplice, for example, have a function to determine possible fusion transcripts [141, 150].

Annotation

Annotation is the process of allocating the aligned sequence reads to known gene model features such as genomic loci, exons or transcripts. This is a prelude to qualitative and quantitative expression analyses. There are multiple annotation databases which contain information on the transcript sequences and splicing of known genes, such as Ensembl, RefSeq and AceView [151–153]. The annotation data used in this process will have significant effects on downstream analyses and should be chosen carefully and consistently [154, 155]. RefSeq is the simplest annotation and appears to give the most consistent gene expression results between samples, whereas AceView is the most comprehensive.

References

1. Gerlinger M, Rowan AJ, Horswell S, Larkin J, Endesfelder D, Gronroos E, Martinez P, Matthews N, Stewart A, Tarpey P et al (2012) Intratumor heterogeneity and branched evolution revealed by multiregion sequencing. N Engl J Med 366:883–892
2. Park SY, Gonen M, Kim HJ, Michor F, Polyak K (2010) Cellular and genetic diversity in the progression of in situ human breast carcinomas to an invasive phenotype. J Clin Invest 120(2):636–644
3. Lawrence MS, Stojanov P, Polak P, Kryukov GV, Cibulskis K, Sivachenko A, Carter SL, Stewart C, Mermel CH, Roberts SA et al (2013) Mutational heterogeneity in cancer and the search for new cancer-associated genes. Nature 499:214–218
4. Kevil CG, Walsh L, Laroux FS, Kalogeris T, Grisham MB, Alexander JS (1997) An improved, rapid Northern protocol. Biochem Biophys Res Commun 238(2):277–279

5. VanGuilder HD, Vrana KE, Freeman WM (2008) Twenty-five years of quantitative PCR for gene expression analysis. Biotechniques 44(5):619–626
6. Becker-Andre M, Hahlbrock K (1989) Absolute mRNA quantification using the polymerase chain reaction (PCR). A novel approach by a PCR aided transcript titration assay (PATTY). Nucleic Acids Res 17(22):9437–9446
7. Noonan KE, Beck C, Holzmayer TA, Chin JE, Wunder JS, Andrulis IL, Gazdar AF, Willman CL, Griffith B, Vonhoff DD et al (1990) Quantitative-analysis of MDR1 (multidrug resistance) gene-expression in human tumors by polymerase chain-reaction. Proc Natl Acad Sci U S A 87(18):7160–7164
8. Wang J, Lin M, Crenshaw A, Hutchinson A, Hicks B, Yeager M, Berndt S, Huang W-Y, Hayes RB, Chanock SJ et al (2009) High-throughput single nucleotide polymorphism genotyping using nanofluidic dynamic arrays. BMC Genomics 10
9. Thiel CT, Kraus C, Rauch A, Ekici AB, Rautenstrauss B, Reis A (2003) A new quantitative PCR multiplex assay for rapid analysis of chromosome 17p11.2-12 duplications and deletions leading to HMSN/HNPP. Eur J Hum Genet 11(2):170–178
10. Velculescu VE, Zhang L, Vogelstein B, Kinzler KW (1995) Serial analysis of gene-expression. Science 270(5235):484–487
11. Schena M, Shalon D, Davis RW, Brown PO (1995) Quantitative monitoring of gene-expression patterns with a complementary-DNA microarray. Science 270(5235):467–470
12. Lashkari DA, DeRisi JL, McCusker JH, Namath AF, Gentile C, Hwang SY, Brown PO, Davis RW (1997) Yeast microarrays for genome wide parallel genetic and gene expression analysis. Proc Natl Acad Sci U S A 94(24):13057–13062
13. Perou C, Sorlie T, Eisen M, van de Rijn M, Jeffrey S, Rees C, Pollack J, Ross D, Johnsen H, Akslen L et al (2000) Molecular portraits of human breast tumours. Nature 406:747–752
14. Sorlie T, Perou CM, Tibshirani R, Aas T, Geisler S, Johnsen H, Hastie T, Eisen MB, van de Rijn M, Jeffrey SS et al (2001) Gene expression patterns of breast carcinomas distinguish tumor subclasses with clinical implications. Proc Natl Acad Sci U S A 98(19):10869–10874
15. Paik S, Shak S, Tang G, Kim C, Baker J, Cronin M, Baehner FL, Walker MG, Watson D, Park T et al (2004) A multigene assay to predict recurrence of tamoxifen-treated, node-negative breast cancer. N Engl J Med 351:2817–2826
16. van 't Veer LJ, Dai HY, van de Vijver MJ, He YDD, Hart AAM, Mao M, Peterse HL, van der Kooy K, Marton MJ, Witteveen AT et al (2002) Gene expression profiling predicts clinical outcome of breast cancer. Nature 415(6871):530–536
17. Bueno-de-Mesquita JM, van Harten WH, Retel VP, van't Veer LJ, van Dam FSAM, Karsenberg K, Douma KFL, van Tinteren H, Peterse JL, Wesseling J et al (2007) Use of 70-gene signature to predict prognosis of patients with node-negative breast cancer: a prospective community-based feasibility study (RASTER). Lancet Oncol 8(12):1079–1087
18. Klang SH, Hammerman A, Liebermann N, Efrat N, Doberne J, Hornberger J (2010) Economic implications of 21-gene breast cancer risk assay from the perspective of an Israeli-managed health-care organization. Value Health 13(4):381–387
19. Partin JF, Mamounas EP (2011) Impact of the 21-gene recurrence score assay compared with standard clinicopathologic guidelines in adjuvant therapy selection for node-negative, estrogen receptor-positive breast cancer. Ann Surg Oncol 18(12):3399–3406
20. Kapronov P, Sementchenko VI, Gingeras TR (2003) Beyond expression profiling: next generation uses of high density oligonucleotide arrays. Brief Funct Genomic Proteomic 2(1):47–56
21. Hacia JG, Collins FS (1999) Mutational analysis using oligonucleotide microarrays. J Med Genet 36(10):730–736
22. Bertone P, Stolc V, Royce TE, Rozowsky JS, Urban AE, Zhu X, Rinn JL, Tongprasit W, Samanta M, Weissman S et al (2004) Global identification of human transcribed sequences with genome tiling arrays. Science 306(5705):2242–2246
23. Manak JR, Dike S, Sementchenko V, Kapranov P, Biemar F, Long J, Cheng J, Bell I, Ghosh S, Piccolboni A et al (2006) Biological function of unannotated transcription during the early development of Drosophila melanogaster. Nat Genet 38(10):1151–1158

24. Ishida H, Yagi T, Tanaka M, Tokuda Y, Kamoi K, Hongo F, Kawauchi A, Nakano M, Miki T, Tashiro K (2013) Identification of a novel gene by whole human genome tiling array. Gene 516(1):33–38
25. Coman D, Gruissem W, Hennig L (2013) Transcript profiling in Arabidopsis with genome tiling microarrays. Methods Mol Biol 1067:35–49
26. Mockler TC, Ecker JR (2005) Applications of DNA tiling arrays for whole-genome analysis. Genomics 85(1):1–15
27. Wong CW, Albert TJ, Vega VB, Norton JE, Cutler DJ, Richmond TA, Stanton LW, Liu ET, Miller LD (2004) Tracking the evolution of the SARS coronavirus using high-throughput, high-density resequencing arrays. Genome Res 14(3):398–405
28. Liu C, Aronow BJ, Jegga AG, Wang N, Miethke A, Mourya R, Bezerra JA (2007) Novel resequencing chip customized to diagnose mutations in patients with inherited syndromes of intrahepatic cholestasis. Gastroenterology 132(1):119–126
29. Kothiyal P, Cox S, Ebert J, Husami A, Kenna MA, Greinwald JH, Aronow BJ, Rehm HL (2010) High-throughput detection of mutations responsible for childhood hearing loss using resequencing microarrays. BMC Biotechnol 10
30. Fokstuen S, Munoz A, Melacini P, Iliceto S, Perrot A, Oezcelik C, Jeanrenaud X, Rieubland C, Farr M, Faber L et al (2011) Rapid detection of genetic variants in hypertrophic cardiomyopathy by custom DNA resequencing array in clinical practice. J Med Genet 48(8):572–576
31. Bertone P, Trifonov V, Rozowsky JS, Schubert F, Emanuelsson O, Karro J, Kao MY, Snyder M, Gerstein M (2006) Design optimization methods for genomic DNA tiling arrays. Genome Res 16(2):271–281
32. Jurka J, Kapitonov VV, Pavlicek A, Klonowski P, Kohany O, Walichiewicz J (2005) Repbase update, a database of eukaryotic repetitive elements. Cytogenet Genome Res 110(1-4):462–467
33. Wang XW, Seed B (2003) Selection of oligonucleotide probes for protein coding sequences. Bioinformatics 19(7):796–802
34. Sorek R, Cossart P (2010) Prokaryotic transcriptomics: a new view on regulation, physiology and pathogenicity. Nat Rev Genet 11(1):9–16
35. Morin RD, Bainbridge M, Fejes A, Hirst M, Krzywinski M, Pugh TJ, McDonald H, Varhol R, Jones SJM, Marra MA (2008) Profiling the HeLa S3 transcriptome using randomly primed cDNA and massively parallel short-read sequencing. Biotechniques 45(1):81–94
36. Shah SP, Roth A, Goya R, Oloumi A, Ha G, Zhao Y, Turashvili G, Ding J, Tse K, Haffari G et al (2012) The clonal and mutational evolution spectrum of primary triple-negative breast cancers. Nature 486:395–399
37. Morin RD, Mendez-Lago M, Mungall AJ, Goya R, Mungall KL, Corbett RD, Johnson NA, Severson TM, Chiu R, Field M et al (2011) Frequent mutation of histone-modifying genes in non-Hodgkin lymphoma. Nature 476(7360):298–303
38. Ehrenreich A (2006) DNA microarray technology for the microbiologist: an overview. Appl Microbiol Biotechnol 73(2):255–273
39. Taton TA, Mirkin CA, Letsinger RL (2000) Scanometric DNA array detection with nanoparticle probes. Science 289(5485):1757–1760
40. Pease AC, Solas D, Sullivan EJ, Cronin MT, Holmes CP, Fodor SPA (1994) Light-generated oligonucleotide arrays for rapid DNA-sequence analysis. Proc Natl Acad Sci U S A 91(11):5022–5026
41. Fodor SPA, Read JL, Pirrung MC, Stryer L, Lu AT, Solas D (1991) Light-directed, spatially addressable parallel chemical synthesis. Science 251(4995):767–773
42. Hughes TR, Mao M, Jones AR, Burchard J, Marton MJ, Shannon KW, Lefkowitz SM, Ziman M, Schelter JM, Meyer MR et al (2001) Expression profiling using microarrays fabricated by an ink-jet oligonucleotide synthesizer. Nat Biotechnol 19(4):342–347
43. Sanchez-Cabo F, Rainer J, Dopazo A, Trajanoski Z, Hackl H (2011) Insights into global mechanism and disease by gene expression profiling. Methods Mol Biol 719:269–298
44. Yang YH, Speed T (2002) Design issues for cDNA microarray experiments. Nat Rev Genet 3(8):579–588

45. Golub TR, Slonim DK, Tamayo P, Huard C, Gaasenbeek M, Mesirov JP, Coller H, Loh ML, Downing JR, Caligiuri MA et al (1999) Molecular classification of cancer: class discovery and class prediction by gene expression monitoring with specific chromosomal translocations. Science 286(286):531–537
46. Dobbin K, Simon R (2002) Comparison of microarray designs for class comparison and class discovery. Bioinformatics 18(11):1438–1445
47. Churchill GA (2002) Fundamentals of experimental design for cDNA microarrays. Nat Genet 32:490–495
48. Rosenzweig BA, Pine PS, Domon OE, Morris SM, Chen JJ, Sistare FD (2004) Dye-bias correction in dual-labeled cDNA microarray gene expression measurements. Environ Health Perspect 112(4):480–487
49. Patterson TA, Lobenhofer EK, Fulmer-Smentek SB, Collins PJ, Chu TM, Bao WJ, Fang H, Kawasaki ES, Hager J, Tikhonova IR et al (2006) Performance comparison of one-color and two-color platforms within the MicroArray Quality Control (MAQC) project. Nat Biotechnol 24(9):1140–1150
50. Peixoto B, Vencio R, Egidio C, Mota-Vieira L, Verjovski-Almeida S, Reis E (2006) Evaluation of reference-based two-color methods for measurement of gene expression ratios using spotted cDNA microarrays. BMC Genomics 7(1):35
51. Kerr M, Churchill G (2001) Experimental design for gene expression microarrays. Biostatistics 2:183–201
52. Teo ZL, McQueen-Miscamble L, Turner K, Martinez G, Madakashira B, Dedhar S, Robinson ML, de Iongh RU (2014) Integrin linked kinase (ILK) is required for lens epithelial cell survival, proliferation and differentiation. Exp Eye Res 121:130–142
53. Bhattacharjee A, Richards WG, Staunton J, Li C, Monti S, Vasa P, Ladd C, Beheshti J, Bueno R, Gillette M et al (2001) Classification of human lung carcinomas by mRNA expression profiling reveals distinct adenocarcinoma subclasses. Proc Natl Acad Sci 98(24):13790–13795
54. Steger D, Berry D, Haider S, Horn M, Wagner M, Stocker R, Loy A (2011) Systematic spatial bias in DNA microarray hybridization is caused by probe spot position-dependent variability in lateral diffusion. PLoS One 6(8), e23727
55. Ritchie ME, Silver J, Oshlack A, Holmes M, Diyagama D, Holloway A, Smyth GK (2007) A comparison of background correction methods for two-colour microarrays. Bioinformatics 23(20):2700–2707
56. Lockhart DJ, Dong HL, Byrne MC, Follettie MT, Gallo MV, Chee MS, Mittmann M, Wang CW, Kobayashi M, Horton H et al (1996) Expression monitoring by hybridization to high-density oligonucleotide arrays. Nat Biotechnol 14(13):1675–1680
57. Irizarry RA, Hobbs B, Collin F, Beazer-Barclay YD, Antonellis KJ, Scherf U, Speed TP (2003) Exploration, normalization, and summaries of high density oligonucleotide array probe level data. Biostatistics 4(2):249–264
58. Naef F, Lim DA, Patil N, Magnasco M (2002) DNA hybridization to mismatched templates: a chip study. Phys Rev E Stat Nonlin Soft Matter Phys 65(4 Pt 1):040902
59. McGee M, Chen Z (2006) Parameter estimation for the exponential-normal convolution model for background correction of affymetrix GeneChip data. Stat Appl Genet Mol Biol 5
60. Neuvial P, Hupe P, Brito I, Liva S, Manie E, Brennetot C, Radvanyi F, Aurias A, Barillot E (2006) Spatial normalization of array-CGH data. BMC Bioinformatics 7:264
61. Suarez-Farinas M, Pellegrino M, Wittkowski KM, Magnasco MO (2005) Harshlight: a "corrective make-up" program for microarray chips. BMC Bioinformatics 6:294
62. Yang Y, Dudoit S, Luu P, Lin D, Peng V, Ngai J, Speed T (2002) Normalization for cDNA microarray data: a robust composite method addressing single and multiple slide systematic variation. Nucleic Acids Res 30, e15
63. Dudoit S, Yang YH, Callow MJ, Speed TP (2002) Statistical methods for identifying differentially expressed genes in replicated cDNA microarray experiments. Stat Sin 12(1):111–139
64. Bolstad BM, Irizarry RA, Åstrand M, Speed TP (2003) A comparison of normalization methods for high density oligonucleotide array data based on variance and bias. Bioinformatics 19(2):185–193

65. Goryachev AB, Macgregor PF, Edwards AM (2001) Unfolding of microarray data. J Comput Biol 8(4):443–461
66. Kerr MK, Martin M, Churchill GA (2000) Analysis of variance for gene expression microarray data. J Comput Biol 7(6):819–837
67. Berger J, Hautaniemi S, Jarvinen A-K, Edgren H, Mitra S, Astola J (2004) Optimized LOWESS normalization parameter selection for DNA microarray data. BMC Bioinformatics 5(1):194
68. Cleveland WS (1979) Robust locally weighted regression and smoothing scatterplots. J Am Stat Assoc 74(368):829–836
69. Cleveland WS (1981) Lowess – a program for smoothing scatterplots by robust locally weighted regression. Am Stat 35(1):54
70. Irizarry RA, Bolstad BM, Collin F, Cope LM, Hobbs B, Speed TP (2003) Summaries of Affymetrix GeneChip probe level data. Nucleic Acids Res 31(4), e15
71. Petri T, Berchtold E, Zimmer R, Friedel C (2012) Detection and correction of probe-level artefacts on microarrays. BMC Bioinformatics 13(1):114
72. Gentleman RC, Carey VJ, Bates DM, Bolstad B, Dettling M, Dudoit S, Ellis B, Gautier L, Ge YC, Gentry J et al (2004) Bioconductor: open software development for computational biology and bioinformatics. Genome Biol 5(10)
73. Wilson CL, Miller CJ (2005) Simpleaffy: a BioConductor package for Affymetrix Quality Control and data analysis. Bioinformatics 21(18):3683–3685
74. GeneSpring GX Software [http://www.chem.agilent.com]
75. Kauffmann A, Gentleman R, Huber W (2009) arrayQualityMetrics—a bioconductor package for quality assessment of microarray data. Bioinformatics 25(3):415–416
76. Cui XQ, Churchill GA (2003) Statistical tests for differential expression in cDNA microarray experiments. Genome Biol 4(4):210
77. Tusher VG, Tibshirani R, Chu G (2001) Significance analysis of microarrays applied to the ionizing radiation response. Proc Natl Acad Sci U S A 98(9):5116–5121
78. Lonnstedt I, Speed T (2002) Replicated microarray data. Stat Sin 12(1):31–46
79. Smyth GK (2005) Limma: linear models for microarray data. In: Gentleman R, Carey VJ, Huber W, Irizarry RA, Dudoit S (eds) Bioinformatics and computational biology solution using R and bioconductor. Springer, New York, pp 397–420
80. Smyth GK (2004) Linear models and empirical Bayes methods for assessing differential expression in microarray experiments. Stat Appl Genet Mol Biol 3(Article 3): Article 3
81. Wu H, Kerr MK, Cui XQ, Churchill GA (2003) MAANOVA: a software package for the analysis of spotted cDNA microarray experiments. In: Parmigiani G, Garret ES, Irizarry RA, Zeger SL (eds) The analysis of gene expression data: an overview of methods and software, 1st edn. Springer, London, pp 313–342
82. Benjamini Y, Hochberg Y (1995) Controlling the false discovery rate – a practical and powerful approach to multiple testing. J R Stat Soc Series B Methodol 57(1):289–300
83. Benjamini Y, Yekutieli D (2001) The control of the false discovery rate in multiple testing under dependency. Ann Stat 29(4):1165–1188
84. Sturn A, Quackenbush J, Trajanoski Z (2002) Genesis: cluster analysis of microarray data. Bioinformatics 18(1):207–208
85. Hartigan JA, Wong MA (1979) A K-means clustering algorithm. Appl Stat 28(1):100–108
86. Shannon W, Culverhouse R, Duncan J (2003) Analyzing microarray data using cluster analysis. Pharmacogenomics 4(1):41–52
87. Kohavi R, John GH (1997) Wrappers for feature subset selection. Artif Intell 97(1–2):273–324
88. Inza I, Larranaga P, Blanco R, Cerrolaza AJ (2004) Filter versus wrapper gene selection approaches in DNA microarray domains. Artif Intell Med 31(2):91–103
89. Cover TM, Hart PE (1967) Nearest neighbor pattern classification. IEEE Trans Inf Theory 13(1):21–27
90. De'ath G, Fabricius KE (2000) Classification and regression trees: a powerful yet simple technique for ecological data analysis. Ecology 81(11):3178–3192

91. Furey TS, Cristianini N, Duffy N, Bednarski DW, Schummer M, Haussler D (2000) Support vector machine classification and validation of cancer tissue samples using microarray expression data. Bioinformatics 16(10):906–914
92. Michaelsen J (1987) Cross-validation in statistical climate forecast models. J Clim Appl Meteorol 26(11):1589–1600
93. Dupuy A, Simon RM (2007) Critical review of published microarray studies for cancer outcome and guidelines on statistical analysis and reporting. J Natl Cancer Inst 99(2):147–157
94. Ashburner M, Ball CA, Blake JA, Botstein D, Butler H, Cherry JM, Davis AP, Dolinski K, Dwight SS, Eppig JT et al (2000) Gene ontology: tool for the unification of biology. Nat Genet 25(1):25–29
95. Kanehisa M, Goto S (2000) KEGG: Kyoto encyclopedia of genes and genomes. Nucleic Acids Res 28:27–30
96. Nishimura D (2001) BioCarta. Biotech Software & Internet Report 2(3):4
97. Mlecnik B, Scheideler M, Hackl H, Hartler J, Sanchez-Cabo F, Trajanoski Z (2005) PathwayExplorer: web service for visualizing high-throughput expression data on biological pathways. Nucleic Acids Res 33:W633–W637
98. Mootha VK, Lindgren CM, Eriksson K-F, Subramanian A, Sihag S, Lehar J, Puigserver P, Carlsson E, Ridderstrale M, Laurila E et al (2003) PGC-1[alpha]-responsive genes involved in oxidative phosphorylation are coordinately downregulated in human diabetes. Nat Genet 34(3):267–273
99. Subramanian A, Tamayo P, Mootha VK, Mukherjee S, Ebert BL, Gillette MA, Paulovich A, Pomeroy SL, Golub TR, Lander ES et al (2005) Gene set enrichment analysis: a knowledge-based approach for interpreting genome-wide expression profiles. Proc Natl Acad Sci U S A 102(43):15545–15550
100. Allison DB, Cui XQ, Page GP, Sabripour M (2006) Microarray data analysis: from disarray to consolidation and consensus. Nat Rev Genet 7(1):55–65
101. Shi L, Reid LH, Jones WD, Shippy R, Warrington JA, Baker SC, Collins PJ, de Longueville F, Kawasaki ES, Lee KY et al (2006) The MicroArray Quality Control (MAQC) project shows inter- and intraplatform reproducibility of gene expression measurements. Nat Biotechnol 24(9):1151–1161
102. 't Hoen PAC, Friedländer MR, Almlöf J, Sammeth M, Pulyakhina I, Anvar SY, Laros JFJ, Buermans HPJ, Karlberg O, Brännvall M et al (2013) Reproducibility of high-throughput mRNA and small RNA sequencing across laboratories. Nat Biotechnol 31:1015–1022
103. Steijger T, Abril JF, Engström PG, Kokocinski F, Akerman M, Alioto T, Ambrosini G, Antonarakis SE, Behr J, Bertone P et al (2013) Assessment of transcript reconstruction methods for RNA-seq. Nat Methods 10:1177–1184
104. Griffith M, Griffith OL, Mwenifumbo J, Goya R, Morrissy AS, Morin RD, Corbett R, Tang MJ, Hou Y-CC, Pugh TJ et al (2010) Alternative expression analysis by RNA sequencing. Nat Methods 7:843–847
105. Maher CA, Kumar-Sinha C, Cao X, Kalyana-Sundaram S, Han B, Jing X, Sam L, Barrette T, Palanisamy N, Chinnaiyan AM (2009) Transcriptome sequencing to detect gene fusions in cancer. Nature 458:97–101
106. Joensuu H, Roberts PJ, Sarlomo-Rikala M, Andersson LC, Tervahartiala P, Tuveson D, Silberman S, Capdeville R, Dimitrijevic S, Druker B et al (2001) Effect of the tyrosine kinase inhibitor STI571 in a patient with a metastatic gastrointestinal stromal tumor. N Engl J Med 344:1052–1056
107. Kwak EL, Bang Y-J, Camidge DR, Shaw AT, Solomon B, Maki RG, Ou S-HI, Dezube BJ, Jänne PA, Costa DB et al (2010) Anaplastic lymphoma kinase inhibition in non-small-cell lung cancer. N Engl J Med 363:1693–1703
108. Jiang L, Schlesinger F, Davis CA, Zhang Y, Li R, Salit M, Gingeras TR, Oliver B (2011) Synthetic spike-in standards for RNA-seq experiments. Genome Res 21:1543–1551
109. Nagalakshmi U, Wang Z, Waern K, Shou C, Raha D, Gerstein M, Snyder M (2008) The transcriptional landscape of the yeast genome defined by RNA sequencing. Science 320:1344–1349

110. Wang Z, Gerstein M, Snyder M (2009) RNA-Seq: a revolutionary tool for transcriptomics. Nat Rev Genet 10:57–63
111. Ojesina AI, Lichtenstein L, Freeman SS, Pedamallu CS, Imaz-Rosshandler I, Pugh TJ, Cherniack AD, Ambrogio L, Cibulskis K, Bertelsen B et al (2013) Landscape of genomic alterations in cervical carcinomas. Nature 506:371–375
112. Kannan K, Wang L, Wang J, Ittmann MM, Li W, Yen L (2011) Recurrent chimeric RNAs enriched in human prostate cancer identified by deep sequencing. Proc Natl Acad Sci U S A 108:9172–9177
113. Eswaran J, Cyanam D, Mudvari P, Reddy SDN, Pakala SB, Nair SS, Florea L, Fuqua SAW, Godbole S, Kumar R (2012) Transcriptomic landscape of breast cancers through mRNA sequencing. Sci Rep 2:264
114. Cronin M, Pho M, Dutta D, Stephans JC, Shak S, Kiefer MC, Esteban JM, Baker JB (2004) Measurement of gene expression in archival paraffin-embedded tissues: development and performance of a 92-gene reverse transcriptase-polymerase chain reaction assay. Am J Pathol 164:35–42
115. Norton N, Sun Z, Asmann YW, Serie DJ, Necela BM, Bhagwate A, Jen J, Eckloff BW, Kalari KR, Thompson KJ et al (2013) Gene expression, single nucleotide variant and fusion transcript discovery in archival material from breast tumors. PLoS One 8, e81925
116. Sinicropi D, Qu K, Collin F, Crager M, Liu M-L, Pelham RJ, Pho M, Dei Rossi A, Jeong J, Scott A et al (2012) Whole transcriptome RNA-Seq analysis of breast cancer recurrence risk using formalin-fixed paraffin-embedded tumor tissue. PLoS One 7, e40092
117. von der Haar T (2008) A quantitative estimation of the global translational activity in logarithmically growing yeast cells. BMC Syst Biol 2:87
118. Warner JR (1999) The economics of ribosome biosynthesis in yeast. Trends Biochem Sci 24:437–440
119. Thore S, Mayer C, Sauter C, Weeks S, Suck D (2003) Crystal structures of the Pyrococcus abyssi Sm core and its complex with RNA. Common features of RNA binding in archaea and eukarya. J Biol Chem 278:1239–1247
120. Kiss T (2001) Small nucleolar RNA-guided post-transcriptional modification of cellular RNAs. EMBO J 20:3617–3622
121. Tsai M-C, Manor O, Wan Y, Mosammaparast N, Wang JK, Lan F, Shi Y, Segal E, Chang HY (2010) Long noncoding RNA as modular scaffold of histone modification complexes. Science 329:689–693
122. Brantl S (2007) Regulatory mechanisms employed by cis-encoded antisense RNAs. Curr Opin Microbiol 10:102–109
123. Lustig AJ (1999) Crisis intervention: the role of telomerase. Proc Natl Acad Sci U S A 96:3339–3341
124. Ahmad K, Henikoff S (2002) Epigenetic consequences of nucleosome dynamics. Cell 111:281–284
125. Tariq MA, Kim HJ, Jejelowo O, Pourmand N (2011) Whole-transcriptome RNAseq analysis from minute amount of total RNA. Nucleic Acids Res 39, e120
126. Morlan JD, Qu K, Sinicropi DV (2012) Selective depletion of rRNA enables whole transcriptome profiling of archival fixed tissue. PLoS One 7, e42882
127. Zhulidov PA, Bogdanova EA, Shcheglov AS, Vagner LL, Khaspekov GL, Kozhemyako VB, Matz MV, Meleshkevitch E, Moroz LL, Lukyanov SA et al (2004) Simple cDNA normalization using kamchatka crab duplex-specific nuclease. Nucleic Acids Res 32, e37
128. Kingston RE (2001) Preparation of poly(A)+RNA. In: Ausubel FM et al (eds) Current protocols in molecular biology. Wiley, New York (Chapter 4, Unit 4.5)
129. Mortazavi A, Williams BA, McCue K, Schaeffer L, Wold B (2008) Mapping and quantifying mammalian transcriptomes by RNA-Seq. Nat Methods 5:621–628
130. Cui P, Lin Q, Ding F, Xin C, Gong W, Zhang L, Geng J, Zhang B, Yu X, Yang J et al (2010) A comparison between ribo-minus RNA-sequencing and polyA-selected RNA-sequencing. Genomics 96:259–265

131. Zeng W, Mortazavi A (2012) Technical considerations for functional sequencing assays. Nat Immunol 13:802–807
132. Adiconis X, Borges-Rivera D, Satija R, DeLuca DS, Busby MA, Berlin AM, Sivachenko A, Thompson DA, Wysoker A, Fennell T et al (2013) Comparative analysis of RNA sequencing methods for degraded or low-input samples. Nat Methods 10:623–629
133. Schroeder A, Mueller O, Stocker S, Salowsky R, Leiber M, Gassmann M, Lightfoot S, Menzel W, Granzow M, Ragg T (2006) The RIN: an RNA integrity number for assigning integrity values to RNA measurements. BMC Mol Biol 7:3
134. Dunn TA, Fedor H, Isaacs WB, De Marzo AM, Luo J (2009) Genome-wide expression analysis of recently processed formalin-fixed paraffin embedded human prostate tissues. Prostate 69:214–218
135. von Ahlfen S, Missel A, Bendrat K, Schlumpberger M (2007) Determinants of RNA quality from FFPE samples. PLoS One 2, e1261
136. Ozsolak F, Platt AR, Jones DR, Reifenberger JG, Sass LE, McInerney P, Thompson JF, Bowers J, Jarosz M, Milos PM (2009) Direct RNA sequencing. Nature 461:814–818
137. Roberts A, Trapnell C, Donaghey J, Rinn JL, Pachter L (2011) Improving RNA-Seq expression estimates by correcting for fragment bias. Genome Biol 12:R22
138. Huang R, Jaritz M, Guenzl P, Vlatkovic I, Sommer A, Tamir IM, Marks H, Klampfl T, Kralovics R, Stunnenberg HG et al (2011) An RNA-Seq strategy to detect the complete coding and non-coding transcriptome including full-length imprinted macro ncRNAs. PLoS One 6, e27288
139. Levin JZ et al (2010) Comprehensive comparative analysis of strand-specific RNA sequencing methods. Nat Methods 7(9):709–715
140. Ozsolak F, Milos PM (2011) RNA sequencing: advances, challenges and opportunities. Nat Rev Genet 12:87–98
141. The Cancer Genome Atlas Research Network (2012) Comprehensive genomic characterization of squamous cell lung cancers. Nature 489:519–525
142. Tilgner H, Grubert F, Sharon D, Snyder MP (2014) Defining a personal, allele-specific, and single-molecule long-read transcriptome. Proc Natl Acad Sci U S A 111:9869–9874
143. Tarazona S, García-Alcalde F, Dopazo J, Ferrer A, Conesa A (2011) Differential expression in RNA-seq: a matter of depth. Genome Res 21:2213–2223
144. Levin JZ, Berger MF, Adiconis X, Rogov P, Melnikov A, Fennell T, Nusbaum C, Garraway LA, Gnirke A (2009) Targeted next-generation sequencing of a cancer transcriptome enhances detection of sequence variants and novel fusion transcripts. Genome Biol 10:R115
145. Mercer TR, Gerhardt DJ, Dinger ME, Crawford J, Trapnell C, Jeddeloh JA, Mattick JS, Rinn JL (2012) Targeted RNA sequencing reveals the deep complexity of the human transcriptome. Nat Biotechnol 30:99–104
146. Halvardson J, Zaghlool A, Feuk L (2013) Exome RNA sequencing reveals rare and novel alternative transcripts. Nucleic Acids Res 41, e6
147. Jiang H, Lei R, Ding S-W, Zhu S (2014) Skewer: a fast and accurate adapter trimmer for next-generation sequencing paired-end reads. BMC Bioinformatics 15:182
148. Engström PG, Steijger T, Sipos B, Grant GR, Kahles A, Alioto T, Behr J, Bertone P, Bohnert R, Campagna D et al (2013) Systematic evaluation of spliced alignment programs for RNA-seq data. Nat Methods 10:1185–1191
149. Fonseca NA, Rung J, Brazma A, Marioni JC (2012) Tools for mapping high-throughput sequencing data. Bioinformatics 28:3169–3177
150. Wang K, Singh D, Zeng Z, Coleman SJ, Huang Y, Savich GL, He X, Mieczkowski P, Grimm SA, Perou CM et al (2010) MapSplice: accurate mapping of RNA-seq reads for splice junction discovery. Nucleic Acids Res 38, e178
151. Flicek P, Amode MR, Barrell D, Beal K, Brent S, Carvalho-Silva D, Clapham P, Coates G, Fairley S, Fitzgerald S (2012) Ensembl 2012. Nucleic Acids Res 40:D84–D90
152. Pruitt KD, Tatusova T, Maglott DR (2007) NCBI reference sequences (RefSeq): a curated non-redundant sequence database of genomes, transcripts and proteins. Nucleic Acids Res 35:D61–D65

153. Thierry-Mieg D, Thierry-Mieg J (2006) AceView: a comprehensive cDNA-supported gene and transcripts annotation. Genome Biol 7:1–14
154. Su Z, Łabaj PP, Li S, Thierry-Mieg J, Thierry-Mieg D, Shi W, Wang C, Schroth GP et al (2014) A comprehensive assessment of RNA-seq accuracy, reproducibility and information content by the Sequencing Quality Control Consortium. Nat Biotechnol
155. Wu P-Y, Phan JH, Wang MD (2013) Assessing the impact of human genome annotation choice on RNA-seq expression estimates. BMC Bioinformatics 14(Suppl 1):S8

Gene Expression Analysis: Applications

Peter Savas, Zhi Ling Teo, and Sherene Loi

DNA Microarray

Introduction

Cancer is a highly variable, heterogeneous disease induced by the accumulation of numerous genetic and environmental factors. Understanding of such a complex system and the intertwining of its multitude of biological functions would require complete deciphering of the human genome [1]. In the post-genomic era, the field of biology has transitioned from detecting differentially expressed single genes to a more systems-based focus, turning to approaches for finding differentially altered pathways. The DNA microarray has emerged as one of the key tools used in gene expression profiling. The power of microarrays, compared with other traditional methods of gene expression analysis (i.e. serial analysis of gene expression and quantitative real time PCR), lies in its ability to quanitify in parallel thousands of genes across multiple samples. The increased availability and affordability of genomic technologies together with the development of information processing technologies has enabled the generation and analysis of copious amounts of data. As a result, gene expression profiling has become a readily used tool, integral to characterising tumour molecular profiles.

P. Savas • Z.L. Teo • S. Loi (✉)
Translational Breast Cancer Genomics Lab, Cancer Therapeutics Program, Division of Research, Peter MacCallum Cancer Centre, East Melbourne, VIC 3002, Australia
e-mail: Sherene.loi@petermac.org

S.R. Lakhani, S.B. Fox (eds.), *Molecular Pathology in Cancer Research*,
DOI 10.1007/978-1-4939-6643-1_7

Classifying Tumour Molecular Profile

Understanding the complex factors that induce tumourigenesis is an arduous task. The process of discovery and development of a therapeutic target necessitates comprehensive knowledge of the functions and effects of the target on cellular activities. The genetic diversity of cancers has alluded to the need for a more tailored medicine, where decisions regarding prevention, diagnosis and treatment of disease are guided by the individual's genetic profile. The specific molecular pathways deregulated in the tumour need to be identified so that each patient receives the optimal targeted therapy. Microarray technology has proven to be a robust yet affordable tool for this purpose by illuminating the differences between healthy and diseased tissues, and identifying key biological pathways that are commonly involved across different cancer types.

Key examples of these pathways are the Wnt signalling pathway that is involved in cell differentiation, migration and polarity [2–5], the p53 signalling pathways involved in genomic stability, cell cycle regulation and inhibition of angiogenesis [6–8], the MAPK growth signalling pathway [2, 9, 10] and the NF-kB pathway that is involved in immune response to infection and cellular responses to free radicals and cytokines [3, 6, 9, 11]. Gene expression profiling through the use of microarrays has also allowed for the identification of potential therapeutic targets [12–16] and depiction of the actions of therapeutic targets at the molecular level, illustrating the engagement of the targeted pathway(s), receptor(s) or network through up- or down-regulated patterns of the intended drug target(s).

The capability of this approach has also been further demonstrated via its use in characterising a variety of cancers including breast, colon, head and neck, liver, lung, ovary, pancreatic, prostate and stomach cancers [17–27]. These cancers have been characterised into different biologically and clinically relevant subgroups based on the relative differences in abundance of certain groups of mRNAs and some of these gene signatures have been put forward as predictors of prognosis or treatment response. The large scale gene expression data sets that have been collected and used to characterise these cancers have been analysed in two fundamentally different ways. One approach is the unsupervised classification or hierarchical clustering approach (class discovery) [28] where similarities in gene expression patterns, which could translate into similarities in biological behaviour or phenotype, can be used to classify a cancer into its subtypes. Additional clinical data is not required for this approach. The other approach is known as supervised classification (class prediction). Samples are first grouped according to different clinical endpoints (i.e. response to therapy) and the analysis identifies the gene expression patterns that best distinguish between the groups.

In the case of breast cancer, microarray analysis, via the unsupervised classification approach, has divided the disease into at least five subtypes with clinical relevance [29, 30]: luminal A, luminal B, basal-like, HER2 enriched and normal-like. The luminal A and B subtypes are oestrogen receptor (ER) positive, luminal B tumours are associated with increased expression of proliferation-related genes and also have poorer outcomes compared with luminal A tumours [31]. Microarray studies show that luminal tumours are associated with high expression of luminal cytokeratins and genetic markers of luminal epithelial cells of normal breast tissue [32]. HER2-enriched tumours show amplification and high expression of the *ERBB2* gene and are associated with high levels of proliferation. Over-expression of the *ERBB2* gene has been found to be associated with low expression levels of ER and numerous other genes shown to be associated with ER expression. This trait was also shown in basal-like tumours [29]. Basal-like breast cancers are ER, progesterone receptor (PR) and HER2 negative and do not express some genes that are typical of myoepithelial cells of normal breast tissue. Instead, basal-like breast cancers, as the name suggests, express basal cell markers such as keratins 5/6 and/or 17 [29].

Prognostic and Predictive Multigene Signatures

The main purpose of the above studies was to establish a molecular classification of tumours and does not serve well in the search for prognostic or predictive classifiers. The supervised approach is normally used for the latter. There are two main strategies involved in the development of prognostic or predictive gene signatures: the 'top-down' approach and the 'bottom-up' approach. In the 'top-down' approach, gene expression data from groups of patients with known clinical outcomes are compared in the search for genes that are associated with prognosis without a priori biological assumptions. The strength of this approach is that it is unbiased as there are no assumptions made about which genes are likely to be involved in the biological pathway of interest. The shortcoming of this approach is that the outcome of the analysis is highly dependent on the quality of the samples and data produced. The 'bottom-up' approach identifies gene expression patterns that are previously known to be associated with a specific disease phenotype or aberrant molecular pathway which are then subsequently correlated with clinical outcome. A drawback to this approach is that the outcome is as good as the state of knowledge: gene expression patterns not previously known to be involved in the process of interest are not incorporated or considered in the analysis [28, 33].

The MammaPrint (Agendia®, The Netherlands) 70-gene prognostic signature uses a supervised 'top-down' approach to predict the risk of breast cancer distant metastases and was the first commercialised DNA microarray predictor. Retrospective information from 78 patients, from the Netherlands Cancer Institute, diagnosed with node-negative breast cancer and who had not received systemic adjuvant therapy was used [34]. It was subsequently validated in a cohort of both node-positive and node-negative patients from the same institution [35]. The prognostic value of MammaPrint was established in patients with node-positive and node-negative breast cancers [35, 36]. It predicts early metastasis in post-menopausal women between the ages of 55–70 years [37] and is also predictive of response to neoadjuvant and adjuvant chemotherapy [38, 39].

Using the 'bottom-up' strategy, a 97 prognostic gene-signature was identified to consistently discriminate between low- and high-grade breast tumours [40]. The set of genes that were commonly overexpressed in grade three tumours and which distinguished between grade three and grade one tumours have been previously associated with cell cycle progression and proliferation. It has also helped to restratify grade two tumours into grade one or grade three tumours. This 97 gene-signature has been shown to be associated with pathological response to primary chemotherapy [41] and has been developed commercially as the MapQuant Dx© Genomic Grade Assay (Ipsogen®, France).

The efficacies of single agent therapeutic targets are usually short-lived either through de novo or acquired resistance. Presence of intra-tumoural heterogeneity could be a factor that contributes to de novo resistance. Acquired drug resistance occurs after prolonged drug treatment and tumour cells develop derangements to bypass dependence on the targeted cell survival/proliferation pathway to overcome the toxic effect of treatment. Gene expression profiling through microarrays has demonstrated the potential to determine response to treatment, identify mechanisms of resistance and further uncover potential therapeutic pathways that can be targeted in combination therapies to overcome resistance. The 'bottom-up' strategy was also used to determine if the various molecular subtypes of breast cancer respond differently to preoperative chemotherapy [42]. Rouzier et al. defined the four molecular classes of breast cancer by clustering and found that the rates of pathologic complete response to preoperative chemotherapy, T/FAC, (paclitaxel followed by 5-fluorouracil, doxorubicin and cyclophosphamide) were significantly different among the four classes. They also identified gene sets within the HER2-enriched and basal-like groups that were associated with pathologic complete response to preoperative T/FAC. Another 74-multigene signature was developed to predict response to preoperative chemotherapy (T/FAC) [43]. The combination of these two gene sets were used to create a 30-gene set to predict complete response to T/FAC chemotherapy and a 200-gene set that predicts tumour recurrence after 5 years of endocrine therapy (NuvoSelect™, Nuvera Bioscience, USA).

A recent pilot study showed that gene expression profiling of tumours that have become treatment-resistant can identify molecular profiles that are associated with sensitivity to certain currently available therapies [16]. The progression free survival (PFS) of a patient using the treatment regimen selected by their molecular profile was compared to the PFS for the most recent regimen on which the same patient had experienced progression. 27 % ($p = 0.007$) of the patients had a longer PFS on the molecular profiling suggested regimen than on the regimen on which the patient experienced progression. New strategies, like that of the pilot study, based on tumour genomic or gene expression profiles will be able to guide patients to early phase trials that are focused on small dedicated populations of patients which could result in theoretically increased efficacy of therapeutics, further translating into shorter and more cost effective trials [44]. Nonetheless, it is important to note that such biopsy-driven therapeutic targeting is limited by the emerging concept of tumoural heterogeneity [45, 46], suggesting that single tumour biopsies might lead to the underestimation of the tumour genomic landscape.

Data Analysis Challenges

While microarray is a robust technology that has allowed significant headway to be gained in our battle against cancer, one of the major hurdles lies in the integration of extensive amounts of dynamic data. The ability to perform robust statistical analyses to generate clinically relevant gene expression signature continues to remain challenging. Despite widespread use of microarray technology in the field of cancer, only very few useful biomarkers have been identified and have been or are being translated into useful clinical assays or companion diagnostics (MammaPrint®, NuvoSelect™). Some of the more salient limitations of the technology have been its lack of reproducibility [47–49] and our ability to tease useful information out of the often noisy and complex amount of data that is generated. The use of DNA microarray for cancer prognosis has been demonstrated [50]. Nonetheless, much more work needs to be done to understand the most accurate way to analyse such complex data and the generalisability of a classifier derived from the data. The Microarray Quality Control project has started a second phase, comparing various approaches to the development and validation of microarray-based classifiers to be used in clinical practices [51].

Conclusion

The capability of microarrays to analyse gene expression patterns of thousands of transcripts in parallel provides us with a unique insight into the biological mechanisms underlying malignancy. There have been successful employments of gene expression signatures in the field of cancer, taking us one step closer to personalised treatment. Appropriate experimental design and robust bioinformatics analysis are required to generate results that are of clinical relevance. Recent advances in the next generation sequencing technologies have made whole exome and whole genome sequencing increasingly affordable and accessible for more comprehensive detection of genetic mutations in tumours. Integrating gene expression patterns with genetic mutation data is an approach that has the potential to allow detailed dissection of the biological pathways underlying the diverse responses to treatments and the mechanisms of resistance that are innate or develop in response to treatment, paving the way for greater clinical efficacy at the population level. Nevertheless, the concept of personalised medicine still faces several challenges and issues before its translation to the clinic. These include the need for new bioinformatics tools to allow quick yet accurate access to genomic/transcriptomic explorations. Furthermore, these high throughput technologies provide us with the tools that can potentially allow us to further identify and characterise new tumour subsets and/or biomarkers. As such, there is further potential to stratify patients into more homogeneous populations.

RNASeq

RNAseq is growing in importance as a research tool. There has been a steadily growing list of applications and associated computational methods. The discussion here relates to the use of RNAseq for analyses of gene expression, and in particular comparing differential expression between samples, time points or experimental conditions.

Following on from the discussion in the Methods chapter, after basic data processing of sequencer reads, RNAseq data is presented as an annotated transcript with an associated read count. Simply the presence or absence of certain transcripts detected via RNAseq has a variety of uses, such as determining if a genomic mutation detected with DNAseq is expressed [52], and as orthogonal validation of mutations detected with whole exome or whole genome DNAseq [53]. RNAseq has also been used to improve detection of low frequency variants detected with DNAseq in low purity tumour samples [54].

Quantitative analysis requires further processing to permit statistical modelling. The nature of short read sequencing requires the read counts to be normalised to enable comparison between different genes in the same sample, or the same gene under different experimental conditions. This has two aspects. Firstly, longer RNA transcripts will produce more sequencer reads than a shorter transcript, despite the two having the same level of expression. The simplest method of correcting for this is to divide the read count for an annotated transcript by the total number of bases in the transcript, which is determined by combining the exons. This will produce a 'reads per kilobase of exon model' measurement for each transcript.

An additional factor requiring normalisation is the inherent variability in sample handling and library preparation which means that the total number of sequencer reads and therefore mapped reads varies between different samples and sequencing runs. To correct for this, the 'reads per kilobase of exon model' is further divided by the number of mapped reads for the sample, in millions of reads. This double normalisation procedure expresses read counts as RPKM or 'reads per kilobase of exon model per million mapped reads'. For paired end reads 'fragments per kilobase of exon model per million mapped reads' or FPKM is used. PKM normalisation was used in the earliest RNAseq experiments, and has been shown to perform well when known quantities of known transcripts, so-called spike ins are added into a sample which is then sequenced under differing conditions [55].

It has since been recognised that normalising by transcript length and mapped reads is overly simplistic [56, 57]. In particular, abundantly expressed genes may take up most of the mapped reads in a sample, thus causing less abundant transcripts to appear under-represented. RPKM remains a useful metric, but more robust normalisation methods such as scaling factors should be used when trying to quantify differential gene expression [58]. A comparative analysis of normalisation methods suggested that the normalisation methods employed by the software edgeR and DESeq were most appropriate [59]. Both of these software packages are available for use with the statistical programming language R.

One of the most commonly used packages for RNAseq data analysis is Cufflinks [60]. Cufflinks models gene expression at the transcript level, and rather than relying on raw read counts it estimates the gene expression level which would produce the detected reads. This requires accurate splice variant detection, which adds another layer of complexity in comparison to read counting.

Differential expression involves applying a statistical test to the observed differences in gene expression between two different states. The output from the analysis is generally a fold change in expression between the experimental groups with an associated *p*-value corrected for multiple comparisons. Some examples of differential expression experiments: comparing the differentially expressed genes between different subtypes of gastric cancer [61], or between cell lines that are variably sensitive to some drug therapy [62]. Statistical approaches to differential expression differ in their flexibility, underlying statistical methodology, sensitivity and computational requirements. The performance of different software packages providing a statistical treatment of differential expression has been compared. In general, edgeR, DESeq or baySeq-based analysis performed well, and were superior to Cuffdiff which is part of the Cufflinks package [63]. Of note, in this study some of the more successful approaches involved applying the robust and mature methods developed for differential expression analysis of microarrays using, for example, the limma package in R [64]. The discrete count-based data produced by RNAseq must be transformed into continuous data reminiscent of microarray gene expression values to enable this approach.

Experimental design is important to achieving sufficient statistical power in a differential expression experiment. The nature of RNAseq is that it is highly sensitive to experimental conditions. In the example of conducting an experiment in cell lines, subtle differences in culture conditions or treatment may introduce variability that will reduce the external validity of the results or obscure a true effect. No analytical method can overcome this limitation, which must be mitigated by using biological replicates—that is, repeating the experiment several times attempting to keep the experimental conditions constant. This provides an estimate of the inherent variability in the gene expression levels, which in turn allows appropriate statistical modelling. In less controlled experimental situations, such as comparing patient samples, much larger sample sizes are required to estimate the background variability [65].

Tools are available to estimate the sample size and power of differential expression experiments [66]. These tools are most useful if pilot data on expression levels can be supplied prior to planning an experiment. Several published analyses have shown that in general, the statistical power of an RNAseq experiment is most easily improved by increasing the number of biological replicates, rather than sequencing more deeply [63, 67]. Increasing replicates and reducing sequencing depth is also a cost effective strategy. If replicates are not possible, then attempts can be made to estimate the background variability of the genes in question using housekeeper genes that are not thought to change expression, or using publicly available RNAseq datasets [65].

Following differential expression analysis, functional annotation of the expressed genes is usually undertaken. This may be conducted using pathway analysis tools such as Gene Ontology [68]. Careful consideration must still be given to the biases inherent to RNAseq when performing functional annotation, however [69]. Normalised

Table 1 Sources of bias in RNASeq

Bias	Source
Overrepresentation of introns	FFPE processing [73]
GC bias: increased coverage of GC rich sequences	Library preparation chemistry [72]
More reads for longer transcripts	Short read platform [55]
Bias towards 3′ transcript sequence reads	Poly(A) enrichment
Bias towards 5′ transcript sequence reads	Random hexamer priming [74]
Variable coverage	Ribosomal RNA depletion [72]

RNAseq expression data may also be used for feature extraction and clustering into molecular subtypes, using the same methods employed for gene expression microarray data such as non-negative matrix factorisation and *k*-means clustering [61, 70].

Limitations

RNAseq remains a new technology, and has important limitations. There are many sources of potential variability in sample acquisition, library preparation and data analysis, which create concerns for the reproducibility of results. A number of biases are known to exist, shown in Table 1. Analytical methods must account for these biases, some of which have only been discovered recently. A consortium led by the United States FDA examined the reproducibility and accuracy of RNAseq performed on multiple platforms and in multiple laboratories [71]. This study found that although absolute quantification of gene expression was unreliable, relative gene expression was consistent on different platforms in different laboratories. It was also noted that there is currently no reliable gold standard reference sample of RNA to test workflow performance—the supposed reference samples in this study showed considerable variability. For this reason, multiple technical replicates—that is, repeating library preparation and sequencing using the same initial sample—were recommended. Another recently published study used transcribed cDNA libraries to test RNAseq performance [72]. This removed the complicating factor of splice variants. Using a number of library preparation methods, unexpected read count variability was seen that appeared dependent on the library preparation protocol and transcript sequence characteristics in various combinations.

Conclusion

Future directions of RNAseq still under development include single cell RNAseq and in situ RNAseq. Single cell RNAseq has already been performed by several groups [75, 76], and efforts are underway to improve the accuracy and throughput of the technique [77, 78]. In situ RNAseq refers to sequencing RNA transcripts in their native cellular location and allows the integration of cellular spatial organisation with the transcriptome providing a direct link between gene expression, microenvironment and phenotype. It has been achieved recently in tissue slides and whole mount embryos [79]. As these technologies improve in efficiency, cost and throughput, an overwhelming amount of transcriptome data will be generated. The success of RNAseq will depend, however, on optimising experimental design, library preparation and analytical methods.

References

1. Han JD (2008) Understanding biological functions through molecular networks. Cell Res 18(2):224–237
2. Nam S, Park T (2012) Pathway-based evaluation in early onset colorectal cancer suggests focal adhesion and immunosuppression along with epithelial-mesenchymal transition. PLoS One 7(4), e31685
3. Ooi CH, Ivanova T, Wu J, Lee M, Tan IB, Tao J, Ward L, Koo JH, Gopalakrishnan V, Zhu Y et al (2009) Oncogenic pathway combinations predict clinical prognosis in gastric cancer. PLoS Genet 5(10), e1000676
4. Krivtsov AV, Twomey D, Feng Z, Stubbs MC, Wang Y, Faber J, Levine JE, Wang J, Hahn WC, Gilliland DG et al (2006) Transformation from committed progenitor to leukaemia stem cell initiated by MLL-AF9. Nature 442(7104):818–822
5. Lehmann BD, Bauer JA, Chen X, Sanders ME, Chakravarthy AB, Shyr Y, Pietenpol JA (2011) Identification of human triple-negative breast cancer subtypes and preclinical models for selection of targeted therapies. J Clin Invest 121:2750–2767
6. Fabbri G, Rasi S, Rossi D, Trifonov V, Khiabanian H, Ma J, Grunn A, Fangazio M, Capello D, Monti S et al (2011) Analysis of the chronic lymphocytic leukemia coding genome: role of NOTCH1 mutational activation. J Exp Med 208(7):1389–1401
7. Shah MA, Khanin R, Tang L, Janjigian YY, Klimstra DS, Gerdes H, Kelsen DP (2011) Molecular classification of gastric cancer: a new paradigm. Clin Cancer Res 17(9):2693–2701
8. Perroud B, Lee J, Valkova N, Dhirapong A, Lin PY, Fiehn O, Kultz D, Weiss RH (2006) Pathway analysis of kidney cancer using proteomics and metabolic profiling. Mol Cancer 5:64
9. Setlur SR, Royce TE, Sboner A, Mosquera J-M, Demichelis F, Hofer MD, Mertz KD, Gerstein M, Rubin MA (2007) Integrative microarray analysis of pathways dysregulated in metastatic prostate cancer. Cancer Res 67(21):10296–10303
10. Nucera C, Porrello A, Antonello ZA, Mekel M, Nehs MA, Giordano TJ, Gerald D, Benjamin LE, Priolo C, Puxeddu E et al (2010) B-Raf(V600E) and thrombospondin-1 promote thyroid cancer progression. Proc Natl Acad Sci U S A 107(23):10649–10654
11. Compagno M, Lim WK, Grunn A, Nandula SV, Brahmachary M, Shen Q, Bertoni F, Ponzoni M, Scandurra M, Califano A et al (2009) Mutations of multiple genes cause deregulation of NF-kappaB in diffuse large B-cell lymphoma. Nature 459(7247):717–721
12. Madhamshettiwar PB, Maetschke SR, Davis MJ, Reverter A, Ragan MA (2012) Gene regulatory network inference: evaluation and application to ovarian cancer allows the prioritization of drug targets. Genome Med 4(5):41
13. Welsh JB, Sapinoso LM, Su AI, Kern SG, Wang-Rodriguez J, Moskaluk CA, Frierson HF, Hampton GM (2001) Analysis of gene expression identifies candidate markers and pharmacological targets in prostate cancer. Cancer Res 61(16):5974–5978
14. Debouck C, Goodfellow PN (1999) DNA microarrays in drug discovery and development. Nat Genet 21:48–50

15. Scherf U, Ross DT, Waltham M, Smith LH, Lee JK, Tanabe L, Kohn KW, Reinhold WC, Myers TG, Andrews DT et al (2000) A gene expression database for the molecular pharmacology of cancer. Nat Genet 24(3):236–244
16. Von Hoff DD, Stephenson JJ, Rosen P, Loesch DM, Borad MJ, Anthony S, Jameson G, Brown S, Cantafio N, Richards DA et al (2010) Pilot study using molecular profiling of patients' tumors to find potential targets and select treatments for their refractory cancers. J Clin Oncol 28:4877–4883
17. Bhattacharjee A, Richards WG, Staunton J, Li C, Monti S, Vasa P, Ladd C, Beheshti J, Bueno R, Gillette M et al (2001) Classification of human lung carcinomas by mRNA expression profiling reveals distinct adenocarcinoma subclasses. Proc Natl Acad Sci 98(24):13790–13795
18. Hedenfalk I, Duggan D, Chen Y, Radmacher M, Bittner M, Simon R, Meltzer P, Gusterson B, Esteller M, Raffeld M et al (2001) Gene-expression profiles in hereditary breast cancer. N Engl J Med 344(8):539–548
19. Chen X, Cheung ST, So S, Fan ST, Barry C, Higgins J, Lai K-M, Ji J, Dudoit S, Ng IOL et al (2002) Gene expression patterns in human liver cancers. Mol Biol Cell 13(6):1929–1939
20. Han H, Bearss DJ, Browne LW, Calaluce R, Nagle RB, Von Hoff DD (2002) Identification of differentially expressed genes in pancreatic cancer cells using cDNA microarray. Cancer Res 62(10):2890–2896
21. Tonin PN, Hudson TJ, Rodier F, Bossolasco M, Lee PD, Novak J, Manderson EN, Provencher D, Mes-Masson AM (2001) Microarray analysis of gene expression mirrors the biology of an ovarian cancer model. Oncogene 20(45):6617–6626
22. Dhanasekaran SM, Barrette TR, Ghosh D, Shah R, Varambally S, Kurachi K, Pienta KJ, Rubin MA, Chinnaiyan AM (2001) Delineation of prognostic biomarkers in prostate cancer. Nature 412(6849):822–826
23. Hippo Y, Taniguchi H, Tsutsumi S, Machida N, Chong JM, Fukayama M, Kodama T, Aburatani H (2002) Global gene expression analysis of gastric cancer by oligonucleotide microarrays. Cancer Res 62(1):233–240
24. Al Moustafa AE, Alaoui-Jamali MA, Batist G, Hernandez-Perez M, Serruya C, Alpert L, Black MJ, Sladek R, Foulkes WD (2002) Identification of genes associated with head and neck carcinogenesis by cDNA microarray comparison between matched primary normal epithelial and squamous carcinoma cells. Oncogene 21(17):2634–2640
25. Kitahara O, Furukawa Y, Tanaka T, Kihara C, Ono K, Yanagawa R, Nita ME, Takagi T, Nakamura Y, Tsunoda T (2001) Alterations of gene expression during colorectal carcinogenesis revealed by cDNA microarrays after laser-capture microdissection of tumor tissues and normal epithelia. Cancer Res 61(9):3544–3549
26. Garber ME, Troyanskaya OG, Schluens K, Petersen S, Thaesler Z, Pacyna-Gengelbach M, van de Rijn M, Rosen GD, Perou CM, Whyte RI et al (2001) Diversity of gene expression in adenocarcinoma of the lung. Proc Natl Acad Sci U S A 98(24):13784–13789
27. Belbin TJ, Singh B, Barber I, Socci N, Wenig B, Smith R, Prystowsky MB, Childs G (2002) Molecular classification of head and neck squamous cell carcinoma using cDNA microarrays. Cancer Res 62(4):1184–1190
28. van 't Veer LJ, Bernards R (2008) Enabling personalized cancer medicine through analysis of gene-expression patterns. Nature 452(7187):564–570
29. Perou C, Sorlie T, Eisen M, van de Rijn M, Jeffrey S, Rees C, Pollack J, Ross D, Johnsen H, Akslen L et al (2000) Molecular portraits of human breast tumours. Nature 406:747–752
30. Sorlie T, Perou CM, Tibshirani R, Aas T, Geisler S, Johnsen H, Hastie T, Eisen MB, van de Rijn M, Jeffrey SS et al (2001) Gene expression patterns of breast carcinomas distinguish tumor subclasses with clinical implications. Proc Natl Acad Sci U S A 98(19):10869–10874
31. Cheang MCU, Chia SK, Voduc D, Gao D, Leung S, Snider J, Watson M, Davies S, Bernard PS, Parker JS et al (2009) Ki67 index, HER2 status, and prognosis of patients with luminal B breast cancer. J Natl Cancer Inst
32. Rakha EA, El-Sayed ME, Green AR, Paish EC, Powe DG, Gee J, Nicholson RI, Lee AHS, Robertson JFR, Ellis IO (2007) Biologic and clinical characteristics of breast cancer with single hormone receptor–positive phenotype. J Clin Oncol 25(30):4772–4778

33. Sotiriou C, Pusztai L (2009) Gene-expression signatures in breast cancer. N Engl J Med 360(8):790–800
34. van 't Veer LJ, Dai HY, van de Vijver MJ, He YDD, Hart AAM, Mao M, Peterse HL, van der Kooy K, Marton MJ, Witteveen AT et al (2002) Gene expression profiling predicts clinical outcome of breast cancer. Nature 415(6871):530–536
35. van de Vijver MJ, He YD, van 't Veer LJ, Dai H, Hart AAM, Voskuil DW, Schreiber GJ, Peterse JL, Roberts C, Marton MJ et al (2009) A gene-expression signature as a predictor of survival in breast cancer. N Engl J Med 347(25):1999–2009
36. Saghatchian M, Mook S, Pruneri G, Viale G, Glas AM, Guerin S, Cardoso F, Piccart M, Tursz T, Delaloge S et al (2013) Additional prognostic value of the 70-gene signature (MammaPrint((R))) among breast cancer patients with 4–9 positive lymph nodes. Breast 22(5):682–690
37. Mook S, Schmidt MK, Weigelt B, Kreike B, Eekhout I, van de Vijver MJ, Glas AM, Floore A, Rutgers EJ, van 't Veer LJ (2010) The 70-gene prognosis signature predicts early metastasis in breast cancer patients between 55 and 70 years of age. Ann Oncol 21(4):717–722
38. Straver ME, Glas AM, Hannemann J, Wesseling J, van de Vijver MJ, Rutgers EJ, Vrancken Peeters MJ, van Tinteren H, Van't Veer LJ, Rodenhuis S (2010) The 70-gene signature as a response predictor for neoadjuvant chemotherapy in breast cancer. Breast Cancer Res Treat 119(3):551–558
39. Knauer M, Mook S, Rutgers EJ, Bender RA, Hauptmann M, van de Vijver MJ, Koornstra RH, Bueno-de-Mesquita JM, Linn SC, van 't Veer LJ (2010) The predictive value of the 70-gene signature for adjuvant chemotherapy in early breast cancer. Breast Cancer Res Treat 120(3):655–661
40. Sotiriou C, Wirapati P, Loi S, Harris A, Fox S, Smeds J, Nordgren H, Farmer P, Praz V, Haibe-Kains B et al (2006) Gene expression profiling in breast cancer: understanding the molecular basis of histologic grade to improve prognosis. J Natl Cancer Inst 98:262–272
41. Liedtke C, Hatzis C, Symmans WF, Desmedt C, Haibe-Kains B, Valero V, Kuerer H, Hortobagyi GN, Piccart-Gebhart M, Sotiriou C et al (2009) Genomic grade index is associated with response to chemotherapy in patients with breast cancer. J Clin Oncol 27(19):3185–3191
42. Rouzier R, Perou CM, Symmans WF, Ibrahim N, Cristofanilli M, Anderson K, Hess KR, Stec J, Ayers M, Wagner P et al (2005) Breast cancer molecular subtypes respond differently to preoperative chemotherapy. Clin Cancer Res 11(16):5678–5685
43. Ayers M, Symmans WF, Stec J, Damokosh AI, Clark E, Hess K, Lecocke M, Metivier J, Booser D, Ibrahim N et al (2004) Gene expression profiles predict complete pathologic response to neoadjuvant paclitaxel and fluorouracil, doxorubicin, and cyclophosphamide chemotherapy in breast cancer. J Clin Oncol 22(12):2284–2293
44. Sabatier R, Gonçalves A, Bertucci F (2014) Personalized medicine: present and future of breast cancer management. Crit Rev Oncol Hematol 91(3):223–233
45. Navin N, Krasnitz A, Rodgers L, Cook K, Meth J, Kendall J, Riggs M, Eberling Y, Troge J, Grubor V et al (2010) Inferring tumor progression from genomic heterogeneity. Genome Res 20(1):68–80
46. Gerlinger M, Rowan AJ, Horswell S, Larkin J, Endesfelder D, Gronroos E, Martinez P, Matthews N, Stewart A, Tarpey P et al (2012) Intratumor heterogeneity and branched evolution revealed by multiregion sequencing. N Engl J Med 366:883–892
47. Marshall E (2004) Getting the noise out of gene arrays. Science 306(5696):630–631
48. Ioannidis JPA (2005) Why most published research findings are false. PLoS Med 2(8), e124
49. Simon R (2006) Development and evaluation of therapeutically relevant predictive classifiers using gene expression profiling. J Natl Cancer Inst 98(17):1169–1171
50. Fan X, Shi L, Fang H, Cheng Y, Perkins R, Tong W (2010) DNA microarrays are predictive of cancer prognosis: a re-evaluation. Clin Cancer Res 16(2):629–636
51. Shi L, Perkins RG, Fang H, Tong W (2008) Reproducible and reliable microarray results through quality control: good laboratory proficiency and appropriate data analysis practices are essential. Curr Opin Biotechnol 19(1):10–18
52. Shah SP, Roth A, Goya R, Oloumi A, Ha G, Zhao Y, Turashvili G, Ding J, Tse K, Haffari G et al (2012) The clonal and mutational evolution spectrum of primary triple-negative breast cancers. Nature 486:395–399

53. Cancer Genome Atlas Research Network (2012) Comprehensive genomic characterization of squamous cell lung cancers. Nature 489:519–525
54. Wilkerson MD, Cabanski CR, Sun W, Hoadley KA, Walter V, Mose LE, Troester MA, Hammerman PS, Parker JS, Perou CM et al (2014) Integrated RNA and DNA sequencing improves mutation detection in low purity tumors. Nucleic Acids Res 42:e107
55. Mortazavi A, Williams BA, McCue K, Schaeffer L, Wold B (2008) Mapping and quantifying mammalian transcriptomes by RNA-Seq. Nat Methods 5:621–628
56. Bullard JH, Purdom E, Hansen KD, Dudoit S (2010) Evaluation of statistical methods for normalization and differential expression in mRNA-Seq experiments. BMC Bioinform 11:94
57. Robinson MD, Oshlack A (2010) A scaling normalization method for differential expression analysis of RNA-seq data. Genome Biol 11:R25
58. Oshlack A, Robinson MD, Young MD (2010) From RNA-seq reads to differential expression results. Genome Biol 11:220
59. Dillies M-A, Rau A, Aubert J, Hennequet-Antier C, Jeanmougin M, Servant N, Keime C, Marot G, Castel D, Estelle J et al (2013) A comprehensive evaluation of normalization methods for Illumina high-throughput RNA sequencing data analysis. Brief Bioinform 14:671–683
60. Trapnell C, Hendrickson DG, Sauvageau M, Goff L, Rinn JL, Pachter L (2013) Differential analysis of gene regulation at transcript resolution with RNA-seq. Nat Biotechnol 31:46–53
61. Bass AJ, Thorsson V, Shmulevich I, Reynolds SM, Miller M, Bernard B, Hinoue T, Laird PW, Curtis C, Shen H et al (2014) Comprehensive molecular characterization of gastric adenocarcinoma. Nature
62. Daemen A, Griffith OL, Heiser LM, Wang NJ, Enache OM, Sanborn Z, Pepin F, Durinck S, Korkola JE, Griffith M et al (2013) Modeling precision treatment of breast cancer. Genome Biol 14:R110
63. Rapaport F, Khanin R, Liang Y, Pirun M, Krek A, Zumbo P, Mason CE, Socci ND, Betel D (2013) Comprehensive evaluation of differential gene expression analysis methods for RNA-seq data. Genome Biol 14:R95
64. Smyth GK (2005) Limma: Linear models for microarray data. In: Gentleman R, Carey VJ, Huber W, Irizarry RA, Dudoit R (eds) Bioinformatics and computational biology solution using R and bioconductor. Springer, New York, pp 397–420
65. Anders S, McCarthy DJ, Chen Y, Okoniewski M, Smyth GK, Huber W, Robinson MD (2013) Count-based differential expression analysis of RNA sequencing data using R and bioconductor. Nat Protoc 8:1765–1786
66. Busby MA, Stewart C, Miller CA, Grzeda KR, Marth GT (2013) Scotty: a web tool for designing RNA-Seq experiments to measure differential gene expression. Bioinformatics 29:656–657
67. Liu Y, Zhou J, White KP (2014) RNA-seq differential expression studies: more sequence or more replication? Bioinformatics 30:301–304
68. Ashburner M, Ball CA, Blake JA, Botstein D, Butler H, Cherry JM, Davis AP, Dolinski K, Dwight SS, Eppig JT et al (2000) Gene ontology: tool for the unification of biology. The Gene Ontology Consortium. Nat Genet 25:25–29
69. Young MD, Wakefield MJ, Smyth GK, Oshlack A (2010) Gene ontology analysis for RNA-seq: accounting for selection bias. Genome Biol 11:R14
70. Kandoth C, Schultz N, Cherniack AD, Akbani R, Liu Y, Shen H, Robertson AG, Pashtan I, Shen R, Benz CC et al (2013) Integrated genomic characterization of endometrial carcinoma. Nature 497:67–73
71. Su Z, Łabaj PP, Li S, Thierry-Mieg J, Thierry-Mieg D, Shi W, Wang C, Schroth GP, Setterquist RA, Thompson JF et al (2014) A comprehensive assessment of RNA-seq accuracy, reproducibility and information content by the Sequencing Quality Control Consortium. Nat Biotechnol
72. Lahens NF, Kavakli IH, Zhang R, Hayer K, Black MB, Dueck H, Pizarro A, Kim J, Irizarry R, Thomas RS et al (2014) IVT-seq reveals extreme bias in RNA-sequencing. Genome Biol 15:R86
73. Adiconis X, Borges-Rivera D, Satija R, DeLuca DS, Busby MA, Berlin AM, Sivachenko A, Thompson DA, Wysoker A, Fennell T et al (2013) Comparative analysis of RNA sequencing methods for degraded or low-input samples. Nat Methods 10:623–629

74. Hansen KD, Brenner SE, Dudoit S (2010) Biases in Illumina transcriptome sequencing caused by random hexamer priming. Nucleic Acids Res 38, e131
75. Hashimshony T, Wagner F, Sher N, Yanai I (2012) CEL-Seq: single-cell RNA-Seq by multiplexed linear amplification. Cell Rep 2:666–673
76. Tang F, Barbacioru C, Nordman E, Li B, Xu N, Bashkirov VI, Lao K, Surani MA (2010) RNA-Seq analysis to capture the transcriptome landscape of a single cell. Nat Protoc 5:516–535
77. Jaitin DA, Kenigsberg E, Keren-Shaul H, Elefant N, Paul F, Zaretsky I, Mildner A, Cohen N, Jung S, Tanay A et al (2014) Massively parallel single-cell RNA-seq for marker-free decomposition of tissues into cell types. Science 343:776–779
78. Wu AR, Neff NF, Kalisky T, Dalerba P, Treutlein B, Rothenberg ME, Mburu FM, Mantalas GL, Sim S, Clarke MF et al (2014) Quantitative assessment of single-cell RNA-sequencing methods. Nat Methods 11:41–46
79. Lee JH, Daugharthy ER, Scheiman J, Kalhor R, Yang JL, Ferrante TC, Terry R, Jeanty SSF, Li C, Amamoto R et al (2014) Highly multiplexed subcellular RNA sequencing in situ. Science 343:1360–1363

Methods Used for Noncoding RNAs Analysis

Marjan E. Askarian-Amiri, Darren J. Korbie, Debina Sarkar, and Graeme Finlay

Background

A fundamental mechanism underlying the etiology of cancer appears to be the imbalanced expression of genes controlling cell growth, resulting in abnormal control of cell proliferation. For more than half a century the dominant paradigm was that tumourigenesis happened as a result of the accumulation of mutations in the regions of key genes that encoded proteins, such as oncogenes and tumor suppressor genes, leading to alteration of the normal cellular signaling processes that governed cellular proliferation and development. However, recent research has revealed that specific classes of RNA molecules—which were previously believed to only convey genetic information from DNA to protein—can also play diverse roles in cellular processes, and these Noncoding RNA molecules (ncRNA) have added a new layer of complexity to our understanding of gene regulation and disease progression [1–5].

Recent genome and transcriptome studies have revealed ncRNAs (excluding ribosomal RNA and transfer RNA) make up the majority of transcribed RNA species and have a wide range of functions in cellular and developmental processes. Moreover, due to their regulatory functions ncRNAs are involved in the development and pathophysiology of many diseases, and represent potential targets for

---

Marjan E. Askarian-Arimi and Darren J. Korbie contributed equally with all other contributors.

M.E. Askarian-Amiri (✉) • D. Sarkar • G. Finlay
Auckland Cancer Society Research Centre, University of Auckland,
85 Park Rd, Grafton, Auckland 1023, New Zealand
e-mail: m.askarian-amiri@auckland.ac.nz; d.sarkar@auckland.ac.nz; g.finlay@auckland.ac.nz

D.J. Korbie
Australian Institute of Bioengineering and Nanotechnology, University of Queensland,
Corner of College and Coopers Roads, Building 75, St Lucia, QLD 4067, Australia
e-mail: d.korbie@uq.edu.au

S.R. Lakhani, S.B. Fox (eds.), *Molecular Pathology in Cancer Research*,
DOI 10.1007/978-1-4939-6643-1_8

therapeutic intervention [1, 2, 4, 6, 7]. Classification of ncRNAs is typically based on their size, with transcripts larger than 200 bases categorized as long ncRNA (lncRNA), while those RNA species smaller than 200 bases are generally referred to as short or small ncRNAs.

Identification of the genetic and epigenetic events that disrupt ncRNA loci provides opportunities for targeting these disruptions for novel diagnostic or therapeutic applications, and can also offer prognostic and predictive possibilities. Here we briefly explain the role of ncRNAs in cellular function before detailing the methods involved in their isolation and analysis.

Small ncRNA

Transcriptome analysis has now profiled a large number of small ncRNAs, and many of them are accepted as powerful regulatory molecules in gene expression. To date, many different types of small ncRNAs such as miRNA [8], PIWI-interacting RNA (piRNAs) [9], small nucleolar RNAs (snoRNAs) [10], or small Cajal body-specific RNA (scaRNAs) [11] have been identified; of these, miRNA have been studied extensively and are arguably the best studied ncRNAs that directly modulate gene expression. A mature miRNA is typically 22 nucleotides in length and functions in the transcriptional and post-transcriptional regulation of gene expression. The biogenesis and post-transcriptional regulatory mechanism of miRNAs are well studied, and involve a complex protein machinery which includes the Argonaute family, RNA polymerase II dependent transcription, and the Drosha and Dicer RNase IIIs. Mature miRNAs and their associated proteins combine to form the RNA-induced silencing complexes (RISC), of which the Argonaute proteins are the catalytic endonuclease components. The translation of mRNA into proteins is then controlled through imperfect pairing between the miRNA and the 3′UTR of the targeted mRNA, and the subsequent translation of mRNA into proteins is then repressed by miRNAs through two key mechanisms: mRNA degradation, or the inhibition of translation initiation (reviewed in [8, 12]). It is estimated that miRNAs regulate more than 60 % of protein coding genes in this fashion (reviewed in [1]).

Long ncRNA

Initially ncRNA research was focused on small regulatory RNAs such as miRNAs, but more recent transcriptome studies have identified a great number of long transcripts—the so-called lncRNAs [13–16]. These studies have now made it clear that mammalian genomes encode numerous lncRNAs that lack open reading frames of significant length (<100 amino acids) [17, 18]. These lncRNAs often serve key regulatory roles that can induce or suppress the expression of protein coding genes, can modulate the activity, localization, and stability of protein(s) which bind to them [19–21], and serve as key structural components [22, 23]. In addition, many lncRNAs also host small RNAs such as snoRNAs or scaRNAs in their introns [24], and some miRNAs can also be derived from lncRNAs via sequential cleavage by Drosha and Dicer, further broadening the complexity of ncRNA processing [25, 26]. Therefore it is becoming increasingly clear that lncRNAs can function via numerous mechanisms and are key regulatory molecules in the cell.

For these reasons ncRNAs have great potential to be used as novel independent biomarkers for early diagnosis, prognosis, and prediction in cancer. For example, *MALAT-1* is a Noncoding transcript highly expressed in the lung, pancreas, and other healthy organs, but in NSCLC (non-small cell lung carcinoma) was identified as prognostic for patient survival in stage I NSCLC [27]. Similarly, an lncRNA termed *HOTAIR* shows increased in expression in primary breast tumors and metastases, and the *HOTAIR* expression level in primary tumors is a powerful predictor of eventual metastasis and death [28]. Together with the identification of a spectrum of small Noncoding transcripts such as miRNAs that are deregulated in cancer [29, 30], the number of Noncoding transcripts suggested as biomarkers in cancer has exploded. As such, both long and short ncRNAs represent new avenues of investigation for drug discovery with several advantages over traditional protein-based targets; however, they come with their own unique set of challenges. Due to the rapidly evolving pace of ncRNA research in cancer pathology, this review will not attempt to catalogue the current state of ncRNAs implicated in cancer etiology and pathogenesis. Similarly, step-by-step methodologies will not be covered, as detailed protocol breakdowns are available elsewhere and are provided in many of the kits that are mentioned below. Rather, this review will focus on the key applications that are used in ncRNA analysis, highlighting critical steps and points, and attempting to provide a general guide to working with Noncoding RNAs based on years of user experience.

Isolation of ncRNA from Different Human Sources

Although ncRNA can be isolated from whole tissue or paraffin sections, there are also other sources from which ncRNAs can be isolated such as urine, saliva, plasma, serum, and Pap smears [31–35]. NcRNAs also can be isolated from exosomes, which are small membrane vesicles [36] that are released to the extracellular environment and are present in different body fluids [37]. Laser capture micro-dissection can also be employed to enrich tissue sections, but this suffers the double disadvantage of providing low yields as well as extremely degraded RNA, which can be problematic for some analysis streams.

Quality of RNA

The first key parameter to address when working with RNA relates to RNA quality. Of the three major classes of biomolecules typically worked with in a research or pathology laboratory (i.e., DNA, RNA, and proteins), RNA is generally the most fragile and sensitive to degradation, due in large part to the ubiquitous presence of exogenous RNases in the environment.

When working with total RNA the 28S and 18S ribosomal bands make up >90 % of the total amount of RNA, and one metric by which to gauge the overall quality of the extracted sample is the relative intensity of these two transcripts, which can be easily visualized on a standard agarose gel. The 28S and 18S ribosomal bands are 5070 bp and 1869 bp in size, respectively [source, NCBI nucleotide website], and when visualized by eye on a standard DNA agarose gel, high quality intact RNA yields two bright bands with the upper 28S band approximately twice as bright as the lower 18S, with minimal signs of smearing.

Alternatively, certain platforms for RNA analysis which perform digital electrophoresis (i.e., Agilent Bioanalyzer, Qiaxcel, and LabChipGX) employ an algorithm to analyze the electrophoretic trace of the RNA sample, including the presence or absence of degradation products. It then uses this spectrum to determine sample integrity and assign an ***RNA Integrity Score*** (RIN) between 1 and 10, where 1 is highly degraded and 10 is completely intact. Advantages of these platforms are the relative sterility when performing lab-on-a-chip analyses, which limits exposure to contaminating RNases; as well as the quantitative interpretation of the RNA electropherogram, which occurs automatically and is not subject to individual interpretation, thereby limiting bias and improving repeatability of experiments [38]. However, it should be noted that some RNA sources such as exosomes and circulating free RNAs do not contain rRNA, and thus RNA integrity cannot be easily determined.

Broadly speaking, superior results will always be obtained when starting with high quality intact RNA, but depending on the application a degraded sample does not necessarily curtail analysis, and high quality data can still be generated even when an RNA sample has experienced substantial degradation.

Methods of Sample Fixation

The ubiquitous use of formalin fixation and the embedding of tissue in paraffin blocks (formalin-fixed paraffin embedded, or FFPE) in most pathology laboratories means that if a sample has undergone fixation, it mostly likely has been treated with formalin. Although the process of formalin fixation preserves tissue morphology for immunohistochemistry, the fixation method is problematic for when RNA and DNA are extracted from the fixed sample. Although several vendors have kits on the market which are optimized for sample extraction from FFPE tissues, in our experience formalin fixation typically results in lower yields and lower nucleic acid quality (both RNA and DNA), as compared with extractions performed with fresh tissue. When high quality RNA is required alternative fixatives such as Paxgene or RNALater (Ambion) can be employed, and we observe Paxgene in particular to preserve RNA integrity quite well. However, while Paxgene or other alternatives may result in higher nucleic acid quality they typically require separate fixation and processing stations, which may not be a feasible option for many pathology labs.

RNA Purification

All analyses related to characterization and quantitation of RNAs are dependent on effective isolation from source material. This will typically involve extraction from fresh tissue, fresh-frozen tissue, or FFPE sections. The two main extraction methods used for RNA purification can be broadly grouped into two applications: column-based isolation methods employing a silica filter, and methods which involve organic solvent/phenol-chloroform.

Phenol/Chloroform Extractions

Commercial phenol-chloroform-based extraction methods (i.e., Trizol from Life Technologies or Qiazol from Qiagen) are readily available on the market. While users are capable of making their own lots of buffered phenol for nucleic acid extractions, the toxicity of phenol, and the need to ensure it is properly buffered typically mean that the convenience and safety in purchasing commercially prepared reagents outweigh any marginal cost-benefits that accrue from making the reagents in-house.

One critical consideration is that phenol is typically buffered at different pHs depending on whether DNA or RNA isolation is intended; DNA isolations usually employ phenol at pH 8.0, which results in both DNA and RNA being isolated from the aqueous phase. In contrast, using phenol buffered at pH 5.2 helps to partition DNA away from the aqueous phase, resulting in superior RNA enrichment. This is a critical parameter, as differences in pH can affect the overall performance and fidelity of the purification.

While phenol-based methods of RNA extraction are extremely effective at purifying and isolating the small RNA fraction, the toxicity and relative hazard of working with buffered phenol means that some laboratories may prefer to explore alternate means. The second most common method of RNA extraction employs chaotropic guanidinium salts to denature proteins in a tissue or cell sample and protect the RNA from degradation, followed by binding of the RNA to silica filters in plastic columns; the RNeasy (Qiagen) and Nucleospin (Macherey-Nagel) kits are good examples of this.

Although both methods of RNA extraction perform well, it is important to note that both protocols can cause carry-over of genomic DNA into the purified RNA. In general, phenol-chloroform extractions are more prone to protein and DNA carry-over, as these contaminants partition at the interphase and are easy to accidently aspirate while attempting to remove the aqueous upper layer. This can be somewhat reduced by removing only a portion of the aqueous layer, which minimizes the chance of accidental aspiration of contaminants, but this sacrifices some of the RNA sample which may not be ideal, particularly for precious clinical samples. The stability of the interphase can be increased by using bromochloropropane (i.e., 1-bromo-2 chloropropane) instead of chloroform, but this offers only modest improvements. After removal of the aqueous supernatant, the RNA is then precipitated with isopropanol or ethanol, followed by washes in 70 % ethanol to remove residual salts and trace phenol. However, it is recommended that for small RNA ethanol washes be performed with 80 % ethanol, as small RNAs may retain some solubility in 70 % concentrations of ethanol. As mentioned, these ethanol washes are mostly used to remove residual salts and solvents from the RNA pellet, but are not effective at removing either contaminating protein or DNA. To remove these contaminants, separate processing steps must be taken; for example, additional rounds of phenol-chloroform purification can be performed to remove protein contamination.

In comparison, the column method is somewhat simpler and more straightforward, relying on the selective binding of RNA to a silica filter and using wash steps to remove residual salts and other contaminants. The wash step in particular is quite effective at removing protein and salt contaminants, and these washes are also somewhat effective at removing contaminating genomic DNA (gDNA), although trace gDNA still remains.

It should be noted that in the absence of DNase treatment both methods of extraction will still have residual gDNA, in our experience typically in the range from 10 to >30 %. This residual gDNA can be problematic in some applications, because while transcripts which encode for proteins or lncRNAs are frequently spliced, small ncRNAs frequently retain the same sequence as the genomic DNA from which they were derived, making it difficult to design PCR primers and probes in such a way as to limit the effects of contaminating DNA. For this reason, DNase treatment is a critical step, particularly when employing RT-qPCR methods of ncRNA detection and quantitation.

In our experience phenol-chloroform-extracted RNA is more prone to DNA contamination, and residual proteins—including RNases—are frequently co-isolated along with the RNA due to the nature of the extraction. As most purified RNA will be resuspended in DEPC, TE, or purified water and kept on ice, the effect of these contaminating RNases tends to be limited. However, care must be taken when DNasing phenol-chloroform-extracted RNA, since upon the addition of 1× DNase buffer and incubation of the sample at 37 °C, contaminating RNases can be activated leading to degradation of RNA, which can impact downstream processes.

Column-based methods of purification are better suited for DNase treatment, as the residual washes tend to remove these contaminating RNases before the addition of DNase buffer. As well, while contaminating DNA is still problematic, the nature of column purifications is such that smaller DNA fragments (i.e., 500 bp and less) tend to be preferentially lost during column purification. Moreover, some vendors such as Qiagen, Ambion, and NucleoSpin have protocol adaptions that allow isolation of the small RNA fraction separate from the larger RNA pool, and this small RNA fraction tends to be highly pure, with most residual DNA contamination co-localizing with the larger RNA fraction. Interestingly, we have observed that column-based DNase treatment frequently increases the RIN quality score of treated RNA, presumably due to the removal of background gDNA fluorescence from the digital electrophoretic trace, which the automated analysis software interprets as degraded RNA.

For the reasons outlined above, column-based methods of RNA purification are generally recommended, except for those instances where RNAs smaller than ~18 nt are being worked with, in which cases phenol-based extraction methods may offer greater enrichment.

Working with Degraded RNA: Enzymatic Verses Physical Degradation of RNA

RNA integrity is frequently referred to when working with RNA, and given the sample source (i.e., FFPE versus fresh tissue biopsies), the integrity of the RNA sample can inform the downstream application. As mentioned, RNA integrity is a metric that refers to the relative intensities of the 28S and 18S ribosomal peaks, when visualized by either agarose-gel electrophoresis, or digitally using the Agilent Bioanalyzer RNA chips (or similar application).

Depending on the application, degraded RNA is not necessarily problematic; although superior results will always be achieved with a more intact starting sample, many applications will function quite well even with substantial degradation of the sample. For example, microarrays and RT-qPCR applications can perform very well with RINs below 5, whereas next-generation sequencing applications will start to suffer due to the saturating effects of degraded rRNA in sequencing libraries, which produce non-informative rRNA reads.

Detection of RNAs/Validation of High-Throughput Data

One of the most common techniques to investigate the expression of candidate genes is the polymerase chain reaction. Here we explain how this technology can be used to determine the expression of ncRNAs and validate the data derived from high-throughput methods such as microarray or deep-sequencing.

Polymerase Chain Reaction

The expression of ncRNAs can be examined by reverse transcription (RT) PCR and/or quantitative real-time PCR. Measuring the level of ncRNAs is a crucial method to understand the roles they play in cells and how they may contribute towards cancer progression. However, despite the widespread use of PCR in diagnostic and research laboratories, there are still frustrations in performing successful PCR, and key parameters which may enhance the success of PCR applications are outlined below.

Preparation of cDNA

The total amount of RNA required for complementary DNA (cDNA) synthesis can vary between 100 and 1000 ng of total RNA. Since the amount of RNA derived from clinical samples is normally limiting, RT-PCR applications depend on the sample quality and quantity. Reverse-transcription can be performed easily using many commercially available kits.

To prepare cDNA to amplify lncRNA and/or precursor and mature miRNA, RNA first needs to be extracted using TRIZOL or another appropriate RNA purification kit, then treated with DNase I. Only after the RNA has been treated with DNAse can it be used in reverse transcription reactions. It has been shown that great numbers of lncRNAs are polyadenylated (poly-A) and their interactions with poly-A binding proteins affect their stability and turnover [39], therefore to detect the expression of lncRNAs, oligo-dT primers can be used to prepare the cDNA library used in PCR. For pre-miRNA or snoRNA and lncRNA lacking a polyadenylated tail gene-specific primers can be used instead, and random hexamers can also be employed for cDNA synthesis of lncRNA transcripts. However, to reverse transcribe stable hairpins, such as those found in pre-miRNA, the use of gene-specific primers at an elevated temperature rather than short primers (i.e., random hexamers) at room temperature is recommended [40]. Publicly available data can be examined to determine the existence of poly-A tail.

Designing Gene-Specific Primers

For lncRNAs containing more than two exons it is best to design the primers that cover exon–intron junctions to eliminate the effect of genomic DNA contamination in PCR reactions, which can induce false positives and reduce sensitivity. To design the primers several software programs such as Primer3, Primer Express, and NCBI primer design are available, and we have used these with excellent success.

For detection and quantification of RNA, the presence of particular RNA isoforms in the sample represent a major issue that needs to be considered carefully. For example, the isoforms of some lncRNAs are tissue specific, and it is important to select the exons

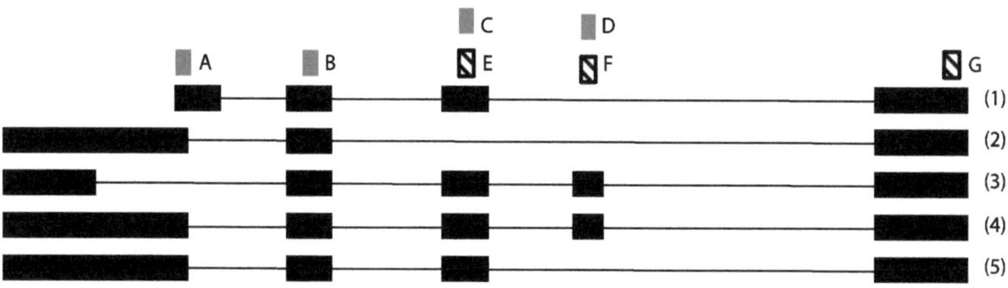

Fig. 1 Schematic of five different isoforms (*1–5*) of a gene and of primer sets used to identify those isoforms by RT-PCR. Different primers are designed to **detect the variety of isoforms**. *Gray* (*A–D*) and *hashed* (*E–G*) boxes are forward and reverse primers, respectively. The combination of each set of primers identifies only one to four isoforms: for instance, primers *A* and *E* identify *1*, *4*, and *5*, primers *B* and *E* identify *1*, *3*, *4*, and *5*, primers *A* and *G* identify isoforms *1*, *2*, *4* and *5* while primers *A* and *F* identify isoform *4*

that are expressed in most of the identified isoforms when designing primers. However, in some cases, to identify the exact sequence and length of isoforms other techniques are required. Figure 1 shows an example of five different isoforms of lncRNA, but each set of primers can distinguish between four isoforms at most. Therefore in this case to identify the 5′ end of the gene 5′ Rapid Amplification of cDNA Ends (5′ RACE) is required; alternately the tissue specific isoform could also potentially be identified by RT-PCR followed by Sanger sequencing or next-generation sequencing [41].

The need for high-stringency reliable PCR methods with reproducibility, selectivity, and sensitivity requires optimization of several conditions such as annealing temperature, elongation time, salt concentration, and can be further optimized by addition of additives (e.g., formamide, DMSO, and BSA) in the reaction mix (reviewed in [42]). The most common cause of unreliable PCR is off-target amplifications, which are normally thought to originate during lower temperatures of annealing, less stringent conditions of sample preparation, and thermal cycler ramping to the initial denaturation temperature of PCR (~94 °C). Therefore a number of methods such as touch-down and hot-start PCR have been developed to reduce the off-target amplification [42, 43].

Real-Time PCR

Real-time RT-PCR or quantitative PCR (qPCR) can easily and reliably quantitate the expression level of ncRNAs, and relies on the amplification of cDNA by gene-specific primers, similar to what was covered above. The recommended length of amplified products is between 80 and 250 nucleotides. The primers are also recommended to cover exon–intron junctions so that the effects of contamination by genomic DNA can be eliminated during the amplification process.

In qPCR the amount of amplified product is measured at the end of each cycle by the use of a fluorescent signal, typically either a dye that is incorporated into the PCR product during amplification, or by a fluorescent probe included into the amplification. In this manner the intensity of the fluorescent signal is positively correlated with the quantity of PCR product produced every cycle, and by monitoring the increase in fluorescence during the exponential phase of amplification the amount of RNA present in the original sample can be determined. The most common method of quantifying specific gene expression is by comparing its expression in relation to another gene(s) called a normalizing gene(s) [44, 45]. Normalized or housekeeping genes are selected for their almost constant rate of expression (e.g., *GAPDH, Actin, Tubulin*, and *HPRT*) and are usually involved in functions related to basic cellular survival. For best practice we recommend the use of more than one normalizer and use the average or geometric mean of gene expression level of multiple housekeeping transcripts as a reference point.

TaqMan PCR

Small ncRNAs such as miRNAs are abundant transcripts which can exhibit differential expression among tissues during development and disease [46]. However, less abundant RNAs can escape detection with some technologies such as cloning, northern hybridization, and microarray analysis, and since many miRNAs also have isoforms almost identical to the mature and precursor sequences, different methods must be used in expression analysis. In these situations, SYBR green detection is not sensitive enough to discriminate between related isoforms if primers are designed to the hairpin. For these reasons the TaqMan RNA assay offers a sensitive alternative which can be used to detect low levels of small RNAs [47]. Furthermore, using TaqMan PCR expands the real-time PCR technology for detecting long transcripts such as mRNA or lncRNA as well as smaller transcripts such as miRNA or snoRNA. It is designed to increase the specificity of quantitative PCR and is used for both diagnostic (e.g., Roche Molecular Diagnostic) and research (e.g., Applied Biosystems) purposes, and for these reasons is the preferred method for detection and quantification.

Microarrays Versus Sequencing for Detecting RNA Expression

With the decrease in cost of next-generation sequencing and the rise of bench-top sequencers such as the Ion Torrent PGM, Proton, and Miseq, there is the question of which application provides the most clinical utility given the pathology question to be addressed. As part of this issue, there is the question as to whether applications such as microarrays remain relevant to the clinic.

One advantage of microarray technology is the fact that it works quite well for samples for which there is limited input, as well as samples which show RNA degradation. Total RNA extractions from small biopsies are typically limiting, and while microgram amounts of total RNA can sometimes be achieved, yields in the hundreds of nanograms are the more likely outcome. Next-generation sequencing

(NGS) library construction becomes more challenging when the total sample amount is in the nanogram range, and in these cases some microarray platform (such as Affymetrix miRNA arrays) can tolerate an input as low as 120 ng of total RNA for miRNA and lncRNA detection; by comparison, most NGS library protocols will struggle to construct quality libraries with such small amounts of starting material. As well, while sequencing will also provide the ability to identify miRNA isoforms (i.e., isomiRs), only one miRNA species will typically dominate, and the identification of uncharacterized and/or lowly expressed ncRNA isoforms will be of lesser concern for many cancer pathologists.

Differential Expression of ncRNAs: Tumor Versus Normal

Similar to protein coding genes, the genes encoding ncRNAs can act as tumor suppressor or oncogenes, and differential expression of those has been reported in different types of cancer [1, 48–50]. The clinical utility of ncRNA expression typically relates to whether or not ncRNA species (or biomarkers) exhibit differential expression as compared with a normal control. This opens up the next topic, which is how to identify differentially expressed species between two sample sets.

There has been a substantial amount of data generated in the literature comparing different methods for ncRNA quantitation (e.g., microarrays, NGS library construction kits, RT-qPCR, etc.). For example, one paper [51] identified clear biases that related to the type of library kit construction method that was used for miRNA sequencing. Although the absolute numbers of miRNA molecules present determined by each method conflicted, one major conclusion from this paper was that the differential expression calls between normal and experimental samples, using the same kits, were highly concordant. Thus, while attempting to infer absolute values for the total number of copies of ncRNA molecules will be affected by the library construction method used, differential expression calls for ncRNA between control and experimental samples (i.e., normal versus tumor) show high concordance in spite of the method used, as long as *the control and experimental samples are prepared in the same manner*. When analysis methods are the same between the two samples being compared, then differential expression calls are typically quite reliable.

Bioinformatic Analysis of ncRNA Expression

When NGS and microarray applications emerged on the market, most analysis was restricted to either vendor-specific software for that particular application, or LINUX-based tools implemented from a command prompt. In both cases a certain amount of bioinformatic and programming experience was required to parse the data and determine expression profiles and statistical significance. However, there are now a variety of user-friendly commercial software solutions which can rapidly assist most pathologists in identifying noncoding transcripts that are differentially expressed between a tumor and normal sample. Two examples of analysis software which can be used with a variety of data (i.e., microarray, NGS, and qPCR) to determine differentially expressed ncRNAs of prognostic or predictive benefit are the Partek and DNASTAR software packages. Although a certain amount of user training is still required to effectively use these software suites, an afternoon or one-day training seminar is typically sufficient to provide the necessary skills to get the answer desired. There is also a rise in vendor-specific application packages, for example, Life Technologies Ion Reporter, which is designed to largely automate the process of analyzing sequencing data.

Detection and Localization of ncRNA

In Situ Hybridization

In situ hybridization (ISH) has become a powerful tool for studying the expression level and localization of specific RNA species within individual cells, or in tissue sections. Identifying the sub-cellular localization of ncRNAs can provide invaluable insights into the physiological and pathological processes in which they are involved, and in principle these applications are very similar to the immunohistochemistry used in standard pathology to determine the localization and expression of key protein biomarkers (i.e., HER2 in breast cancer). Several laboratories have successfully used ISH to detect and localize ncRNA in different cell types or tissue sections [22, 52, 53], and since the expression of ncRNA can occur at key points in disease progression, its detection and localization can be used as a diagnostic marker.

In situ hybrization utilizes the principle of Watson–Crick base pairing, in which a labeled RNA sequence (the probe) can be used to quantify the expression and define the sub-cellular localization of target RNA molecules in fixed tissues/cells, followed by visualization of target transcripts via a fluorescence dye incorporated into the probe. This method can be employed to detect the presence of the target of interest, and gain information which relates to the intracellular spatial localization of lncRNAs and small RNAs. Determining the spatial expression of these ncRNAs may provide an insight to whether they are involved in post-transcriptional regulation of gene expression [54].

The key requirements for successful ISH are preparation of samples, making the right template for the probe, labeling the probe appropriately, permeability of the cells, defining optimum hybridization conditions, and imaging techniques.

Probe Labeling

The technique of probe labeling has undergone substantial changes since its conception in 1968. Initially probes were labeled using radioactive isotopes like ^{32}P, ^{35}S, and ^{3}H [55] which were soon replaced by enzyme linked probes which catalyzed chromogenic or fluorogenic reactions [56]. Amidst all this fluorescence-labeled probes were introduced for the first time in 1980 by Bauman et al. [57], and since then fluorescence labeled probes have been engineered and developed to increase efficiency/sensitivity, specificity and facilitate easy detection of target RNA transcripts.

Probe Design

The design of the probes has also evolved substantially from their inception. Initially probes were produced from clones and were quite large and sparsely labeled to allow specific hybridization. However, using long probes resulted in high background fluorescence, non-specific sites of binding, and non-specific signal detection; to improve the technique these large probes were then chopped into smaller pieces (<200 bases) to obtain more specific signal and improve specificity [58]. Femino et al. used several short 50 nucleotide long probes which were complementary to sequential parts of target mRNAs and each probe was conjugated to five fluorescent moieties at predefined positions, which increased the sensitivity of the technique to detect the single mRNA molecules [59].

More recently Raj et al. developed a technique by designing probes to image individual mRNA molecules in fixed cells. In this method each mRNA is targeted with ≤48 singly labeled oligonucleotide probes which enabled visualization of each individual RNA target by fluorescence microscopy [60]; designing the probes in this way enables simultaneous detection, localization, and quantification of individual RNA molecules at the cellular level [61]. The fluorescence ISH (FISH) technique developed by Raj and colleagues has enabled direct visualization and quantification of lncRNAs (e.g., *TINCR*) [62]. Here we explain this method in more detail.

FISH Technique Using Multiple Singly Labeled Probes

The technique requires designing a set of 48 oligonucleotide probes (20 bases long), with each of these oligonucleotides complementary to different regions of the target RNA molecule (with at least 2 bases separating any two oligonucleotides). The GC content of each individual oligonucleotide also needs to be taken into consideration, to ensure uniform hybridization potential of all oligonucleotide probes, which also ensures the binding of the maximum number of probes at a given hybridization stringency [63]. The oligonucleotide probes are then conjugated with a fluorophore of choice at the 3′ end.

This protocol involves fixation of cells/tissue, permeabilization of cells, and hybridization with probes followed by washing and visualization using a fluorescent microscope [60]. The cells are fixed with 3.7 % formaldehyde in PBS, followed by permeabilization of cells with 70 % ethanol. The hybridization step is very critical and certain parameters should be considered carefully, such as the concentration of probes (normally between 50 nM and 1.25 μM), hybridization temperature (the temperature at which probes hybridize to the RNA of interest), and concentration of formamide (which is a component of the hybridization and wash buffers to enhance specificity); the last two parameters are related and concerned with the stringency of the hybridization itself [63]. The concentration of formamide used is 10 % in both hybridization and wash buffer, with a higher concentration of formamide leading to higher stringency [63]. Hybridization is carried out normally at 37 °C, and the use of higher temperatures can also lead to higher stringency, since fewer probes will bind.

Imaging is done using a wide-field fluorescence microscope using 63×/100× oil objectives and *z*-stack images are obtained. The signals obtained in this case are very weak and need a longer exposure time, as compared with standard fluorescence microscopy applications. However, one disadvantage of using wide-field fluorescence microscopes is that they place a tight limitation on sample thickness, as thicker samples lead to more out-of-focus light that may obscure the relatively weak target RNA signals. Confocal microscopy stands at a disadvantage in this case as the high intensity of the laser excitation can result in rapid photo-bleaching of the sample [63].

This technique is applicable to wide variety of samples, such as adherent mammalian cell lines, cells in suspension, frozen specimens, and FFPE tissue [61]. In addition to this, the technique can be combined with immunofluorescence to get a complete transcription-translation picture, and detect possible interactions between the ncRNA of interest and proteins [64].

LNA-FISH/CISH (Colorimetric in situ DNA probe hybridization)

This is a variation of the FISH technique described in the previous section and enables localization of small ncRNA like miRNAs in cultured cells and FFPE tissue samples/sections [65].

These techniques utilize locked nucleic acid (LNA)-modified oligonucleotide probes which have a greater hybridization affinity towards their complementary/target RNAs, which enables the use of more stringent hybridization conditions to increase specificity and sensitivity. This is particularly important because the small size of miRNAs may lead to non-specific binding of the probes. The length of the probes is restricted to 17–21 nucleotides for miRNA and LNA modification to 20–25 % of the probe (i.e., 1–2 bases). Binding of the probes can be detected either colorimetrically by antibodies specific for DIG (digoxigenin), which are coupled with AP (alkaline phosphatase) or by the use of fluorophores. The sensitivity of the probes can be increased further by labeling both 5′ and 3′ ends [65].

RNA-Protein Interactions

The functional characterization of many ncRNAs has revealed that they are interacting with proteins, and those interactions are known to be vital for ncRNA function. NcRNA-protein interactions have been studied, for example, through classical biochemical techniques developed for studying translation and RNA-processing complexes, as well as more recent technological advances using RNA immunoprecipitation (RIP) [66] and cross-linking followed by immunoprecipitation [67, 68]. These experiments have demonstrated that ncRNAs specifically interact with proteins in the RISC complex, in chromatin and in the chromatin-modifying machinery such as polycomb repressor complexes and trithorax proteins as well as transcription factors, and promoter- or enhancer-associated proteins [28, 69, 70]. Long ncRNAs may also act as scaffolds for multiple complexes [71]. Of note, RNA molecules can also directly interact with DNA through canonical Watson–Crick base pairing, and via non-canonical structures such as triple helices [72] and by indirect interactions mediated through another RNA or a protein molecule [73]. Most common methods for analyzing RNA-protein interactions, such as RIP or chromatin immunoprecipitation (ChIP), rely on using known protein(s) to isolate and identify unknown RNA binding partners. The most common methods for studying protein-ncRNA interactions are described below.

RNA Immunoprecipitation

It is known that mRNA molecules interact with several RNA binding proteins (RBPs) and those interactions are critical for mRNA stability and function. For this reason high-throughput technologies to analyze the entire subset of mRNAs associated with a particular RBP are required [74–76]. RIP uses an antibody specific to the RBP of interest to capture endogenously formed mRNA-proteins complexes, followed by purification of associated mRNAs; the purified RNA can then be used for quantitative analysis using microarrays, RT-PCR, or deep-sequencing. This technology has been used to identify mRNA bound directly or indirectly as parts of the larger mRNA-protein (mRNP) complexes to RBPs. Currently RIP is widely used to identify the short and long ncRNAs that are bound to particular RBPs.

The RIP-Chip (also known as RIP on Chip or RIP-SEQ) method is used to identify discrete subsets of RNAs associated with multiple RNA targets of RBPs globally. This method can be applied to both small and long ncRNAs. Here we describe this procedure for analysis of small and long ncRNAs (procedure summarized in Fig. 2).

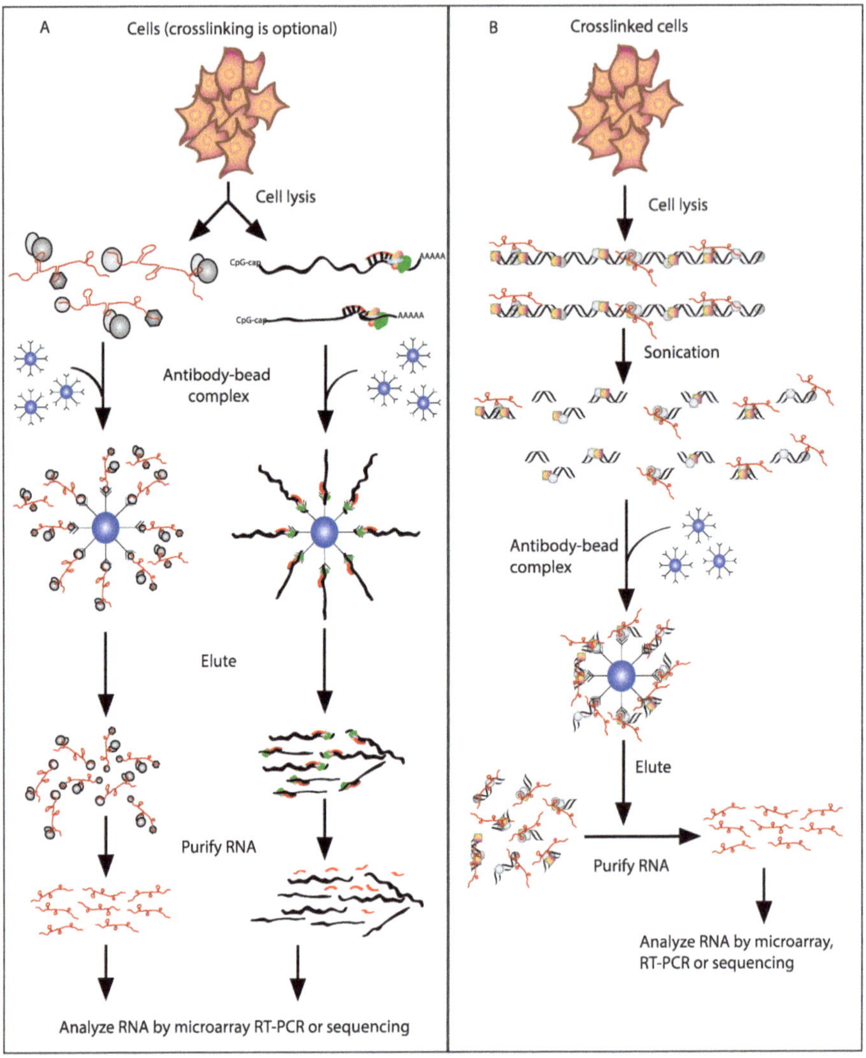

Fig. 2 RNA immunoprecipitation assay (**a**) illustrating the RIP assay for long (*left*) and short (*right*) ncRNA. The cells are lysed and total lysate incubated with an antibody-bead complex. Protein-RNA complexes bind to the antibody-bead and after several washes the bound complex is eluted and total RNA extracted for further sequence analysis. (**b**) The ChIP assay. Cells are lysed after being fixed and total cell lysate is sonicated or enzymatically digested to fragment chromatin. The sample is then incubated with an antibody-bead complex to capture bound RNA and/or DNA-protein complexes. The captured materials are washed and eluted. The nucleic acids recovered from this process are further analyzed

Purification and Identification of miRNA by RIP-Chip Assay

To target miRNA present in the cytoplasmic fraction a mild lysis buffer is used, which leaves nuclei essentially intact and minimizes inappropriate exchange of mRNAs during subsequent immunoprecipitation [77]. Also modifications that include the nuclear fraction use stronger detergents that lyse nuclear membrane [77, 78]. In this method, total, nuclear or cytoplasmic extracts are immunoprecipitated, the pellets washed extensively, the RNPs released and dissociated into RNA and protein components and total RNA extracted [77]. The purified RNA can then be analyzed by various methods including RT-PCR, microarray analysis, or high-throughput sequencing.

An alternate method is to purify the miRNA bound to its target mRNA, which involves using an antibody against a protein in the RISC-complex (such as Argonaute proteins) and immunoprecipitating mRNA-miRNA complexes. In this case the cytoplasmic fraction is incubated with anti-Argonaute antibody-coated beads that bind to mRNA-miRNA complexes and capture all RISC complexes. The precipitate then can be extensively washed and total RNA from captured complexes can be purified by the appropriate method, and further analyzed using various applications (Fig. 2a). This method can be used for any other class of small ncRNAs simply by modifying the specific antibody that binds the RBP.

Purification and Identification of lncRNA by RIP

For purification and identification of lncRNAs, the same procedure as above can be applied by selecting an antibody against a protein that binds to the lncRNA of interest. For example, proteins such as poly-A binding proteins and m7GpppG cap-binding proteins are known to interact with lncRNAs, and antibodies against those proteins are commonly used to immunoprecipitate RNA-protein complexes (Fig. 2a).

It should be noted that several papers report controversial results when cross-linking prior to cell lysis. Some found reversible formaldehyde cross-linking led to high non-specific binding [79], while others recommend using RNA-protein cross-linking in RIP-Chip [68, 78, 80]. Such disparities suggest that an optimization step may be required to compare cross-linked to non-cross-linked materials, so as to determine which yields superior results. If no major differences are obtained one should avoid the cross-linking step as this step may introduce artifacts, reduce cell lysis efficiency, introduce sequence biases, and increase background binding [77, 79]. However, using 0.1 % formaldehyde rather than 1 % formaldehyde for cross-linking RNA to protein is recommended to improve the quality and recovery of bound RNA [78].

Chromatin Immunoprecipitation

It has been shown that ncRNAs play important regulatory roles in chromatin remodeling, and there are several examples of ncRNAs which influence chromatin dynamics and function [28, 81–85]. Chromatin immunoprecipitation (ChIP) is a powerful method that allows us to probe specific protein-DNA/RNA interactions in vivo, and reveals whether a protein-nucleic acid interaction is present at a certain location within the genome. It also shows the density of protein or nucleic acid in that region, and when combined with other techniques [86, 87] can uncover an extraordinarily rich and dynamic chromatin environment [88, 89].

For chromatin immunoprecipitation it is advisable to cross-link the nucleic acids to the proteins with formaldehyde prior to cell lysis. The cells then are lysed followed by sonication or enzymatic digestion to shear chromatin into fragment sizes ranging from 200 to 1000 bp. Complexes containing the factor of interest are then immunoprecipitated using an antibody specific to that protein, the immunoprecipitate is washed, cross-links are reversed, and RNA is purified from the isolated chromatin (Fig. 2b) [90, 91].

Since this technology is extensively used a number of companies such as Cell Signaling, Millipore, and Active Motif have developed kits that can be easily used to perform ChIP.

RNA Pull-Down Assays

The advancement of sequencing technologies has led to the discovery of many new ncRNAs, and has revealed their important physiological functions in cells. Functional characterization of some ncRNAs has demonstrated that they interact with other components in the cell, such as other RNA molecules, proteins, or even specific regions of a chromosome. Identification of these interacting components can be informative, and as explained in the previous section RIP and ChIP can be used for known proteins to identify bound RNAs. Conversely, to identify novel proteins bound to an RNA species of interest different technology is required. To this end the RNA pull-down assay was developed, which uses labeled RNA to capture the proteins, RNA or DNA molecules bound to it.

RNA Pull-Down Assays Using Labeled RNA

This method selectively extracts protein-RNA complexes from a sample by taking advantage of high affinity tags, such as biotin. The template to prepare the probe can be provided by cloning the region of interest into an appropriate vector, which is then used as a template for in vitro transcription using the necessary kits and biotinylated nucleic acids. The resulting in vitro transcribed RNA is then purified and used in pull-down assays [92]. Biotinylated RNA probes are then incubated with protein from total cell lysate, cytoplasmic or nuclear fractions, and purified using agarose or magnetic beads. After washing the beads several times the complex is dissociated from the beads and the total protein is precipitated and analyzed by an appropriate method, such as mass spectrometry (Fig. 3a). Other RNA species or DNA from complexes can also be analyzed by RT-PCR or sequencing. One recommended kit is the Pierce Magnetic RNA-Protein Pull-down Kit which can be used for labeling RNA and pull down. The pulled-down proteins can be detected by Western blotting or used in mass spectrometry. The other RNA species or DNA from such complexes can be analyzed by RT-PCR and sequencing.

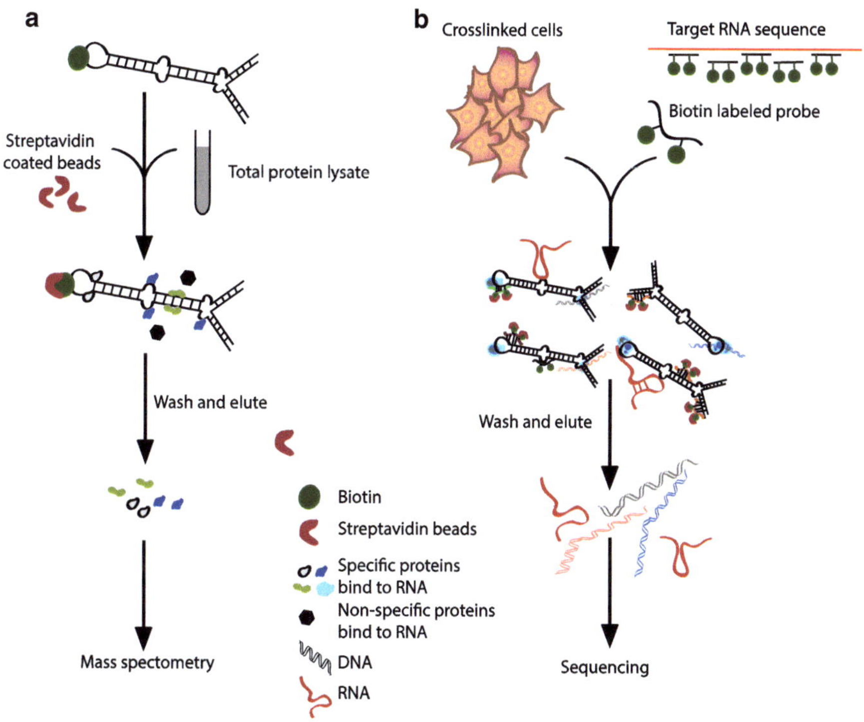

Fig. 3 RNA pull-down assay (**a**) in vitro labeled RNA is incubated with total cell lysate. The labeled RNA binds to protein, and the complex is captured by coated beads with affinity for the labeled RNA (e.g., biotin-labeled RNA and streptavidin-coated beads). Captured complexes are washed and eluted. Total protein from RNA-protein complexes is recovered and analyzed by mass spectrometry. (**b**) RNA antisense purification (RAP). In this method labeled tiled probes specific for the RNA of interest are prepared and incubated with total cell lysate. The probes bind to lncRNA in complex with protein(s), DNA, or other RNA molecule(s). The complexes are captured by beads coated to bind to labeled probe (e.g., biotin-labeled RNA and streptavidin-coated beads). After several washes the complexes are eluted and total RNA and/or DNA extracted for further sequence analysis

RNA Antisense Purification (RAP)

This is a newly developed method that uses labeled antisense probes to hybridize to a target RNA to purify the endogenous RNA and its associated proteins, RNA, and genomic DNA from cross-linked cell lysate. To achieve high specificity, RAP utilizes 120-nucleotide antisense RNA probes complementary to the RNA of interest that will hybridize and capture the target RNA. Strong hybridization conditions provide the opportunity to purify the cellular components bound to RNA molecules under denaturing conditions, and non-specific RNA-protein interactions and non-specific hybridization with RNAs or genomic DNA are limited using these conditions. This technology was initially used to purify *XIST* lncRNA and its associated cellular components. DNase I was used to digest genomic DNA to ~150 bp fragments to achieve high resolution mapping of binding sites, and the DNA bound to *XIST* was then sequenced [53]. However, the DNAse digestion step can be skipped if the ncRNA is derived from cytoplasmic fractions.

RAP uses a pool of overlapping probes tiled across the entire length of the target RNA to capture lncRNA. The tiling of probes across the whole length of RNA ensures efficient capture of the RNA molecules even in the case of extensive protein-RNA interactions, RNA secondary structure, or partial RNA degradation [53]. Isolated protein bound to RNA molecules can then be analyzed by Western blotting or mass spectrometry. The co-localization of RNA and proteins by this method can be examined by ISH and immunofluorescence of identified proteins. The RNA and DNA bound to target RNA can be further analyzed by microarray or sequencing (Fig. 3b).

Prospects

In recent years significant developments in genome and transcriptome analysis has provided us enormous amounts of data which have led to an appreciation of the significant number of ncRNAs present in the human genome. Today there is no debate concerning the functionality of ncRNAs, whereas the mechanisms of action for those are yet to be answered. Due to ease of access and reduced cost of technologies for these analyses, the techniques for ncRNA analysis are being used routinely in many research and clinical laboratories. Currently, many scientists continue to develop methods that will unveil the complex characteristics of ncRNAs. In this chapter we have discussed some of the methods which have been developed recently, or have been adopted by previous studies to explore the function of ncRNA and explore its role in cancer progression. We believe ncRNA research will provide an exciting opportunity for scientists and clinicians to collaborate to discover the hidden layer of complexity that is the noncoding transcriptome in cancer.

Acknowledgement MEA-A is recipient of the Rodney and Elaine Davies Cancer Research Fellowship and funded by the Auckland Medical Research Foundation. Authors are greatly thankful to Professor Bruce Baguley for reading the manuscript and his critical comments. DJK is funded by the National Breast Cancer Foundation, Australia.

References

1. Esteller M (2011) Non coding RNAs in human disease. Nat Rev Genet 12(12):861–874
2. Mattick J (2005) The functional genomics of Non coding RNA. Science 309(5740):1527–1528
3. Mattick J, Makunin IV (2006) Non coding RNA. Hum Mol Genet 15 Spec No 1:R17–R29
4. Prensner J, Chinnaiyan AM (2011) The emergence of lncRNAs in cancer biology. Cancer Discov 1(5):391–407
5. Zhang H, Chen Z, Wang X, Huang Z, He Z, Chen Y (2013) Long Non coding RNA: a new player in cancer. J Hematol Oncol 6(1):37
6. Guttman M, Rinn JL (2012) Modular regulatory principles of large Non coding RNAs. Nature 482(7385):339–346
7. Kim E, Sung S (2012) Long Non coding RNA: unveiling hidden layer of gene regulatory networks. Trends Plant Sci 17(1):16–21
8. Lee R, Feinbaum RL, Ambros V (1993) The C. elegans heterochronic gene lin-4 encodes small RNAs with antisense complementarity to lin-14. Cell 75(5):843–854
9. Aravin A, Naumova NM, Tulin AV, Vagin VV, Rozovsky YM, Gvozdev VA (2011) Double-stranded RNA-mediated silencing of genomic tandem repeats and transposable elements in the D. melanogaster germline. Curr Biol 11(13):1017–1027
10. Bachellerie J, Cavaille J, Huttenhofer A (2002) The expanding snoRNA world. Biochimie 84(8):775–790
11. Gerard M, Myslinski E, Chylak N, Baudrcy S, Krol A, Carbon P (2010) The scaRNA2 is produced by an independent transcription unit and its processing is directed by the encoding region. Nucleic Acids Res 38(2):370–381
12. Mendell J (2005) MicroRNAs: critical regulators of development, cellular physiology and malignancy. Cell Cycle 4(9):1179–1184
13. Bernstein B, Birney E, Dunham I, Green ED, Gunter C, Snyder M (2012) An integrated encyclopedia of DNA elements in the human genome. Nature 489(7414):57–74
14. Carninci P, Kasukawa T, Katayama S, Gough J, Frith MC, Maeda N, Oyama R et al (2005) The transcriptional landscape of the mammalian genome. Science 311(5740):1559–1563
15. Carninci P (2007) Constructing the landscape of the mammalian transcriptome. J Exp Biol 210:1497–1506
16. Djebali S, Davis CA, Merkel A, Dobin A, Lassmann T, Mortazavi A, Tanzer A, Lagarde J et al (2012) Landscape of transcription in human cells. Nature 489(7414):101–108
17. Dinger M, Pang KC, Mercer TR, Mattick JS (2008) Differentiating protein-coding and Non coding RNA: challenges and ambiguities. PLoS Comput Biol 4, e1000176
18. Ponting C, Oliver PL, Reik W (2009) Evolution and functions of long Non coding RNAs. Cell 136(4):629–641
19. Feng J, Bi C, Clark BS, Mady R, Shah P, Kohtz JD (2006) The Evf-2 Non coding RNA is transcribed from the Dlx-5/6 ultraconserved region and functions as a Dlx-2 transcriptional coactivator. Genes Dev 20:1470–1484
20. Shamovsky I, Ivannikov M, Kandel ES, Gershon D, Nudler E (2006) RNA-mediated response to heat shock in mammalian cells. Nature 440(7083):556–560
21. Wang X, Arai S, Song X, Reichart D, Du K, Pascual G, Tempst P, Rosenfeld MG, Glass CK, Kurokawa R (2008) Induced ncRNAs allosterically modify RNA-binding proteins in cis to inhibit transcription. Nature 454(7200):126–130
22. Clemson C, Hutchinson JN, Sara SA, Ensminger AW, Fox AH, Chess A, Lawrence JB (2009) An architectural role for a nuclear Non coding RNA: NEAT1 RNA is essential for the structure of paraspeckles. Mol Cell 33(6):717–726
23. Sunwoo H, Dinger ME, Wilusz JE, Amaral PP, Mattick JS, Spector DL (2009) MEN epsilon/beta nuclear-retained Non coding RNAs are up-regulated upon muscle differentiation and are essential components of paraspeckles. Genome Res 19(3):347–359

24. Dieci G, Preti M, Montanini B (2009) Eukaryotic snoRNAs: a paradigm for gene expression flexibility. Genomics 94(2):83–88
25. Cai X, Hagedorn CH, Cullen BR (2004) Human microRNAs are processed from capped, polyadenylated transcripts that can also function as mRNAs. RNA 10(12):1957–1966
26. Lee Y, Kim M, Han J, Yeom KH, Lee S, Baek SH, Kim VN (2004) MicroRNA genes are transcribed by RNA polymerase II. EMBO J 23(20):4051–4060
27. Ji P et al (2003) MALAT-1, a novel Non coding RNA, and thymosin beta4 predict metastasis and survival in early-stage non-small cell lung cancer. Oncogene 22(39):8031–8041
28. Gupta R, Shah N, Wang KC, Kim J, Horlings HM, Wong DJ, Tsai MC, Hung T, Argani P, Rinn JL, Wang Y, Brzoska P, Kong B, Li R, West RB, van de Vijver MJ, Sukumar S, Chang HY (2010) Long Non coding RNA HOTAIR reprograms chromatin state to promote cancer metastasis. Nature 464(7291):1071–1076
29. Calin G, Croce CM (2006) MicroRNA signatures in human cancers. Nat Rev Cancer 6(11):857–866
30. Iorio M, Ferracin M, Liu CG, Veronese A, Spizzo R, Sabbioni S, Magri E, Pedriali M, Fabbri M et al (2005) MicroRNA gene expression deregulation in human breast cancer. Cancer Res 65(16):7065–7070
31. Hildebrandt E, Lee JR, Crosby JH, Ferris DG, Anderson MG (2003) Liquid based pap smears as a source of RNA for analysis of gene expression. Appl Immunohistochem Mol Morphol 11(4):345–351
32. Lodde M, Fradet Y (2008) The detection of genetic markers of bladder cancer in urine and serum. Curr Opin Urol 18(5):499–503
33. Menke T, Warnecke JM (2004) Improved conditions for isolation and quantification of RNA in urine specimens. Ann N Y Acad Sci 1022:185–189
34. Taback B, Hoon DS (2004) Circulating nucleic acids and proteomics of plasma/serum: clinical utility. Ann N Y Acad Sci 1022:1–8
35. Zhang L, Farrell JJ, Zhou H, Elashoff D, Akin D, Park NH, Chia D, Wong DT (2010) Salivary transcriptomic biomarkers for detection of resectable pancreatic cancer. Gastroenterology 138(3):949–57 e1–7
36. Thery C, Zitvogel L, Amigorena S (2002) Exosomes: composition, biogenesis and function. Nat Rev Immunol 2(8):569–579
37. Lasser C, Alikhani VS, Ekstrom K, Eldh M, Paredes PT, Bossios A, Sjostrand M, Gabrielsson S, Lotvall J, Valadi H (2011) Human saliva, plasma and breast milk exosomes contain RNA: uptake by macrophages. J Transl Med 9:9
38. Borgna S, Armellin M, di Gennaro A, Maestro R, Santarosa M (2012) Mesenchymal traits are selected along with stem features in breast cancer cells grown as mammospheres. Cell Cycle 11(22):4242–4251
39. Beaulieu Y, Kleinman CL, Landry-Voyer AM, Majewski J, Bachand F (2012) Polyadenylation-dependent control of long Non coding RNA expression by the poly(A)-binding protein nuclear 1. PLoS Genet 8(11):e1003078
40. Schmittgen T, Jiang J, Liu Q, Yang L (2004) A high-throughput method to monitor the expression of microRNA precursors. Nucleic Acids Res 32(4), e43
41. Dave R, Dinger ME, Andrew M, Askarian-Amiri M, Hume DA, Kellie S (2013) Regulated expression of PTPRJ/CD148 and an antisense long Non coding RNA in macrophages by pro-inflammatory stimuli. PLoS One 8, e68306
42. Korbie D, Mattick J (2008) Touchdown PCR for increased specificity and sensitivity in PCR amplification. Nat Protoc 3(9):1452–1456
43. Paul N, Shum J, Le T (2010) Hot start PCR. Methods Mol Biol 630:301–318
44. Gibson U, Heid CA, Williams PM (1996) A novel method for real time quantitative RT-PCR. Genome Res 6(10):995–1001
45. Heid C, Stevens J, Livak KJ, Williams PM (1996) Real time quantitative PCR. Genome Res 6(10):986–994

46. Kim J, Krichevsky A, Grad Y, Hayes GD, Kosik KS, Church GM, Ruvkun G (2004) Identification of many microRNAs that copurify with polyribosomes in mammalian neurons. Proc Natl Acad Sci U S A 101(1):360–365
47. Chen C, Ridzon DA, Broomer AJ, Zhou Z, Lee DH, Nguyen JT, Barbisin M, Xu NL, Mahuvakar VR, Andersen MR, Lao KQ, Livak KJ, Guegler KJ (2005) Real-time quantification of microRNAs by stem-loop RT-PCR. Nucleic Acids Res 33(20), e179
48. Gibb E, Brown CJ, Lam WL (2011) The functional role of long Non coding RNA in human carcinomas. Mol Cancer 10:38
49. Gutschner T, Diederichs S (2012) The hallmarks of cancer: a long Non coding RNA point of view. RNA Biol 9(6):703–719
50. Huarte M, Rinn JL (2010) Large Non coding RNAs: missing links in cancer? Hum Mol Genet 19((R2)):R152–R161
51. Sanada Y, Yoshida K, Ohara M, Oeda M, Konishi K, Tsutani Y (2006) Histopathologic evaluation of stepwise progression of pancreatic carcinoma with immunohistochemical analysis of gastric epithelial transcription factor SOX2: comparison of expression patterns between invasive components and cancerous or nonneoplastic intraductal components. Pancreas 32(2):164–170
52. Askarian-Amiri ME, Crawford J, French JD, Smith MA, Smart CE, Ru K, Mercer TR, Thompson ER, Lakhani SR, Vargas AC, Campbell IG, Brown MA, Dinger ME, Mattick JS (2011) SNORD-host RNA Znfx1-as is a regulator of mammary development and a potential marker for breast cancer. RNA 17:878–891
53. Engreitz J, Pandya-Jones A, McDonel P, Shishkin A, Sirokman K, Surka C, Kadri S, Xing J, Goren A, Lander ES, Plath K, Guttman M (2013) The Xist lncRNA exploits three-dimensional genome architecture to spread across the X chromosome. Science 341(6147):1237973
54. Kwon S (2013) Single-molecule fluorescence in situ hybridiz: quantitative imaging of single RNA molecules. BMB Rep 46(2):65–72
55. Gall J, Pardue ML (1969) Formation and detection of RNA-DNA hybrid molecules in cytological preparations. Proc Natl Acad Sci U S A 63(2):378–383
56. Raap A, van de Corput MP, Vervenne RA, van Gijlswijk RP, Tanke HJ, Wiegant J (1995) Ultrasensitive FISH using peroxidase-mediated deposition of biotin- or fluorochrome tyramides. Hum Mol Genet 4(4):529–534
57. Bauman J, Wiegant J, Borst P, van Duijn P (1980) A new method for fluorescence microscopical localization of specific DNA sequences by in situ hybridization of fluorochromelabelled RNA. Exp Cell Res 128(2):485–490
58. Levsky J, Singer RH (2003) Fluorescence in situ hybridization: past, present and future. J Cell Sci 116(Pt 14):2833–2838
59. Femino A, Fay FS, Fogarty K, Singer RH (1998) Visualization of single RNA transcripts in situ. Science 280(5363):585–590
60. Raj A, van den Bogaard P, Rifkin SA, van Oudenaarden A, Tyagi S (2008) Imaging individual mRNA molecules using multiple singly labeled probes. Nat Methods 10:877–879
61. Orjalo Jr A, Johansson HE, Ruth JL (2011) Stellaris[trade] fluorescence in situ hybridization (FISH) probes: a powerful tool for mRNA detection. Nat Methods 8
62. Kretz M, Siprashvili Z, Chu C, Webster DE, Zehnder A, Qu K, Lee CS, Flockhart RJ, Groff AF, Chow J, Johnston D, Kim GE, Spitale RC, Flynn RA, Zheng GX, Aiyer S, Raj A, Rinn JL, Chang HY, Khavari PA (2013) Control of somatic tissue differentiation by the long Non coding RNA TINCR. Nature 493(7431):231–235
63. Raj A, Tyagi S (2010) Chapter 17 – Detection of individual endogenous RNA transcripts in situ using multiple singly labeled probes. In: Nils GW (ed) Methods in enzymology. Academic Press, pp 365–386
64. Shih J, Waks Z, Kedersha N, Silver PA (2011) Visualization of single mRNAs reveals temporal association of proteins with microRNA-regulated mRNA. Nucleic Acids Res 39(17):7740–7749
65. de Planell-Saguer M, Rodicio MC, Mourelatos Z (2010) Rapid in situ codetection of Non coding RNAs and proteins in cells and formalin-fixed paraffin-embedded tissue sections without protease treatment. Nat Protoc 5(6):1061–1073

66. Gilbert C, Svejstrup JQ (2006) RNA immunoprecipitation for determining RNA-protein associations in vivo (Chapter 27, Unit 27.4). In: Ausubel FM et al (eds) Curr Protoc in Mol Biol. doi: 10.1002/0471142727.mb2704s75
67. Ule J, Jensen KB, Ruggiu M, Mele A, Ule A, Darnell RB (2003) CLIP identifies Nova-regulated RNA networks in the brain. Science 302(5648):1212–1215
68. Ule J, Jensen K, Mele A, Darnell RB (2005) CLIP: a method for identifying protein-RNA interaction sites in living cells. Methods 37(4):376–386
69. Cabianca D, Casa V, Bodega B, Xynos A, Ginelli E, Tanaka Y, Gabellini D (2012) A long ncRNA links copy number variation to a polycomb/trithorax epigenetic switch in FSHD muscular dystrophy. Cell 149(4):819–831
70. Orom UA, Derrien T, Beringer M, Gumireddy K, Gardini A, Bussotti G, Lai F, Zytnicki M, Notredame C, Huang Q, Guigo R, Shiekhattar R (2010) Long Non coding RNAs with enhancer-like function in human cells. Cell 143(1):46–58
71. Tsai M, Manor O, Wan Y, Mosammaparast N, Wang JK, Lan F, Shi Y, Segal E, Chang HY (2010) Long Non coding RNA as modular scaffold of histone modification complexes. Science 329(5992):689–693
72. Duca M, Vekhoff P, Oussedik K, Halby L, Arimondo PB (2008) The triple helix: 50 years later, the outcome. Nucleic Acids Res 36(16):5123–5138
73. Koziol MJ, Rinn JL (2010) RNA traffic control of chromatin complexes. Curr Opin Genet Dev 20(2):142–148
74. Pedace L, De Simone P, Castori M, Sperduti I, Silipo V, Eibenschutz L, De Bernardo C, Buccini P, Moscarella E, Panetta C, Ferrari A, Grammatico P, Catricala C (2011) Clinical features predicting identification of CDKN2A mutations in Italian patients with familial cutaneous melanoma. Cancer Epidemiol 35:e116–e120
75. Tenenbaum S, Carson CC, Lager PJ, Keene JD (2000) Identifying mRNA subsets in messenger ribonucleoprotein complexes by using cDNA arrays. Proc Natl Acad Sci U S A 97(26): 14085–14090
76. Tenenbaum S, Lager PJ, Carson CC, Keene JD (2002) Ribonomics: identifying mRNA subsets in mRNP complexes using antibodies to RNA-binding proteins and genomic arrays. Methods 26(2):191–198
77. Keene J, Komisarow JM, Friedersdorf MB (2006) RIP-Chip: the isolation and identification of mRNAs, microRNAs and protein components of ribonucleoprotein complexes from cell extracts. Nat Protoc 1(1):302–307
78. Kaneko S, Manley JL (2005) The mammalian RNA polymerase II C-terminal domain interacts with RNA to suppress transcription-coupled 3′ end formation. Mol Cell 20(1):91–103
79. Penalva L, Tenenbaum SA, Keene JD (2004) Gene expression analysis of messenger RNP complexes. Methods Mol Biol 257:125–134
80. Kunitomo H, Uesugi H, Kohara Y, Iino Y (2005) Identification of ciliated sensory neuron-expressed genes in Caenorhabditis elegans using targeted pull-down of poly(A) tails. Genome Biol 6(2):R17
81. Guil S, Soler M, Portela A, Carrere J, Fonalleras E, Gomez A, Villanueva A, Esteller M (2012) Intronic RNAs mediate EZH2 regulation of epigenetic targets. Nat Struct Mol Biol 19:664–670
82. Guttman M, Amit I, Garber M, French C, Lin MF, Feldser D, Huarte M, Zuk O, Carey BW, Cassady JP, Cabili MN, Jaenisch R, Mikkelsen TS, Jacks T, Hacohen N, Bernstein BE, Kellis M, Regev A, Rinn JL, Lander ES (2009) Chromatin signature reveals over a thousand highly conserved large Non coding RNAs in mammals. Nature 458(7235):223–227
83. Magistri M, Faghihi MA, St Laurent G 3rd, Wahlestedt C (2012) Regulation of chromatin structure by long Non coding RNAs: focus on natural antisense transcripts. Trends Genet 28:389–396
84. Mikkelsen T, Ku M, Jaffe DB, Issac B, Lieberman E, Giannoukos G, Alvarez P, Brockman W, Kim TK, Koche RP, Lee W, Mendenhall E, O'Donovan A, Presser A, Russ C, Xie X, Meissner A, Wernig M, Jaenisch R, Nusbaum C, Lander ES, Bernstein BE (2007) Genome-wide maps of chromatin state in pluripotent and lineage-committed cells. Nature 448:553–560

85. Pandey R, Mondal T, Mohammad F, Enroth S, Redrup L, Komorowski J, Nagano T, Mancini-Dinardo D, Kanduri C (2008) Kcnq1ot1 antisense Non coding RNA mediates lineage-specific transcriptional silencing through chromatin-level regulation. Mol Cell 32(2):232–246
86. Dundr M, Hoffmann-Rohrer U, Hu Q, Grummt I, Rothblum LI, Phair RD, Misteli T (2002) A kinetic framework for a mammalian RNA polymerase in vivo. Science 298(5598):1623–1626
87. Cheutin T, McNairn AJ, Jenuwein T, Gilbert DM, Singh PB, Misteli T (2003) Maintenance of stable heterochromatin domains by dynamic HP1 binding. Science 299(5607):721–725
88. Koyanagi M, Baguet A, Martens J, Margueron R, Jenuwein T, Bix M (2005) EZH2 and histone 3 trimethyl lysine 27 associated with Il4 and Il13 gene silencing in Th1 cells. J Biol Chem 280(36):31470–31477
89. Metivier R, Penot G, Carmouche RP, Hubner MR, Reid G, Denger S, Manu D, Brand H, Kos M, Benes V, Gannon F (2004) Transcriptional complexes engaged by apo-estrogen receptor-alpha isoforms have divergent outcomes. EMBO J 23(18):3653–3666
90. Nelson J, Denisenko O, Bomsztyk K (2006) Protocol for the fast chromatin immunoprecipitation (ChIP) method. Nat Protoc 1(1):179–185
91. Nelson J, Denisenko O, Sova P, Bomsztyk K (2006) Fast chromatin immunoprecipitation assay. Nucleic Acids Res 34(1), e2
92. Topisirovic I, Siddiqui N, Lapointe VL, Trost M, Thibault P, Bangeranye C, Pinol-Roma S, Borden KL (2009) Molecular dissection of the eukaryotic initiation factor 4E (eIF4E) export-competent RNP. EMBO J 28(8):1087–1098

Applications of Non-coding RNA in the Molecular Pathology of Cancer

Keerthana Krishnan and Nicole Cloonan

Introduction

One of the puzzling discoveries to arise from the first sequencing of the human genome was that the number of protein-coding genes (~20,000–25,000) was not significantly larger than that of the roundworm *Caenorhabditis elegans* (~20,000) [1]. Whilst the complexity of an organism is clearly not reflected by the number of genes in its genome, it does appear to correlate with the size of its genome [2]. Recent studies by the *Enc*yclopedia *of* *D*NA *E*lements (ENCODE) have found that the vast majority of the genome is transcribed [3, 4], but the transcripts lack an open reading frame of a substantial size. This raises the possibility that some of the information required for a correctly functioning human cell lies in transcripts that do not code for proteins.

The term non-coding RNAs (ncRNAs) is usually used to describe these ORF-less transcripts; however, it should be regarded as a name of convenience only. A large number of transcripts annotated as ncRNAs appear to be translated as short peptides [5], and could therefore be considered as coding transcripts that were not detected computationally due to an arbitrary definition of the minimum length of an ORF [6]. Complicating the story, several transcripts are now known to be bifunctional—encoding both coding and non-coding functions [7–12]—and if this is a common biological theme, the designation of "non-coding" may become more arbitrary and ambiguous than first envisaged.

The function and biological relevance is still hotly debated for many types of ncRNAs [3, 13–22], and there has been some concern that many of the transcripts could be non-functional, or incidental by products of transcription [14, 18, 21, 22].

K. Krishnan
Forsyth Institute, 245 First Street, Cambridge, MA 02142, USA

N. Cloonan (✉)
QIMR Berghofer Medical Research Institute, 300 Herston Rd, Herston, QLD 4006, Australia
e-mail: nicole.cloonan@qimrberghofer.edu.au

S.R. Lakhani, S.B. Fox (eds.), *Molecular Pathology in Cancer Research*,
DOI 10.1007/978-1-4939-6643-1_9

Given that noise is an integral part of complex systems [23, 24], it does seem likely that at least a portion of these transcripts could be considered to be transcriptional noise [22]. Nevertheless, a large number of ncRNAs have been demonstrated to have regulatory roles in normal and pathological states, including cancer initiation, progression, and response to therapy (as discussed below). This evidence, combined with the versatility of RNA, makes ncRNAs a rich source of biomarkers that could be exploited for the diagnosis, prognosis, and prediction of cancer. This chapter describes the different categories of ncRNAs and their association with cancer, explores the relevance and practical use of ncRNAs in molecular pathology, and examines the major limitations and future challenges.

Association of Non-coding RNAs with Cancer

The two major categories of ncRNA are separated on size: those less than 200 nt long are designated as small ncRNAs, whereas those that are 200 nt or longer are classed as long ncRNAs. This distinction is arbitrary, and is based largely on the failure of the original protocols based on silica-matrix spin-column purification to retain RNA species <200 nt [25]. The major classes within each of these categories of ncRNAs are shown in Fig. 1, and their distinguishing features and association with cancer are detailed below.

Small Non-coding RNAs

MicroRNAs

By far the most well-studied class of non-coding RNAs whether long or short are the microRNAs (miRNAs). These were first identified in 1993 [26], but not recognized as a major class of regulators until 2000 [27, 28]. MicroRNAs vary slightly in length from 18 nt to 25 nt, but are predominantly 22 nt long [29, 30].

Function and Biogenesis of microRNAs

MicroRNAs negatively regulate the translation of mRNAs to protein through multiple mechanisms: direct translational inhibition [31–36], mRNA sequestration [37], and mRNA degradation [38–40]. Although the majority of protein loss after miRNA activity comes from a reduction of mRNA [41], this is a secondary effect of miRNAs targeting transcription factors [42], and the most common result of miRNA–mRNA interactions is translational inhibition [43].

Maturation of miRNAs involves multiple enzymatic steps to convert secondary structures in long RNAs into 80 nt hairpins, and then into 22 nt miRNA duplexes

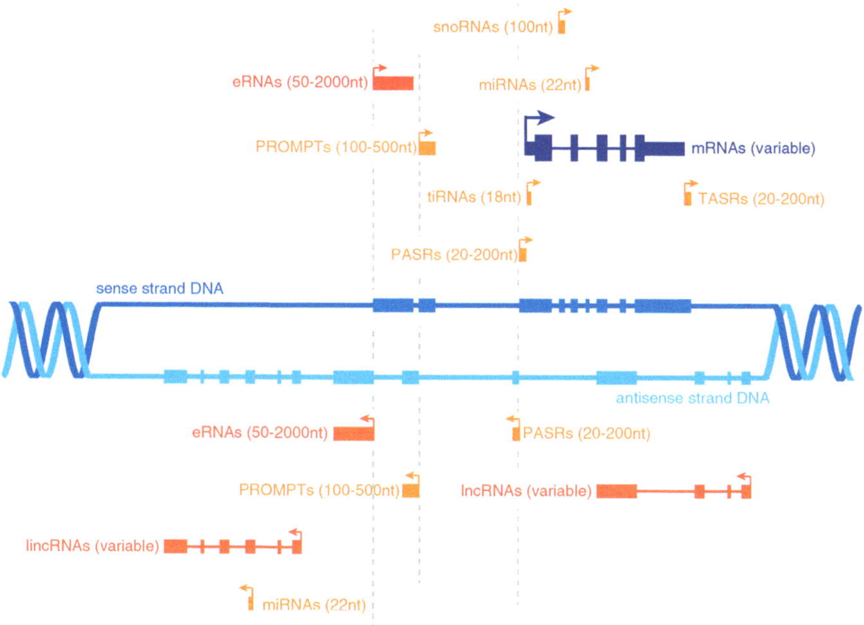

Fig. 1 Most of the genome is transcribed; some of the genome is functional. The sense and antisense strands of DNA are represented by the *helix* and *lines* in *blue*. Exons are depicted by *rectangles*, introns or intergenic sequences are represented by *lines*. Coding sequences are represented by *thick rectangles*, non-coding sequences by *thinner rectangles*. Coding transcripts are colored *blue*, small non-coding RNAs are *orange*, and long non-coding RNAs are *red*. The transcription start site (TSS) and the direction of transcription (5′ to 3′) are represented by *arrows*. *Dashed lines* represent points of bidirectional transcription

(Fig. 2). A nuclear primary miRNA transcript (pri-miRNA) is cleaved into hairpins by multi-protein enzyme complexes containing Drosha [44]. Introns of protein-coding genes can also be processed to hairpins by mRNA splicing machinery [45, 46]. Dicer-containing complexes were originally thought to be essential for the maturation of hairpins into miRNA duplexes [44]; however, recent work has revealed many miRNAs undergo maturation through an alternative Argonaut-dependent cleavage [30, 47–49].

Notes on microRNA Nomenclature

As our understanding of miRNAs and their function has evolved over time, so has their nomenclature, which can lead to headaches for those studying these molecules individually or collectively. Initially, mature miRNA sequences were named using the prefix "miR" (note the uppercase "R") followed by a hyphen and then a unique numeral identifier, and the loci from which they were derived were named "mir"(note the lowercase "r") followed by the same number (e.g., miR-1 and mir-1). The

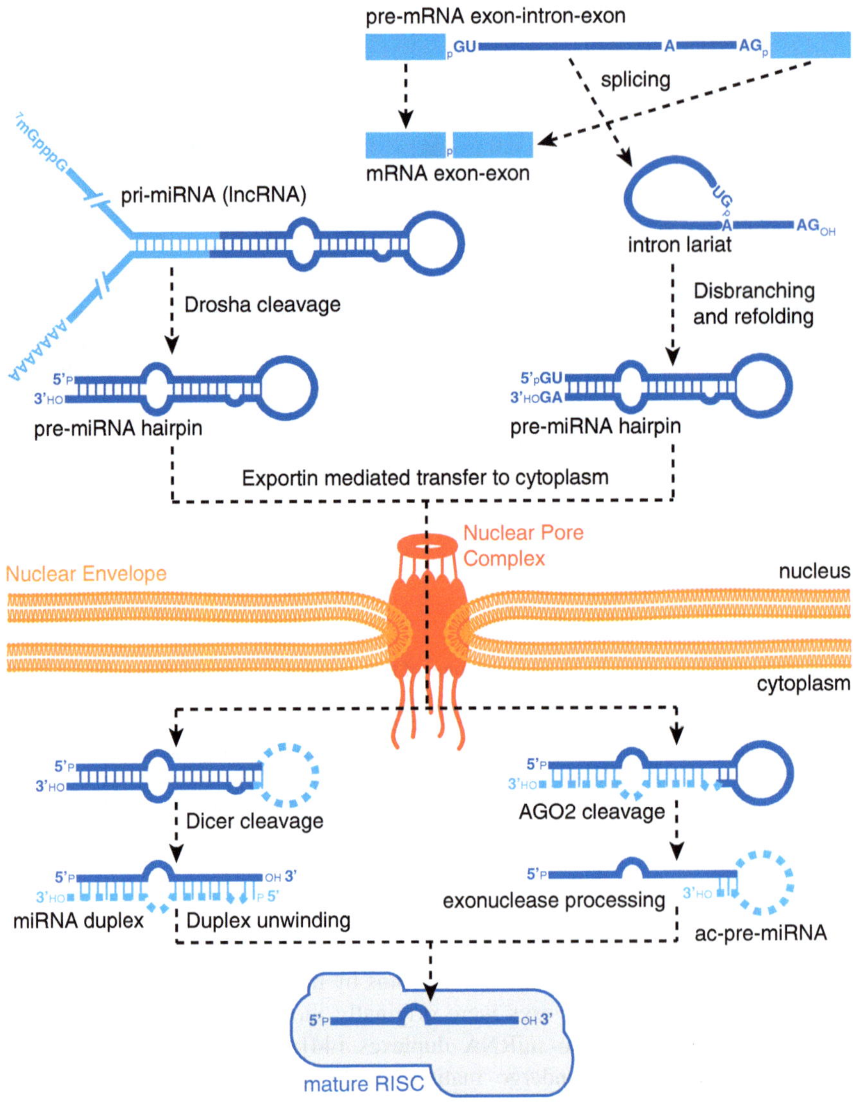

Fig. 2 Biogenesis of miRNAs through multiple redundant pathways. Pre-miRNA hairpins can be generated by DROSHA cleavage of non-coding primary miRNA transcripts (pri-miRNA), or as a byproduct of splicing mRNA and disbranching and refolding of the lariat. Pre-miRNA hairpins are exported to the cytoplasm through the action of Exportin 5 (XPO5). Most of the time, DICER cleaves the pre-miRNA hairpins to form mature miRNA duplexes which then unwind and incorporate into the RNA Induced Silencing Complex (RISC). In an alternative pathway, Argonaute 2 (AGO2) cleaves one arm of the hairpin to create AGO2-cleaved pre-miRNA (ac-pre-miRNA) which undergoes exonuclease processing to generate the mature miRNA. Complimentary base-pairing between the miRNA and mRNA is used to guide RISC to its mRNA targets

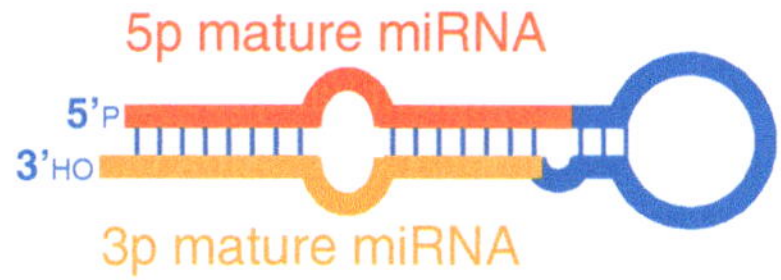

Fig. 3 5p and 3p annotations of mature miRNAs relative to the pre-miRNA hairpin

indentifying numbers are allocated sequentially based on the order of publication or by sequence similarity with miRNAs from another species (now handled by miRBase.org). This last usage required the addition of a three letter species prefix (hsa for *Homo sapiens*, mmu for *Mus musculus*, etc.) to the beginning of the miRNA name to distinguish these orthologous miRNAs as in many cases, the sequence similarity was very close (drosophila miR-1 differs from hsa-miR-1 by a single nucleotide). Within a single species, letter suffixes to the numeral identifier are also used to denote sequence similarity, such as hsa-miR-19a and hsa-miR-19b which derive from different genomic loci, but are only different by one nucleotide. For the miRNA loci, a second hyphenated number is appended if there are multiple genomic loci in the genome that produce the same mature sequence. Two loci on human chromosomes 17 (hsa-mir-196a-1) and 12 (hsa-mir-196a-2) produce different stem-loop structures (110 nt and 70 nt, respectively) but the same major miRNA product (hsa-miR-196a-5p).

Initially, it was thought that miRNAs were only ever derived from one arm of a hairpin (the "mature" or "sense" strand), and the other arm of the hairpin was degraded and non-functional (the "star" or "anti-sense" strand) [50]. However, it appears that only the most highly expressed miRNAs have such striking arm-expression bias, and there are a number of examples of equal expression from both arms (such as miR-199-5p/3p). In these cases, miRNAs were annotated as either 5p or 3p (short for 5 prime or 3 prime, respectively), referring to which arm of the hairpin the mature miRNA derives from in the direction of transcription (Fig. 3). However the mature/star annotation remained to annotate which arm was dominant in expression.

From an annotation perspective, this turned out to be troublesome, as more and more evidence accumulated for arm-switching [30, 51, 52], where different arms of the miRNA would be dominantly expressed in different tissues. As there was no clear relationship between the mature/star and the 5p/3p annotations, the community moved to annotate based on a static feature of the miRNA, and abandoned the mature/star annotations. Naturally, this leaves footprints in the existing literature; however, this chapter has used the most recent nomenclature accepted by the miRNA community, not the names under which they were originally published.

MicroRNAs and Cancer

The first evidence that miRNAs were associated with cancer was based on a study looking for tumor suppressors within the 13q14 locus, a region of the genome which was frequently deleted in chronic lymphocytic leukemia (CLL). This locus contained two very similar pre-miRNA hairpins encoded on the same pri-miRNA [53], and are now called hsa-miR-15a-5p and hsa-miR-16-5p. Although there was no causal link established, these two miRNAs were deleted or not expressed in 41 of 60 (68.3 %) patient samples tested [53]. Subsequent studies showed that these two miRNAs inhibit progression of the cell cycle [54–56], promote apoptosis [57], and suppress tumorigenesis both in vivo [58] and in vitro [57], confirming the strong association originally discovered as causal.

The first oncogenic miRNA locus was the miR-17-92 cluster, which encodes six pre-miRNA hairpins and at least seven mature and functional miRNAs [59]. In a mouse model of B-cell lymphoma, the introduction of the miR-17-92 cluster promoted both initiation and progression of the disease, and these miRNAs were found to be amplified or over-expressed in 30 of 46 (65 %) human patient samples [59]. Some of this effect has been attributed to hsa-miR-17-5p, which regulates progression from G1 to S phase of the cell cycle [60] and inhibits apoptosis in many cell types [61, 62], although other miRNAs are also required for this phenotype [63–65].

Genomics evidence also supports major regulatory roles for miRNAs in cancer. MicroRNAs repressing oncogenes are often located in regions of deletions or mutations (so-called fragile loci), whereas miRNAs repressing tumor suppressors are frequently found in regions of amplification [66–68]. Large scale expression profiling of miRNAs shows that change in miRNA profiles between normal tissue and cancerous samples is the norm, not the exception [69, 70]. Gain of function and loss of function experiments have also been critical in demonstrating that miRNAs are involved in each of the 10 major hallmarks of cancer (Fig. 4; [71]). So much evidence now exists for the role of specific miRNAs in every aspect of cancer that it would be impractical to list it all. More than 26,000 articles have been published on miRNAs and cancer, detailing the association between 236 miRNAs and 79 types of cancer [72]). Instead, the following sections detail some of the key examples for each hallmark.

Proliferative Signaling and miRNAs

One of the major hallmarks of cancer cells is that they stimulate their own growth regardless of exogenous signals [73]. A number of miRNAs have been directly linked to this facet of cancer biology. KRAS is a particularly common and potent oncogene, and is usually targeted for suppression by hsa-miR-143-3p [74]. However this miRNA is lost in a wide variety of tumors (including bladder [75–77]; breast [78], cervical [79]; colorectal [80–85]; esophageal [86]; glioma [87]; nasopharyngeal [88]; osteosarcoma [89, 90]; prostate [91–93]; and renal [94]), leading to its over-expression and activation of proliferative pathways. The G_1/S cell cycle checkpoint usually stops cells from proliferating in the absence of mitogenic signals. MicroRNA hsa-miR-17-5p suppresses this checkpoint through a highly connected network of genes [95], leading to growth factor independence. Another miRNA, hsa-miR-15b-5p, also acts at the G_1/S checkpoint, but this time by suppressing Cyclin E1 (CCNE1), a protein required for progression through the cell cycle [55]. Loss of this miRNA leads to CCNE1 over-expression and proliferation in glioma cells [96].

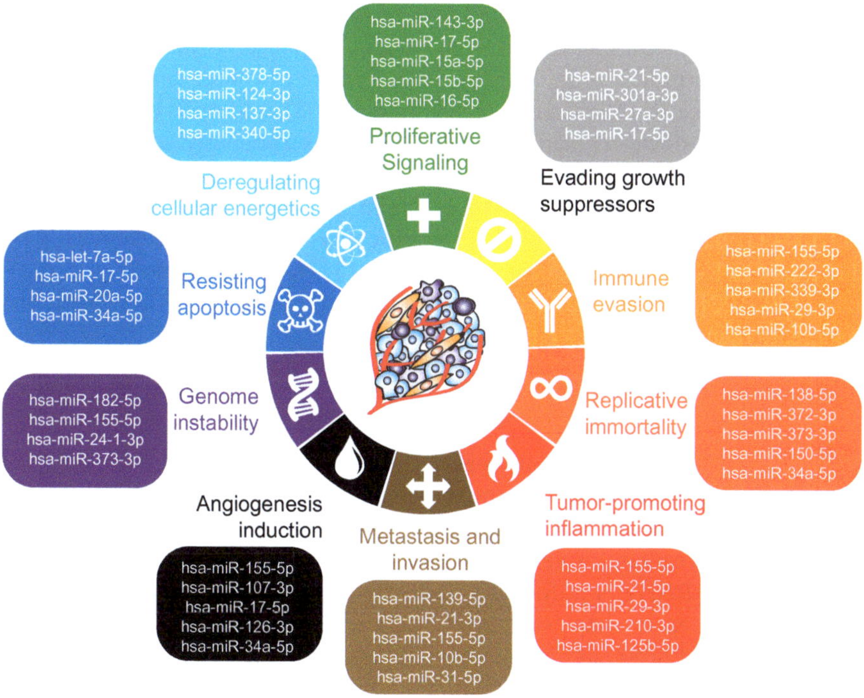

Fig. 4 MicroRNA involvement in the 10 hallmarks of cancer as defined by Hanahan and Weinberg [71]. The miRNAs presented here do not represent a comprehensive list, and there are many miRNAs associated with each hallmark. As miRNAs can have hundreds of *bona fide* targets, individual miRNAs can be associated with more than one hallmark

Evading Growth Suppressors and miRNAs

The second hallmark of cancer is the ability to resist endogenous and exogenous signals that may otherwise prevent growth [73]. By repressing the tumor suppressor PTEN, hsa-miR-21-5p promotes the growth of cancer [97], and is the most commonly up-regulated miRNA in all cancers [98]. Another miRNA targeting PTEN is hsa-miR-301a-3p, where over-expression leads to proliferation in breast cancer cells [99]. Prohibitin (PHB) is a different tumor suppressor that inhibits DNA synthesis [100], and is targeted by hsa-miR-27a-3p [101]. This miRNA is up-regulated in breast cancer [102], gastric cancer [101, 103, 104], and ovarian cancer [105], leading to cellular proliferation.

Avoiding Immune Destruction and miRNAs

Surveillance by the immune system is thought to play a substantial role in recognizing and eliminating the vast majority of cancer cells, and therefore it is believed that tumors that do grow must be able to avoid or reduce the impact of immunological clearance [71]. Transient expression of miR-155 is required for normal immune function [106], and loss of this miRNA prevents the immune system from mounting an effective anti-tumor response [107]. Both hsa-miR-222-3p and hsa-miR-339-3p enhance the resistance of cancer cells to immunological clearance by down-regulating ICAM-1 [108], and miR-10b performs a similar role by down-regulating MICB [109]. Unsurprisingly, as their mechanism of action is very generalized, these miRNAs have been associated with a wide range of cancer types: hsa-miR-10b-5p is up-regulated in glioblastoma [110, 111], pancreatic cancer [112], breast cancer [113], and leukemias [114, 115]; hsa-miR-222-3p is associated with breast [116–118], bladder [119], prostate [120–122], gastric [123], colorectal [124], lung [125], ovarian [126], and pancreatic cancers [127].

Enabling Replicative Immortality and miRNAs

Mammalian cells are limited in their ability to replicate based on the length of their telomeres. Cancer cells escape this limit to become immortalized, and are capable of indefinite growth and division [73]. The length of telomeres is regulated by the telomerase protein [128], and several miRNAs regulate its expression. The microRNA hsa-miR-138-5p directly targets telomerase, and expression is down-regulated in anaplastic thyroid carcinoma [129], hepatocellular carcinoma [130], and head and neck squamous cell carcinoma cell lines [131]. The vitamin-D inducible hsa-miR-498-5p also directly targets telomerase and is down-regulated in ovarian tumors [132]. Similarly, loss of hsa-miR-150-5p leads indirectly to the up-regulation of telomerase through loss of AKT targeting [133]. However it's not only loss of miRNAs that lead to telomerase activation. Loss of p53 function also leads to an increase in telomerase [134], and over-expression of hsa-miR-372-3p or hsa-miR-373-3p leads to deactivation of p53 and an increase in telomerase activity [135].

Tumor-Promoting Inflammation and miRNAs

Although the immune system attempts to clear the tumor before it takes hold, the inflammation created by immune attack can have the paradoxical effect of enhancing tumorigenesis and progression, and assisting the cancer cells to acquire other cancer hallmarks [71]. Some miRNAs that are involved with evasion of the immune system (hsa-miR-29-3p and hsa-miR-155-5p, see section "Avoiding Immune Destruction and miRNAs") also play roles in tumor-promoting inflammation. During the macrophage inflammatory response, expression of hsa-miR-155-5p is induced [136], and this leads to an increased frequency of mutations through the suppression of targets that safeguard the genome [137]. Cancerous cells also secrete hsa-miR-21-5p and hsa-miR-29-3p which bind to Toll-like receptors [138]. This induces inflammatory cytokines which not only enhances metastasis [138] and inhibits apoptosis [139], but also creates a positive feedback loop as inflammation induces the expression of hsa-miR-21-5p [140–142].

Invasion, Metastasis, and miRNAs

At least 90% of the mortality from human cancer is due to metastasis [143]. The ability of pioneering cells to migrate from the site of the primary tumor, and colonize to distant sites in the body is acquired during tumor development [73], and several miRNAs have been shown to regulate this process. hsa-miR-21-5p is the most well-known miRNA to promote metastasis and is over-expressed in a broad range of cancer types [144, 145]. The key metastatic genes ITGAV, RDX, and RHOA are all regulated by hsa-miR-31-5p [146]. Over-expression of hsa-miR-139-5p correlates with reduced metastatic activity in hepatocellular carcinoma and gastric cancer cells [147–149], while loss of hsa-miR-139-5p expression is associated with increased metastatic disease in patients with invasive squamous cell carcinoma [150]. This miRNA also increases the migration and invasion of breast cancer cell lines, and directly targets the TGFβ, Wnt, Rho, and MAPK/PI3K signaling cascades [151].

Angiogenesis and miRNAs

Angiogenesis, the process by which new blood vessels are formed, is a process required by developing cancers to ensure a continuous supply of oxygen and nutrients [73]. The association of miRNAs with fine-tuning angiogenesis has been reviewed [152, 153], but in a cancer context, it is the response to hypoxia that stimulates angiogenesis, and this process is augmented through hsa-miR-107-3p targeting of HIF1 [154]. The miR-17-92 cluster (see MicroRNAs and Cancer) also appears to promote angiogenesis when over-expressed [155], and this may be due directly to hsa-miR-17-5p also targeting HIF1 [95]. However over-expression of hsa-miR-92a-1-3p (the final member of the miR-17-92 cluster) appears to inhibit angiogenesis, by repressing the expression of pro-angiogenic proteins [156].

Genome Instability and miRNAs

The hypothesis of multi-step tumor formation relies on the chance acquisition of mutations that confer a selective advantage to those cells. Usually the rates of spontaneous mutations are extremely low, and therefore some level of genomic instability is required to drive these acquisitions [71]. A number of miRNAs target DNA repair pathways, such as hsa-miR-24-3p and hsa-miR-182-5p which target the homologous repair pathway of double stranded breaks [157, 158], and hsa-miR-373-3p which targets the nucleotide excision repair pathway [159]. However so far only a single miRNA (hsa-miR-155-5p) has been demonstrated to be sufficient to directly increase the mutation burden in cells [137]. Intriguingly, as hsa-miR-155-5p also is involved with the inflammatory response (see tumor-promoting inflammation and miRNAs), this molecule may provide the much sought after link between inflammation and cancer.

Resisting Apoptosis and miRNAs

A primary safeguard of cellular integrity is apoptosis, programmed cell death, a naturally occurring process that prevents cells with minor or major defects from proliferating. Cancer cells need to resist apoptotic signals [73], and miRNAs such as hsa-miR-17-5p and hsa-miR-20a-5p are likely to be involved in that process, given that their knock-down in cancer cell lines using results in increased apoptosis [61]. This aspect of miRNA regulation has been studied extensively, and many miRNAs have been identified as regulating all aspects of apoptosis [160].

Deregulating Cellular Energetics and miRNAs

Cancer cells are able to reprogram their energy generation from oxidative phosphorylation to glycolysis, even in the presence of oxygen. This switch (known as the Warburg Effect [161]) is as widespread as the other hallmarks, and forms the basis for positron emission tomography (PET) using radio-labeled glucose as the reporter. Although the functional rational for this is not yet clear, the Warburg Effect may provide the cancer with sufficient macromolecules for other metabolic pathways [71]. In breast cancer cells, hsa-miR-378-5p targets GABPA and ESRRG to induce the Warburg Effect, and the expression of this miRNA correlates with tumor progression in breast cancer patients [162]. Some miRNAs have been shown to inhibit the Warburg Effect, essentially acting as tumor suppressors. These include hsa-miR-124-3p, hsa-miR-137-3p, and hsa-miR-340-5p [163].

Bifunctional microRNAs

Although the above examples almost always tell a story of miRNAs targeting specific genes to induce an effect, it's important to remember that these relationships are but one part of a large and integrated cellular network. In recent years, it has become clear that miRNAs do not typically exert their effect through one or two key genes, but instead target biological pathways to achieve specific outcomes [30, 95, 164–167]. Sometimes, hundreds or thousands of mRNAs are directly targeted by a single miRNA [30], and this lies behind the ability of some of the miRNAs listed here to target multiple facets of cancer biology. For example, the loss of hsa-miR-143-3p in cancers not only activates proliferation pathways through one set of targets [74], but the loss also promotes invasion and migration through another set of targets [168]. These dual functions are coherent, but not all targets of miRNAs make such intuitive sense.

There are a number of miRNAs that appear to have contradictory effects on cancer initiation and progression depending on the tissue under study. In endometrial, lung, prostate, breast and colon cancers, hsa-miR-182-5p has an oncogenic-like role [169–173], however, in some lung and gastric cancers its role appears to be tumor suppressor-like [174–176]. Similarly, hsa-miR-17-5p was originally found to promote proliferation in B-cell lymphoma, bladder, breast, gastric, liver, lung, and pancreatic cancers [59, 61, 177–188], but in some breast, ovarian, and cervical cell lines expression of this miRNA inhibited proliferation [189, 190]. This may be a common occurrence. Using miRCancer, a database of miRNA association with cancer [72], 46 of the 236

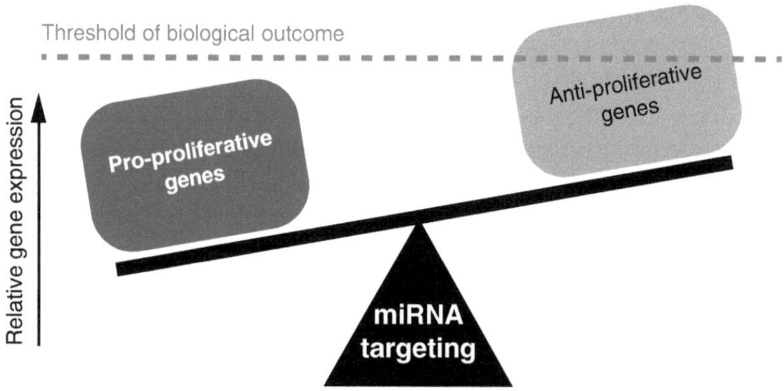

Fig. 5 The biological outcome of bifunctional miRNAs depends upon the expression level of the groups of genes targeted by that miRNA. In the example of cellular proliferation, if the anti-proliferative targets of an miRNA were expressed higher than the pro-proliferative targets in a particular cell line or cancer type, then we would expect a net anti-proliferative outcome from over-expression of the bifunctional miRNA. This model explains how miRNAs targeting identical genes can have different biological outcomes in different cancer types

(19.5 %) cancer associated miRNAs had apparently conflicting biological roles where at least two independent studies supported these roles. When only one study was required, this number doubles to 100 of 236 (42.4 %). Although these results can only be considered to be indicative, not conclusive, it does suggest that these phenomena may be more common than first appreciated.

In the case of hsa-miR-17-5p, the molecular mechanism was demonstrated by systematic identification of targets associated with the cell cycle [95]. This work, later confirmed by alternative techniques [30], demonstrated that miRNAs can simultaneously target a large number of genes with opposing functions, and the final outcome is dependent on the stoichiometry between the miRNA and its target sequences, and how that ratio affects the cellular network (Fig. 5). Therefore the extrapolation of the effect of an miRNA from one biological context to another is not wise.

Small Interfering RNAs

Like miRNAs, small (or short) interfering RNAs (siRNAs) are ~21 nt single-stranded RNA molecules that incorporate into the RNA induced silencing complexes (RISC) [191]. The term siRNA is usually used to describe synthetic or exogenous molecules that promote mRNA cleavage of complementary RNA targets through perfect Watson–Crick base-pairing [192–194], however, endogenous siRNAs also exist in mammals [195]. The distinction between endogenous siRNAs and miRNAs is based largely on biogenesis. Whereas miRNAs are derived from the short secondary structure in a single-stranded RNA molecule, siRNAs are cleaved from long secondary structures, or double-stranded RNAs that can derive from inverted repeat structures and sense/anti-sense transcripts [191, 196].

As the siRNA RISC is biochemically identical to miRNA RISC, miRNAs can function as siRNAS [191], and siRNAs can function as miRNAs [197]. The primary biological role of endogenous siRNAs is proposed to be the silencing of repetitive elements that would otherwise activate or alter transcription of nearby genes [198]. While there are no direct studies linking endogenous siRNAs to cancer, there are now tentative links between retrotransposon activity and cancer [199], so a link between siRNAs and cancer may emerge with further research.

PIWI-Interacting RNAs

The PIWI-interacting RNAs (piRNAs) are typically 24–32 nt long and were originally thought to only be present in germ cells, where they are involved in maintaining genome stability [200]. Subsequent studies have found they are expressed far more widely, and are in fact present in all somatic tissues [201]. Although the role for most piRNAs is not clear, there appears to be some functional redundancy between these molecules and endogenous siRNAs (see above) in silencing retrotransposon activity [202]. Links between piRNAs and cancer are still preliminary and based largely on differential expression. Certain piRNAs such as piR-651 [203] and piR-20365 [204] have been shown to be highly expressed in certain cancer types whereas others such as piR-823 are down-regulated in other cancers [205]. The links between piRNAs and cancer have been recently reviewed elsewhere [206].

Small Nucleolar RNAs

Another well-studied small RNA family are the small nucleolar RNAs (snoRNAs) which are best known for directing chemical modifications of other non-coding RNAs, such as transfer RNA (tRNA) and ribosomal RNA (rRNA) [207], however, can also be processed to regulate the splicing of mRNAs [208]. There are two well-defined categories of snoRNAs that are based on their sequence and structural characteristics (Fig. 6). The first, C/D box snoRNAs mediate 2′-*O*-methylation of RNA residues in complementary RNA molecules [209], while H/ACA box RNAs pseudouridylate their target RNAs [210]. Like piRNAs, snoRNAs have been linked to cancer through primarily through differential expression studies, and functional links are few and preliminary [211]. However lack of evidence for a causal link to cancer does not preclude these molecules from use as a biomarker, as only correlation is required. The snoRNA U50 is under-expressed in prostate cancer, and breast cancer and re-expressing in prostate cancer cells inhibits colony formation [212, 213]. A second example, SNORA42, is frequently over-expressed in lung cancer and is associated with decreased survival [214, 215]. Knocking down SNORA42 in lung cancer cell lines induced apoptosis and reduced tumor formation in mice [215].

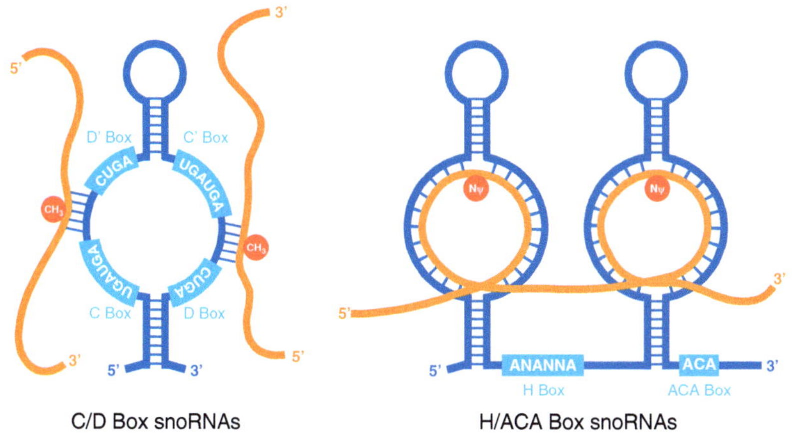

Fig. 6 C/D Box and H/ACA Box snoRNAs enable template-mediated 2′-*O*-Methylation (CH_3) and pseudouridylation (Nψ) of target RNAs, respectively

Small Nuclear RNAs

Small nuclear RNAs are key players in the splicing and maturation of messenger RNA (mRNA) [216]. Their other common name, U-RNAs, refers to their high uridine composition. They are the very longest of the "small" RNA classes (with lengths not far under 200 nt), and serve as templates to recognize the 5′ and 3′ intron/exon boundaries. Since they are involved in regulation of the spliceosome, it is unsurprising that links between snRNAs and cancer have been uncovered. Examples include the U1 snRNA with pro-apoptotic functions in cervical cancer [217], and the tumor suppressor p53 regulation of RNA polymerase II and III-dependent snRNA gene transcription [218].

Y RNAs

Y-RNAs are 80–120 nt non-coding RNAs so named in order to distinguish them from U-RNAs (see snRNAs above). These molecules form stem-loop secondary structures (analogous to pre-miRNA hairpins) that are proposed to stimulate DNA replication during the cell cycle [219], as well as marking misfolded RNAs for degradation [220]. It is likely to be the former biological role that leads to a proliferative advantage in the pathogenesis of cancer when over-expressed. All four human Y-RNAs have been shown to be over-expressed in different carcinomas, with their inhibition leading to decrease in cell proliferation [221].

Vault RNAs

Vault RNAs are another stem-loop structure associated with the eukaryotic vault organelle [222]. 8–16 vault RNAs are complexed with many copies of the major vault protein (MVP) and two types of minor vault proteins (VPARP and TEP1) which together form the vault ribonucleoprotein complex [223]. The vault complex is known to shuttle between the nucleus and cytoplasm and has hence been implicated in intracellular transport (10087261) what is this?. Its major association with cancer comes from the role in multi-drug resistance where knock-down of the vault RNA re-sensitized cancer cells to chemotherapeutics [224].

Other Small RNAs Not Yet Associated with Cancer

Many other classes of small RNAs have been described (Fig. 1), including: promoter associated short RNAs (PASRs)) and termini associated short RNAs (TASRs) [225], transcription initiation RNAs (tiRNAs) [226], promoter upstream transcripts (PROMPTs) [227], transcription start site associated RNAs (TSSa-RNAs) [228], and splice-site RNAs (spliRNAs) [229]. In all these cases, the classification is based around the genomic location of transcriptional products rather than a specific biological function. Indeed, there has been substantial controversy as to whether the wealth of RNA expression that we can now detect using massively parallel sequencing technologies are merely byproducts of transcriptional processing or are functional molecules in their own right (see Introduction). For the purposes of molecular pathology, this is not a debate that needs to be resolved, as a molecule does not need to be causally associated with a disease to be a useful biomarker, however, in the case of these small RNA classes, there has not yet been an association with any disease or pathological state, including cancer.

Long Non-coding RNAs

Structure and Function of lncRNAs

Unlike small RNAs, no functionally distinct classes based on location or sizes have emerged for long non-coding RNAs (lncRNAs). The term lncRNA tends to exclude RNAs that make up more than 80 % of total cellular RNA (ribosomal RNAs), and some people distinguish between the lncRNAs found within protein-coding loci and long intergenic non-coding RNAs (lincRNAs) [230]. This distinction is arbitrary, and given the variable lengths and sheer number different non-coding transcripts [231], it seems likely that the individual biological functions of lncRNAs could be as diverse as the functions of protein-coding RNAs. Even so, several functional themes are emerging, and these include: (1) regulating the expression of other genes [232]; (2) interfering with miRNA function by competitive binding [233]; (3) interactions with chromatin remodeling complexes [234]; and (4) as an intermediate product in processing to smaller RNA species [235].

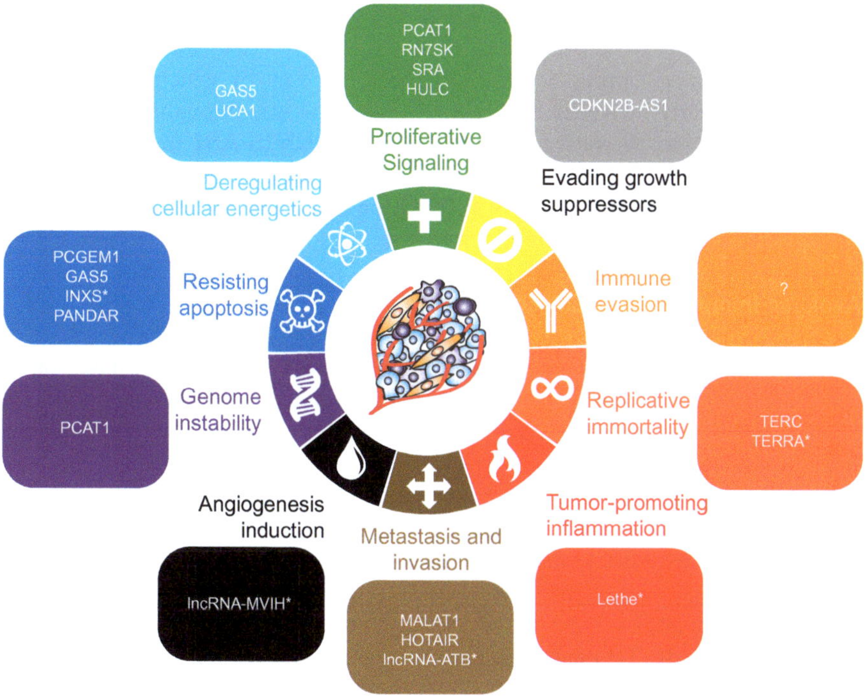

Fig. 7 MicroRNA involvement in the 10 hallmarks of cancer as defined by Hanahan and Weinberg [71]. *Asterisks* represent lncRNA names that do not yet have an official gene symbol. The lncRNAs presented here do not represent a comprehensive list. Just like protein-coding genes, non-coding genes can have multiple biological roles in different pathways

Long Non-coding RNAs and Cancer

Although the characterization of cancer associated lncRNAs has taken place at a substantially slower rate than miRNAs due to the lack of a well studied and common mechanism of action, there are still lncRNA candidates identified for most of the 10 hallmarks of cancer (Fig. 7).

Proliferative Signaling and ncRNAs

Through RNA-seq of prostate tissue and cell lines, PCAT1 was identified to be as associated with prostate cancer, and shown to regulate the proliferative ability of cells [236]. Other lncRNAs that regulate cell proliferation in cancer include RN7SK which regulates transcriptional elongation via pTEFb [237] and SRA which acts as a coactivator for the estrogen, progesterone, and glucocorticoid receptors [11]. The lncRNA HULC was also recently reported to regulate the proliferation of gastric cancers [238].

Evading Growth Suppressors and ncRNAs

Silencing of tumor suppressor genes has been one way by which tumorigenesis proceeds. Similarly lncRNAs also promote tumorigenesis through the evasion of growth suppression. A well-studied example is CDKN2B-AS1 (more commonly known as ANRIL), which interacts with SUZ12, recruiting the PRC2 complex to repress the expression of p15 (INK4B) [239], a well-documented tumor suppressor.

Avoiding Immune Destruction and ncRNAs

Although lncRNAs that play a direct role in immune evasion of tumors have not yet been identified, examples of immune response regulators can be found [240]. It seems reasonable to assume that discovery of lncRNAs for this category is only a matter of time.

Enabling Replicative Immortality and ncRNAs

Replicative immortality is a hallmark of tumor cells which is achieved through the action of telomerase. Telomerase is a holoenzyme consisting of an lncRNA component (TERC) which is amplified in several human cancers [241]. Other lncRNAs associated with telomerase include the TERRA family members which are believed to be negative regulators of telomerase [242].

Tumor-Promoting Inflammation and ncRNAs

Rapicavoli et al. reported the identification of a novel lncRNA, Lethe, induced by TNF and IL-1β in response to inflammation [243]. Lethe negatively regulates nuclear factor-kB signaling, which is a classical regulator of inflammation in cancer [244].

Invasion, Metastasis, and ncRNAs

The MALAT1 lncRNA (Metastasis-Associated Lung Adenocarcinoma Transcript 1) was first reported to be able to predict metastasis in non-small cell lung cancers [245]. It has since been shown to enhance motility of lung cancer cells [232] and also leads to increased invasive ability of cervical cancer cells [246]. Another lncRNA, HOTAIR, has been shown to enhance tumor cell invasiveness by reprogramming the chromatin state of the cells in breast cancer [234] and gastric cancer [247]. More recently, lncRNA-ATB was shown to be activated by TGFbeta promoting an epithelial–mesenchymal transition in hepatocellular carcinomas, promoting the invasion-metastasis cascade [248].

Angiogenesis and ncRNAs

lncRNA-MVIH is over-expressed in hepatocellular carcinomas and activates tumor-inducing angiogenesis, potentially serving as a predictor of recurrence-free survival [249]. A natural anti-sense transcript, aHIF, has been shown to inhibit the expression of HIF1 which is a key regulator of angiogenesis [250].

Genome Instability and ncRNAs

PCAT1, originally identified to regulate cell proliferation in prostate cancers, has more recently been shown to interfere with homologous recombination by repression of BRCA2 [251]. Significantly, active PCAT1 has also been shown to impart high sensitivity to PARP inhibitors, and could therefore mark a cohort of tumors sensitive to this chemotherapy.

Resisting Apoptosis and ncRNAs

PCGEM1, a prostate cancer associated lncRNA plays an anti-apoptotic role preventing cell death following treatment with chemotherapeutic drugs such as doxorubicin [252]. In contrast, the lncRNA GAS5 has been shown to promote a pro-apoptotic function in prostate cancers [253]. The lncRNA INXS was shown to be a critical regulator of BCL-XS-induced apoptosis in tumor cell lines from various tissue types [254]. PANDAR, an lncRNA induced upon onset of DNA damage, has been shown to limit the expression of pro-apoptotic genes, its loss leading to greater sensitivity to doxorubicin [255].

Deregulating Cellular Energetic and ncRNAs

In addition to playing a role in apoptosis, GAS5 has also been shown to modulate cellular metabolism by antagonizing the glucocorticoid receptor (GR) by repressing GR-induced genes [256]. lncRNA UCA1 was also shown to promote glycolysis through the induction of mTOR and hexokinase 2, playing a role in cancer cell glucose metabolism [257].

Applications of Non-coding RNAs in Cancer Pathology

Nucleic acids in general and RNAs in particular have a number of advantages when used as a clinical tool for diagnosis and prognosis. Unlike proteins, all nucleic acids are easily amplifiable for exquisite sensitivity, which means that small sample sizes can be used without compromising the accuracy of the tests. Depending on the detection method used, assays for nucleic acids can be less open to misinterpretation than colorimetric or other visual assays. For RNA specifically, although it is not usually as robust as DNA due to the high levels of environmental RNAses, RNA profiling allows a snapshot of cellular activity that static DNA information cannot provide.

Just as the vast majority of basic research on non-coding RNAs has been performed on miRNAs, the majority of research on ncRNA biomarkers has also been directed towards miRNAs. This has been driven primarily by the discovery that despite the high levels of nucleases in the blood [258], miRNAs are easily and stably detected in whole blood [259], plasma [260], and sera [261]. This stability is mediated by either association with protein partners including RISC [262, 263], or inclusion of miRNAs in exosomes [264], both of which shield the miRNA from nuclease attack. MicroRNAs have also been identified in all remaining bodily fluids [265], sputum [266], and stool samples [267], making less-invasive and non-invasive assay development possible. Other ncRNAs have also been identified in body fluids [268], although it is less clear whether their presence is due to active or passive export. Providing that robust differences between disease fluids and healthy fluids are identified, this distinction may not be necessary.

Tests that use next-generation sequencing methods to study the DNA or RNA of all genes simultaneously are being developed (e.g., acute myeloid leukemia [269] and an adenocarcinoma [270]), and are in common use as research tools [271], but none are currently approved as diagnostic or prognostic tools. Most current clinical assays for ncRNAs rely on either quantitative reverse-transcriptase PCR (qRT-PCR) or microarrays for signal detection. The following sections outline the applications for which non-coding RNAs are currently being used in the clinic, are in clinical trials, or are on the research horizon.

Non-coding RNAs as Diagnostic Tools

The first major use of non-coding RNAs in the clinic has been for diagnostic purposes, exploiting the highly tissue specific nature of both short and long non-coding RNAs. Assays in this category are more often in clinical use, although some of the applications in this category remain highly speculative.

Early Detection

It is generally accepted that the earlier a cancer is detected, the more effective and more successful subsequent treatment is. Using melanoma as an example, the 5-year survival rate of stage I disease is 98 %. However, survival drops to approximately 16 % by stage IV [272]. Only certain cancers have screening tests that reduce mortality (including breast [273], cervical [274], lung [275], and colon cancer [276]), and there is still debate as to whether the screening tests available for others are effective at preventing deaths [277]. Nevertheless, non-coding RNAs have already contributed to this particular diagnostic use, and the potential for more widespread application is large.

Whole Population Screening

Routine screening for cancer in healthy subjects is generally considered an unviable option, both because of the low incidence of most specific cancer types within the general population (as individual cancer types would need to be screened for individually) and the lack of highly sensitive and specific markers [268, 278, 279]. Given the higher tissue specificity in the expression of non-coding RNAs, it seems unlikely that candidates for general cancer screening will be discovered. However as the population ages, and the incidence of cancer rises [280], it may become viable to screen for specific prevalent cancers, and ncRNAs would be ideally suited for this role.

Top non-coding RNA candidates for early detection of cancer would include a group of three miRNAs (hsa-miR-205-5p, hsa-miR-210-3p, and hsa-miR-708-5p) which can distinguish between sputum from healthy patients and those with squamous cell lung carcinoma with 73 % sensitivity and 96 % specificity [281]. A second example, hsa-miR-378, appears to discriminate the serum of healthy controls from patients with gastric cancer with 87.5 % sensitivity and 70.73 % specificity [282].

Targeted Population Screening

Where whole population screening is not economically viable, screening subpopulations of at-risk individuals may still provide substantial social benefits. The best known example of a non-coding RNA in this role is the use of PCA3 as a marker for prostate cancer [283]. Marketed as PROGENSA PCA3, this FDA approved urine test is used to determine whether men with elevated PSA in their serum but a negative biopsy should be referred for a second biopsy. The strong pathological-specificity of PCA3 expression [284, 285] means that this assay has a specificity of 76 % for the prediction of prostate cancer at follow-up biopsy, compared to PSA alone with a specificity of 47 % [268]. Importantly, the assay appears not to be affected by important clinical variables such as age, inflammation, or prostate volume [268].

Genetic Risk

Inherited DNA polymorphisms can increase the risk of developing cancer, and genetic screening individuals with family history of cancer can influence the management of their future medical care. An example of this are the polymorphisms in the BRCA1 and BRCA2 genes which increase the likelihood of developing breast [286, 287], ovarian [287, 288], and other types of cancers [289]. A test result confirming the presence of a risk polymorphism allows a patient to take action to reduce their risk of cancer, such as changing personal behavior, medication, preventative surgery, or earlier and more frequent screening [290].

Obviously, DNA is largely identical in most cells of a body, whereas the RNA expression levels can vary dramatically, and therefore RNA would not usually be considered as a potential screening molecule for genetic risk. However, it is conceivable that for clearly familial cancers, where the DNA test has not revealed a known polymorphism, non-coding RNAs that phenocopy a particular set of polymorphisms could be examined. For the BRCA1/BRCA2 genes, the leading candidate would be hsa-miR-182-5p, which has been shown to repress BRCA1 and other members of the HR DNA repair pathway [157, 291]. Not only does over-expression of this miRNA replicate the sensitivity of cells to PARP inhibitor therapy, but it may also explain why some BRCA-like familial cancers may not have obvious BRCA1/2 polymorphisms [292]. Such an application is still highly speculative, and would require substantial basic and confirmatory studies before even exploratory clinical use could be considered.

Tumors of Unknown Origin

Cancers that present as metastatic lesions where the primary tumor site cannot be identified with certainty prevent adequate management of treatment which is increasingly based on the tissue of origin [293]. Failure to identify the tissue of origin can also be a psychological burden on the patient, although it is thought to be unlikely to impact the prognosis of metastatic disease [294]. Leveraging the tissue specificity of miRNAs, Rosetta Genomics developed a microarray based assay (originally qRT-PCR based [295]) that is able to distinguish between 42 different tumor types (covering 92 % of all solid tumors) using the expression levels of 64 microRNAs [296]. The reported sensitivity of the test was 85 %, and the specificity >99 %. Given the very high tissue specificity of long non-coding RNAs, it is conceivable that a test developed with lncRNAs could show even higher sensitivity.

Classification of Molecular Subtypes

Even when the primary cancer can be identified, highly heterogeneous tumor types will have subtypes with different prognoses and different management strategies. For example, breast cancer can be classified into approximately 20 major and 18 minor subtypes primarily by histology and morphological properties of the tumor at the time of diagnosis which correlates with response to treatment and clinical outcome [297].

Several assays based on miRNAs are currently being used clinically. In 2012, Rosetta Genomics was given FDA approval to use their Lung Cancer Test as a diagnostic to differentiate between four major types of lung cancer: (1) small cell lung cancer (SCLC); (2) squamous non-small cell lung cancer (NSCLC); (3) non-squamous NSCLC; and (4) carcinoid [298, 299]. This is a qRT-PCR based assay for 8 miRNAs which is performed on FFPE samples. Specificity and sensitivity are 98 % and 97.3 %, respectively.

Two more clinical assays, also produced by Rosetta Genomics, distinguish the subtypes of kidney cancers and mesothelioma. The first test uses a custom microarray of miRNAs to differentiate between benign oncocytoma and the three most common subtypes of renal cell carcinoma (RCC): clear cell, papillary, and chromophobe [300]. The second test uses qRT-PCR on 3 miRNAs to differentiate between mesothelioma and other primary and metastatic lung cancers [301, 302].

Non-coding RNAs as Prognostic or Predictive Tools

The second major use for non-coding RNAs has been to understand their prognostic or predictive power in a clinical setting. Fewer examples of commercial clinical assays exist in this category; however, there is far more research in the pipeline reflecting the enormous potential of these molecules to impact upon patient outcomes.

Patient Stratification

Underscored by the genomics research demonstrating that no two patients have the same spectrum of mutations within the same cancer subtype [303, 304] there has been a desire to move towards personalized or precision medicine—treating and understanding the individual patient rather than understanding a population of patients. This has driven a desire to stratify patient therapies, prognosis, and predictions on recurrence and metastasis based on meaningful biological markers.

Assignment of Therapeutic Intervention

The classification of cancers into subtypes helps in the making of therapeutic decisions, providing guidelines on how individual tumors may respond to certain kinds of therapy. For example, in breast cancer, classification of tumors into different molecular subtypes [305] has ensured straightforward means of identifying HER2 positive tumors which will respond to treatment with trastuzumab [306]. Similar molecular approaches have also been used to classify lung cancers [307], melanoma [308], and glioblastoma [309] amongst others.

Traditional methods of molecular classification have often relied on identifying genomic alterations and measuring the expression levels of protein-coding genes. Often classification of tumors into different subtypes requires large efforts in expression profiling and analyses, however, are beneficial in diagnosis, predicting prognosis, and directing therapy. More recently, the diagnostic and prognostic power of using non-coding RNA has been uncovered. MicroRNA signatures have been used to segregate molecular subtypes of breast cancer as well as differentiate between the DCIS and IDC stages of tumor progression [310, 311]. miRNA expression has been associated with clinicopathological features and to specifically identify ER, PR, and HER2 status of breast cancers [312]. Therefore it is possible that the identification of a small number of miRNAs that are more predictive of tumor prognosis could substitute for having to perform extensive molecular profiling and analyses.

Survival Prognosis

Patients and clinicians alike need to understand the likely impact of cancer on the patient to make informed choices about therapies or palliative care. MicroRNA signatures have been used to stratify TNBCs based on predictability of overall survival and metastasis-free survival [313], and a great many more examples exist in the literature.

Recurrence Prediction

Even when a treatment appears to be successful (where the tumor burden drops below the sensitivity limits of the available tests), a recurrence of that cancer can occur months or years later. While it is desirable to remove any unnecessary chemotherapy treatments from consideration, it is also important that those most likely to have a recurrence are treated more aggressively. Expression of hsa-miR-221-3p was been shown to be inversely correlated with the likelihood of recurrence of prostate cancer [314], and a small panel of less than ten miRNAs was observed to predict the recurrence of non-small cell lung cancer after surgical resection [315].

Metastatic Potential

Metastasis is the leading cause of mortality in patients suffering from cancer, accounting for over 90 % of cancer-related deaths [316]. This is indicative of both our inability to detect cancers at an early enough stage to prevent metastasis, and our inability to predict which tumors are more likely to metastasize, implying the need for quicker and more reliable diagnostics. Since miRNAs have been shown to be able to segregate tumors based on their grades and aggressiveness [310, 311], they might present with unique tools to detect tumors that have a higher propensity to metastasize.

Monitoring Tumor Burden

It is beneficial to measure the burden of cancer in the patient in order to monitor the effectiveness (or otherwise) of a prescribed treatment. This style of monitoring has been described for chromosomal rearrangements that are unique to a tumor (so-called personalized diagnostics [317, 318]), however, non-coding RNAs also have potential to impact in this space. Although the purpose behind the cancer detection may be different (post-diagnosis rather than pre-diagnosis), all the potential and caveats of early detection and screening apply to this application too.

A Note on Non-coding RNAs as Therapeutics

Although not strictly within the bounds of molecular pathology, it is useful to understand that where functional contribution to disease has been demonstrated (as is the case for most of the miRNAs described here), the molecule being detected is the same molecule that can be targeted for replacement or inhibition with some variety of anti-sense molecule. While using non-coding RNAs as therapeutics has a number of challenges that have been reviewed extensively elsewhere [319–321], the scope for precision medicine in this context is exciting given that the disease specific expression of many non-coding RNAs could lead to treatments with very few side effects.

Challenges and Limitations

So far this chapter has highlighted many of the potential benefits of using non-coding RNAs for diagnosis, prognosis, and prediction in a clinical setting, but this strategy is not without its challenges. Some of these issues are described in the following sections.

Validation of Non-coding RNA Biomarkers

The preceding sections have sampled the wide range of studies associating non-coding RNAs with cancer, many of which report on the possibility of using ncRNAs as biomarkers for diagnostic and prognostic purposes. A large number also report the potential of using miRNAs as therapeutics for several cancer types [319]. Despite this enormous interest, there are currently no phase II or III clinical trials that involve either long or short ncRNAs, and the numbers of markers available to the clinic are very small. There are a fair number of trials that are investigating the possibility of using miRNAs as potential biomarkers in different cancers (Table 1), and some that are being considered as diagnostics (Table 2), however, none have

Table 1 Non-coding RNAs currently being explored for use as prognostic or predictive tumor biomarkers in registered clinical trials

Cancer type	ncRNA	Title	Trial ID
B-ALL	Multiple miRNAs	Studying biomarkers in samples from patients with B-call acute lymphoblastic leukemia	NCT01505699
Breast	Multiple miRNAs	Identifying circulating miRNA to identify responders to adjuvant therapy	NCT01722851
Breast	Multiple miRNAs	miRNA profiling of breast cancer in patients undergoing neoadjuvant or adjuvant treatment	NCT01231386
Breast	Multiple miRNAs	Circulating miRNAs as biomarkers of hormone sensitivity in breast cancer	NCT01612871
Endometrial	Multiple miRNAs	miRNAs associated with lymph node metastasis in endometrial cancer	NCT01119573
Glioma	hsa-miR-10b	Evaluating the expression levels of MicroRNA-10b in patients with gliomas	NCT01849952
Leukemias, lymphomas, CNS tumors	Multiple miRNAs	Circulating microRNAs as disease markers in pediatric cancers	NCT01541800
Leukemia	Multiple miRNAs	Pediatric myeloid leukemia-specific miRNA expression profiles induced by the leukemic stem cell niche	NCT01298414
Neurofibromatosis, glioma	Multiple miRNAs	MicroRNAs as disease markers for central nervous system tumors in patients with neurofibromatosis type 1	NCT01595139
Oral SCC	hsa-miR-29b	To test the prognostic value of miR-29b in oral cancer	NCT02009852
Ovarian, fallopian tube, peritoneal serous-type	Multiple miRNAs	Predictors in therapeutic response in patients with ovarian cancer	NCT01391351
Prostate	Multiple miRNAs	To determine whether specific miRNA expression profiles are related to prostate cancer outcome	NCT01220427
Prostate	Multiple miRNAs	Analysis and quantification of microRNAs in prostate tumors	ISRCTN67055660
Pancreatic	Multiple miRNAs	miRNA analysis study using a peptide vaccine for pancreatic cancer after surgery	JPRN-UMIN000013053

Table 2 Non-coding RNAs currently being explored for use as diagnostic tumor biomarkers in registered clinical trials

Cancer type	ncRNA	Title	Trial ID
Colorectal	Multiple miRNAs	Analysis of microRNA profiles in blood samples of patients with colorectal cancer	DRKS00005982
Gastric		MicroRNA as novel markers of gastric adenoma	JPRN-UMIN000005902
Glioma	hsa-miR-10b	Evaluating the expression levels of microRNA-10b in patients with gliomas	NCT01849952
Leukemia	Multiple miRNAs	Studying RNA biomarkers in tissue samples from infants with acute myeloid leukemia	NCT01229124
Neurofibromatosis, glioma	Multiple miRNAs	MicroRNAs as disease markers for central nervous system tumors in patients with neurofibromatosis type 1	NCT01595139
Prostate	Multiple miRNAs	Analysis and quantification of microRNAs in prostate tumors	ISRCTN67055660
Prostate		Serum microRNAs in the diagnosis of prostate cancer	DRKS00003155
Rhabdoid	Multiple miRNAs	Biomarkers in samples from patients with rhabdoid tumor of the kidney and atypical teratoid rhabdoid tumor	NCT01453465
Thyroid	Multiple miRNAs	Use of a microRNA panel to identify thyroid malignancy in FNA leftover cells	NCT01964508

gone past the initial discovery stage. This may indicate that a large number of potential candidates are about to be validated in the near future, or, it may reflect on the failure in the translation of pre-clinical studies to the human context [322, 323]. The following sections discuss some of the validation that will be critical to adoption of non-coding RNAs for clinical use.

Analytic Validation

Any clinical assay using non-coding RNAs would have to undergo analytical validation to ensure its reproducibility and quality. Because of differences in both practice and skill sets between research and clinical laboratories, techniques that are routine in research labs may not be robust enough for inclusion in a clinical lab repertoire. For example, assays such as quantitative real-time PCR are routinely carried out in research laboratories and the results obtained at different times may vary depending on a number of factors [324]. Such variability may hinder the use of this technique in a clinical setting for diagnostic purposes as these differences may greatly impact diagnostics leading to differences in results and interpretation.

To a certain extent, issues of reproducibility can be overcome using quality control initiatives that, when implemented, could reduce inter- and intra-lab variability. An example of such an initiative is MACQ (Microarray Quality Control)—initiated to improve microarray technology and foster its appropriate application in discovery, development, and review of FDA-regulated products [324]. This initiative has also progressed to include next-generation sequencing (SEQC), which aims to increase accuracy and reproducibility in RNA-seq [95, 325], a tool that is increasingly becoming pertinent to clinical diagnosis [326]. Such initiatives may enable the use of tools such as sequencing for clinical applications, although it seems unlikely until automated and robust workflow robotics can ease the burden of these labor intensive research techniques.

Additionally, while a patient's genotype does not change, the RNA expression profile varies substantially with cell type and biological state. Blood cells in particular are poised and ready to respond to small changes in the extracellular environment, and small variation in protocol to collect or process blood samples can lead to substantial changes in gene expression [327], an obvious challenge to overcome for clinical assays reliant upon non-coding RNAs.

The variation in expression of both non-coding and protein-coding genes is also challenging for biomarker development. Protein expression is considered to be less variable than mRNA expression due to canalization—the tolerance of living systems to slight fluctuations to both endogenous and exogenous signals [328]. Such tolerance is important to ensure the stability of phenotypes, but leads to variability of RNA profiles that are not reflected in protein profiles from the same samples [329]. MicroRNAs appear to be an exception to this rule. Indeed, a biological role proposed for these molecules is to stabilize the expression from mRNAs and assist in canalization [330, 331]. If true, this could be an argument for preferring miRNAs over ncRNAs for more robust biomarker development.

Finally, care needs to be taken when deciding upon the method of RNA extraction, especially for the small non-coding RNAs, as non-random loss of certain sequences can occur when comparing one set of extraction methods with another [332]. This may mean that the optimal extraction technique may depend upon the transcript being detected.

Clinical Validation

The clinical validation of an assay refers to its ability to differentiate a positive and negative outcome, and assays to be used for diagnostic purposes require a large multi-institutional effort to ensure that bias from small sample sizes and single institution studies do not affect the evaluation. Several bodies such as Consolidated Standard Randomized Trials (CONSORT) movement and the Standard for Reporting Diagnostic Accuracy (STARD) group [333, 334] have initiated quality criteria that should be fulfilled before a diagnostic can be approved. Additionally, the Statistics Subcommittee of the NCI-EORTC Working Group on Cancer Diagnostics released reporting recommendation for tumor marker prognostic studies to address the "pitifully small" proportion of markers that have emerged as clinically useful from the thousands of publish articles [335]. Although issued before the bulk of non-coding RNA research has been conducted, the guidelines are no less applicable.

A factor that may complicate the clinical validation of miRNAs is their ability to play dual oncogenic and tumor suppressive functions across different tumors types (see Bifunctional microRNAs). While there have not yet been reports of differences in miRNA function between subtypes, if the function of an miRNA is largely dependent on the mRNAs that it targets, then it is conceivable that its phenotypic effects could vary significantly based on whether or not its targets are expressed. This begs the question, how important is it to understand the target genes for a given miRNA? And how important is it to incorporate the expression of target genes into the diagnostic/prognostic algorithm? It is possible that a detailed understanding of the mechanism by which the miRNA acts may be essential to prevent misinterpretation of miRNA based assays.

Clinical Utility

The development of new applications for non-coding RNA tools should be dictated by the benefit to patients, and should additionally offer an independent benefit over existing clinical assays. So far it is unclear whether there would be any wide-reaching benefit to using lncRNAs over protein-coding RNAs for clinical applications based on multi-gene expression profiles. Whilst ncRNAs are often more tissue specific than protein-coding genes, and can therefore offer benefits in classifying poorly differentiated tumors, the expression of lncRNAs in particular is often substantially lower [3]. It is still unknown whether this lower level of expression is due to high base-expression in a small sub-population of cells, or low-level expression in all cells. The answer has substantial importance for its use as a biomarker. Although highly cell type specific expression may be of use diagnostically and therapeutically, a poorly expressed marker may lead into a trade-off between sensitivity and specificity.

Conclusions

It's clear that non-coding RNAs are an extremely active area of both basic and applied research in the detection and management of cancer, and that the results generated so far are undeniably in support of non-coding RNAs playing a substantial role in every aspect of cancer biology. Regarding their application to molecular pathology, prospects are high for their widespread use in many areas of diagnostics and prognostics, provided that issues surrounding the analytical validation, clinical validation, and clinical utility can be addressed.

Acknowledgments NC is supported by an Australian Research Council Future Fellowship (FT120100453) and the Cancer Council of Queensland (APP1063119). We would like to thank our colleagues, students, collaborators, and family for their patience and assistance during the preparation and writing of this chapter.

References

1. Lander ES et al (2001) Initial sequencing and analysis of the human genome. Nature 409(6822):860–921
2. Taft RJ, Pheasant M, Mattick JS (2007) The relationship between non-protein-coding DNA and eukaryotic complexity. Bioessays 29(3):288–299
3. Djebali S et al (2012) Landscape of transcription in human cells. Nature 489(7414): 101–108
4. Harrow J et al (2012) GENCODE: the reference human genome annotation for The ENCODE Project. Genome Res 22(9):1760–1774
5. Gascoigne DK et al (2012) Pinstripe: a suite of programs for integrating transcriptomic and proteomic datasets identifies novel proteins and improves differentiation of protein-coding and non-coding genes. Bioinformatics 28(23):3042–3050
6. Dinger ME et al (2008) Differentiating protein-coding and noncoding RNA: challenges and ambiguities. PLoS Comput Biol 4(11):e1000176
7. Chooniedass-Kothari S et al (2004) The steroid receptor RNA activator is the first functional RNA encoding a protein. FEBS Lett 566(1-3):43–47
8. Ephrussi A, Lehmann R (1992) Induction of germ cell formation by oskar. Nature 358: 387–392
9. Jenny A et al (2006) A translation-independent role of oskar RNA in early Drosophila oogenesis. Development 133(15):2827–2833
10. Kloc M et al (2005) Potential structural role of non-coding and coding RNAs in the organization of the cytoskeleton at the vegetal cortex of Xenopus oocytes. Development 132(15): 3445–3457
11. Lanz RB et al (1999) A steroid receptor coactivator, SRA, functions as an RNA and is present in an SRC-1 complex. Cell 97(1):17–27
12. Zhang J et al (1998) The role of maternal VegT in establishing the primary germ layers in Xenopus embryos. Cell 94(4):515–524
13. Doolittle WF (2013) Is junk DNA bunk? A critique of ENCODE. Proc Natl Acad Sci U S A 110(14):5294–5300
14. Ebisuya M et al (2008) Ripples from neighbouring transcription. Nat Cell Biol 10(9): 1106–1113
15. Eddy SR (2012) The C-value paradox, junk DNA and ENCODE. Curr Biol 22(21): R898–R899
16. Graur D et al (2013) On the immortality of television sets: "function" in the human genome according to the evolution-free gospel of ENCODE. Genome Biol Evol 5(3):578–590
17. Guttman M, Rinn J (2012) Modular regulatory principles of large non-coding RNAs. Nature 482(7385):339–346
18. Kim T-K et al (2010) Widespread transcription at neuronal activity-regulated enhancers. Nature 465(7295):182–187

19. Niu D-K, Jiang L (2013) Can ENCODE tell us how much junk DNA we carry in our genome? Biochem Biophys Res Commun 430(4):1340–1343
20. Nobrega M et al (2004) Megabase deletions of gene deserts result in viable mice. Nature 431(October):988–993
21. De Santa F et al (2010) A large fraction of extragenic RNA pol II transcription sites overlap enhancers. PLoS Biol 8(5):e1000384
22. Struhl K (2007) Transcriptional noise and the fidelity of initiation by RNA polymerase II. Nat Struct Mol Biol 14(2):103–105
23. Hänggi P (2002) Stochastic resonance in biology how noise can enhance detection of weak signals and help improve biological information processing. Chemphyschem 3(3):285–290
24. Ozbudak EM et al (2002) Regulation of noise in the expression of a single gene. Nat Genet 31(1):69–73
25. Baker M (2011) Long noncoding RNAs: the search for function. Nat Methods 8(5): 379–383
26. Lee RC, Feinbaum RL, Ambros V (1993) The C. elegans heterochronic gene lin-4 encodes small RNAs with antisense complementarity to lin-14. Cell 75(5):843–854
27. Pasquinelli AE et al (2000) Conservation of the sequence and temporal expression of let-7 heterochronic regulatory RNA. Nature 408(6808):86–89
28. Reinhart BJ et al (2000) The 21-nucleotide let-7 RNA regulates developmental timing in Caenorhabditis elegans. Nature 403(6772):901–906
29. Bartel DP (2009) MicroRNAs: target recognition and regulatory functions. Cell 136(2): 215–233
30. Cloonan N et al (2011) MicroRNAs and their isomiRs function cooperatively to target common biological pathways. Genome Biol 12(12):R126
31. Humphreys DT et al (2005) MicroRNAs control translation initiation by inhibiting eukaryotic initiation factor 4E/cap and poly(A) tail function. Proc Natl Acad Sci U S A 102(47): 16961–16966
32. Maroney PA et al (2006) Evidence that microRNAs are associated with translating messenger RNAs in human cells. Nat Struct Mol Biol 13(12):1102–1107
33. Mathonnet G et al (2007) MicroRNA inhibition of translation initiation in vitro by targeting the cap-binding complex eIF4F. Science 317(5845):1764–1767
34. Nottrott S, Simard MJ, Richter JD (2006) Human let-7a miRNA blocks protein production on actively translating polyribosomes. Nat Struct Mol Biol 13(12):1108–1114
35. Petersen CP et al (2006) Short RNAs repress translation after initiation in mammalian cells. Mol Cell 21(4):533–542
36. Pillai RS et al (2005) Inhibition of translational initiation by Let-7 microRNA in human cells. Science 309(5740):1573–1576
37. Liu J et al (2005) MicroRNA-dependent localization of targeted mRNAs to mammalian P-bodies. Nat Cell Biol 7(7):719–723
38. Giraldez AJ et al (2006) Zebrafish MiR-430 promotes deadenylation and clearance of maternal mRNAs. Science 312(5770):75–79
39. Valencia-Sanchez MA et al (2006) Control of translation and mRNA degradation by miRNAs and siRNAs. Genes Dev 20(5):515–524
40. Wu L, Fan J, Belasco JG (2006) MicroRNAs direct rapid deadenylation of mRNA. Proc Natl Acad Sci U S A 103(11):4034–4039
41. Guo H et al (2010) Mammalian microRNAs predominantly act to decrease target mRNA levels. Nature 466(7308):835–840
42. Cui Q et al (2006) Principles of microRNA regulation of a human cellular signaling network. Mol Syst Biol 2:46
43. Clancy JL et al (2011) mRNA isoform diversity can obscure detection of miRNA-mediated control of translation. RNA 17(6):1025–1031
44. Krol J, Loedige I, Filipowicz W (2010) The widespread regulation of microRNA biogenesis, function and decay. Nat Rev Genet 11(9):597–610
45. Berezikov E et al (2007) Mammalian mirtron genes. Mol Cell 28(2):328–336

46. Okamura K et al (2007) The mirtron pathway generates microRNA-class regulatory RNAs in Drosophila. Cell 130(1):89–100
47. Cheloufi S et al (2010) A dicer-independent miRNA biogenesis pathway that requires Ago catalysis. Nature 465(7298):584–589
48. Cifuentes D et al (2010) A novel miRNA processing pathway independent of Dicer requires Argonaute2 catalytic activity. Science 328(5986):1694–1698
49. Yang J-S et al (2010) Conserved vertebrate mir-451 provides a platform for Dicer-independent, Ago2-mediated microRNA biogenesis. Proc Natl Acad Sci U S A 107(34):15163–15168
50. Matranga C et al (2005) Passenger-strand cleavage facilitates assembly of siRNA into Ago2-containing RNAi enzyme complexes. Cell 123(4):607–620
51. Griffiths-Jones S et al (2011) MicroRNA evolution by arm switching. EMBO Rep 12(2): 172–177
52. Humphreys DT et al (2012) Complexity of murine cardiomyocyte miRNA biogenesis, sequence variant expression and function. PLoS One 7(2):e30933
53. Calin GA et al (2002) Frequent deletions and down-regulation of micro- RNA genes miR15 and miR16 at 13q14 in chronic lymphocytic leukemia. Proc Natl Acad Sci U S A 99(24): 15524–15529
54. Bandi N et al (2009) miR-15a and miR-16 are implicated in cell cycle regulation in a Rb-dependent manner and are frequently deleted or down-regulated in non-small cell lung cancer. Cancer Res 69(13):5553–5559
55. Linsley PS et al (2007) Transcripts targeted by the microRNA-16 family cooperatively regulate cell cycle progression. Mol Cell Biol 27(6):2240–2252
56. Liu Q et al (2008) miR-16 family induces cell cycle arrest by regulating multiple cell cycle genes. Nucleic Acids Res 36(16):5391–5404
57. Cimmino A et al (2005) miR-15 and miR-16 induce apoptosis by targeting BCL2. Proc Natl Acad Sci U S A 102(39):13944–13949
58. Calin GA et al (2008) MiR-15a and miR-16-1 cluster functions in human leukemia. Proc Natl Acad Sci U S A 105(13):5166–5171
59. He L et al (2005) A microRNA polycistron as a potential human oncogene. Nature 435(7043):828–833
60. Cloonan N et al (2008) The miR-17-5p microRNA is a key regulator of the G1/S phase cell cycle transition. Genome Biol 9(8):R127
61. Matsubara H et al (2007) Apoptosis induction by antisense oligonucleotides against miR-17-5p and miR-20a in lung cancers overexpressing miR-17-92. Oncogene 26(41):6099–6105
62. Yan H et al (2009) Repression of the miR-17-92 cluster by p53 has an important function in hypoxia-induced apoptosis. EMBO J 28(18):2719–2732
63. Mu P et al (2009) Genetic dissection of the miR-17~92 cluster of microRNAs in Myc-induced B-cell lymphomas. Genes Dev 23(24):2806–2811
64. Olive V et al (2009) miR-19 is a key oncogenic component of mir-17-92. Genes Dev 23(24):2839–2849
65. Ventura A et al (2008) Targeted deletion reveals essential and overlapping functions of the miR-17 through 92 family of miRNA clusters. Cell 132(5):875–886
66. Calin GA, Sevignani C et al (2004) Human microRNA genes are frequently located at fragile sites and genomic regions involved in cancers. Proc Natl Acad Sci U S A 101(9):2999–3004
67. Tagawa H, Seto M (2005) A microRNA cluster as a target of genomic amplification in malignant lymphoma. Leukemia 19(11):2013–2016
68. Zhang L et al (2006) microRNAs exhibit high frequency genomic alterations in human cancer. Proc Natl Acad Sci U S A 103(24):9136–9141
69. Croce CM (2009) Causes and consequences of microRNA dysregulation in cancer. Nat Rev Genet 10(10):704–714
70. Iorio MV, Croce CM (2012) MicroRNA dysregulation in cancer: diagnostics, monitoring and therapeutics. A comprehensive review. EMBO Mol Med 4(3):143–159
71. Hanahan D, Weinberg RA (2011) Hallmarks of cancer: the next generation. Cell 144(5): 646–674

72. Xie B et al (2013) miRCancer: a microRNA-cancer association database constructed by text mining on literature. Bioinformatics 29(5):638–644
73. Hanahan D, Weinberg R (2000) The hallmarks of cancer. Cell 100:57–70
74. Chen X et al (2009) Role of miR-143 targeting KRAS in colorectal tumorigenesis. Oncogene 28(10):1385–1392
75. Lin T et al (2009) MicroRNA-143 as a tumor suppressor for bladder cancer. J Urol 181(3):1372–1380
76. Noguchi S et al (2011) MicroRNA-143 functions as a tumor suppressor in human bladder cancer T24 cells. Cancer Lett 307(2):211–220
77. Noguchi S et al (2013) Replacement treatment with microRNA-143 and -145 induces synergistic inhibition of the growth of human bladder cancer cells by regulating PI3K/Akt and MAPK signaling pathways. Cancer Lett 328(2):353–361
78. Ng EKO et al (2014) MicroRNA-143 is downregulated in breast cancer and regulates DNA methyltransferases 3A in breast cancer cells. Tumour Biol 35:2591–2598
79. Liu L et al (2012) miR-143 is downregulated in cervical cancer and promotes apoptosis and inhibits tumor formation by targeting Bcl-2. Mol Med Rep 5(3):753–760
80. Borralho PM et al (2011) miR-143 overexpression impairs growth of human colon carcinoma xenografts in mice with induction of apoptosis and inhibition of proliferation. PLoS One 6(8):e23787
81. Ng EKO et al (2009) MicroRNA-143 targets DNA methyltransferases 3A in colorectal cancer. Br J Cancer 101(4):699–706
82. Pichler M et al (2012) Down-regulation of KRAS-interacting miRNA-143 predicts poor prognosis but not response to EGFR-targeted agents in colorectal cancer. Br J Cancer 106(11):1826–1832
83. Qian X et al (2013) MicroRNA-143 inhibits tumor growth and angiogenesis and sensitizes chemosensitivity to oxaliplatin in colorectal cancers. Cell Cycle 12(9):1385–1394
84. Slaby O et al (2007) Altered expression of miR-21, miR-31, miR-143 and miR-145 is related to clinicopathologic features of colorectal cancer. Oncology 72(5-6):397–402
85. Takaoka Y et al (2012) Forced expression of miR-143 represses ERK5/c-Myc and p68/p72 signaling in concert with miR-145 in gut tumors of Apc(Min) mice. PLoS One 7(8):e42137
86. Ni Y et al (2013) MicroRNA-143 functions as a tumor suppressor in human esophageal squamous cell carcinoma. Gene 517(2):197–204
87. Zhao S et al (2013) miR-143 inhibits glycolysis and depletes stemness of glioblastoma stem-like cells. Cancer Lett 333(2):253–260
88. Chen H-C et al (2009) MicroRNA deregulation and pathway alterations in nasopharyngeal carcinoma. Br J Cancer 100(6):1002–1011
89. Ouyang L et al (2013) A three-plasma miRNA signature serves as novel biomarkers for osteosarcoma. Med Oncol 30(1):340
90. Zhang H et al (2010) microRNA-143, down-regulated in osteosarcoma, promotes apoptosis and suppresses tumorigenicity by targeting Bcl-2. Oncol Rep 24:1363–1369
91. Clapé C et al (2009) miR-143 interferes with ERK5 signaling, and abrogates prostate cancer progression in mice. PLoS One 4(10):e7542
92. Kojima S et al (2014) The tumor-suppressive microRNA-143/145 cluster inhibits cell migration and invasion by targeting GOLM1 in prostate cancer. J Hum Genet 59:78–87
93. Xu B et al (2011) miR-143 decreases prostate cancer cells proliferation and migration and enhances their sensitivity to docetaxel through suppression of KRAS. Mol Cell Biochem 350(1-2):207–213
94. Yoshino H et al (2013) The tumor-suppressive microRNA-143/145 cluster targets hexokinase-2 in renal cell carcinoma. Cancer Sci 104(12):1567–1574
95. Cloonan N et al (2008) Stem cell transcriptome profiling via massive-scale mRNA sequencing. Nat Methods 5(7):613–619
96. Xia H et al (2009) MicroRNA-15b regulates cell cycle progression by targeting cyclins in glioma cells. Biochem Biophys Res Commun 380(2):205–210

97. Talotta F et al (2009) An autoregulatory loop mediated by miR-21 and PDCD4 controls the AP-1 activity in RAS transformation. Oncogene 28(1):73–84
98. Volinia S et al (2010) Reprogramming of miRNA networks in cancer and leukemia. Genome Res 20(5):589–599
99. Shi W et al (2011) MicroRNA-301 mediates proliferation and invasion in human breast cancer. Cancer Res 71(8):2926–2937
100. Coates PJ et al (2001) Mammalian prohibitin proteins respond to mitochondrial stress and decrease during cellular senescence. Exp Cell Res 265(2):262–273
101. Liu T et al (2009) MicroRNA-27a functions as an oncogene in gastric adenocarcinoma by targeting prohibitin. Cancer Lett 273(2):233–242
102. Li X et al (2013) c-MYC-regulated miR-23a/24-2/27a cluster promotes mammary carcinoma cell invasion and hepatic metastasis by targeting Sprouty2. J Biol Chem 288(25): 18121–18133
103. Zhang Z et al (2011) miR-27 promotes human gastric cancer cell metastasis by inducing epithelial-to-mesenchymal transition. Cancer Genet 204(9):486–491
104. Zhao X, Yang L, Hu J (2011) Down-regulation of miR-27a might inhibit proliferation and drug resistance of gastric cancer cells. J Exp Clin Cancer Res 30(1):55
105. Park YT et al (2013) MicroRNAs overexpressed in ovarian ALDH1-positive cells are associated with chemoresistance. J Ovarian Res 6(1):18
106. Rodriguez A et al (2007) Requirement of bic/microRNA-155 for normal immune function. Science 316(5824):608–611
107. Zonari E et al (2013) A role for miR-155 in enabling tumor-infiltrating innate immune cells to mount effective antitumor responses in mice. Blood 122(2):243–252
108. Ueda R et al (2009) Dicer-regulated microRNAs 222 and 339 promote resistance of cancer cells to cytotoxic T-lymphocytes by down-regulation of ICAM-1. Proc Natl Acad Sci U S A 106(26):10746–10751
109. Tsukerman P et al (2012) MiR-10b downregulates the stress-induced cell surface molecule MICB, a critical ligand for cancer cell recognition by natural killer cells. Cancer Res 72(21):5463–5472
110. Ciafrè SA et al (2005) Extensive modulation of a set of microRNAs in primary glioblastoma. Biochem Biophys Res Commun 334(4):1351–1358
111. Silber J et al (2008) miR-124 and miR-137 inhibit proliferation of glioblastoma multiforme cells and induce differentiation of brain tumor stem cells. BMC Med 6:14
112. Bloomston M et al (2007) MicroRNA expression patterns to differentiate pancreatic adenocarcinoma from normal pancreas and chronic pancreatitis. JAMA 297(17):1901–1908
113. Ma L, Teruya-Feldstein J, Weinberg RA (2007) Tumour invasion and metastasis initiated by microRNA-10b in breast cancer. Nature 449(7163):682–688
114. Calin GA, Liu C-G et al (2004) MicroRNA profiling reveals distinct signatures in B cell chronic lymphocytic leukemias. Proc Natl Acad Sci U S A 101(32):11755–11760
115. Garzon R et al (2008) Distinctive microRNA signature of acute myeloid leukemia bearing cytoplasmic mutated nucleophosmin. Proc Natl Acad Sci U S A 105(10):3945–3950
116. Hwang MS et al (2013) miR-221/222 targets adiponectin receptor 1 to promote the epithelial-to-mesenchymal transition in breast cancer. PLoS One 8(6):e66502
117. Shah MY, Calin GA (2011) MicroRNAs miR-221 and miR-222: a new level of regulation in aggressive breast cancer. Genome Med 3(8):56
118. Stinson S et al (2011) TRPS1 targeting by miR-221/222 promotes the epithelial-to-mesenchymal transition in breast cancer. Sci Signal 4(177):ra41
119. Puerta-Gil P et al (2012) miR-143, miR-222, and miR-452 are useful as tumor stratification and noninvasive diagnostic biomarkers for bladder cancer. Am J Pathol 180(5):1808–1815
120. Amankwah EK et al (2013) miR-21, miR-221 and miR-222 expression and prostate cancer recurrence among obese and non-obese cases. Asian J Androl 15(2):226–230

121. Fuse M et al (2012) Tumor suppressive microRNAs (miR-222 and miR-31) regulate molecular pathways based on microRNA expression signature in prostate cancer. J Hum Genet 57(11):691–699
122. Sun T et al (2012) The altered expression of MiR-221/-222 and MiR-23b/-27b is associated with the development of human castration resistant prostate cancer. Prostate 72(10):1093–1103
123. Li N et al (2012) Increased miR-222 in H. pylori-associated gastric cancer correlated with tumor progression by promoting cancer cell proliferation and targeting RECK. FEBS Lett 586(6):722–728
124. Tsunoda T et al (2011) Oncogenic KRAS regulates miR-200c and miR-221/222 in a 3D-specific manner in colorectal cancer cells. Anticancer Res 31(7):2453–2459
125. Zhang Y et al (2011) High-mobility group A1 proteins enhance the expression of the oncogenic miR-222 in lung cancer cells. Mol Cell Biochem 357(1-2):363–371
126. Sun C et al (2013) miR-222 is upregulated in epithelial ovarian cancer and promotes cell proliferation by downregulating P27(kip1.). Oncol Lett 6(2):507–512
127. Lee C et al (2013) Elevated expression of tumor miR-222 in pancreatic cancer is associated with Ki67 and poor prognosis. Med Oncol 30(4):700
128. Harley CB (2008) Telomerase and cancer therapeutics. Nat Rev Cancer 8(3):167–179
129. Mitomo S et al (2008) Downregulation of miR-138 is associated with overexpression of human telomerase reverse transcriptase protein in human anaplastic thyroid carcinoma cell lines. Cancer Sci 99(2):280–286
130. Wang W et al (2012) MiR-138 induces cell cycle arrest by targeting cyclin D3 in hepatocellular carcinoma. Carcinogenesis 33(5):1113–1120
131. Liu X et al (2009) MicroRNA-138 suppresses invasion and promotes apoptosis in head and neck squamous cell carcinoma cell lines. Cancer Lett 286(2):217–222
132. Kasiappan R et al (2012) 1,25-Dihydroxyvitamin D3 suppresses telomerase expression and human cancer growth through microRNA-498. J Biol Chem 287(49):41297–41309
133. Watanabe A et al (2011) The role of microRNA-150 as a tumor suppressor in malignant lymphoma. Leukemia 25(8):1324–1334
134. Stampfer MR et al (2003) Loss of p53 function accelerates acquisition of telomerase activity in indefinite lifespan human mammary epithelial cell lines. Oncogene 22(34):5238–5251
135. Voorhoeve PM et al (2006) A genetic screen implicates miRNA-372 and miRNA-373 as oncogenes in testicular germ cell tumors. Cell 124(6):1169–1181
136. O'Connell RM et al (2007) MicroRNA-155 is induced during the macrophage inflammatory response. Proc Natl Acad Sci U S A 104(5):1604–1609
137. Tili E et al (2011) Mutator activity induced by microRNA-155 (miR-155) links inflammation and cancer. Proc Natl Acad Sci U S A 108(12):4908–4913
138. Fabbri M, Paone A, Calore F (2012) MicroRNAs bind to Toll-like receptors to induce prometastatic inflammatory response. Proc Natl Acad Sci U S A 109(31):E2110–E2116
139. Yang CH et al (2010) IFN induces miR-21 through a signal transducer and activator of transcription 3-dependent pathway as a suppressive negative feedback on IFN-induced apoptosis. Cancer Res 70(20):8108–8116
140. Lu TX, Munitz A, Rothenberg ME (2009) MicroRNA-21 is up-regulated in allergic airway inflammation and regulates IL-12p35 expression. J Immunol 182(8):4994–5002
141. Moschos SA et al (2007) Expression profiling in vivo demonstrates rapid changes in lung microRNA levels following lipopolysaccharide-induced inflammation but not in the anti-inflammatory action of glucocorticoids. BMC Genomics 8:240
142. Sheedy FJ et al (2010) Negative regulation of TLR4 via targeting of the proinflammatory tumor suppressor PDCD4 by the microRNA miR-21. Nat Immunol 11(2):141–147
143. Weigelt B, Peterse JL, & van't Veer LJ (2005) Breast cancer metastasis: markers and models. Nat Rev Cancer 5(8):591–602
144. Asangani IA et al (2008) MicroRNA-21 (miR-21) post-transcriptionally downregulates tumor suppressor Pdcd4 and stimulates invasion, intravasation and metastasis in colorectal cancer. Oncogene 27(15):2128–2136

145. Zhu S et al (2008) MicroRNA-21 targets tumor suppressor genes in invasion and metastasis. Cell Res 18(3):350–359
146. Valastyan S et al (2009) A pleiotropically acting microRNA, miR-31, inhibits breast cancer metastasis. Cell 137(6):1032–1046
147. Bao W et al (2011) HER2 interacts with CD44 to up-regulate CXCR4 via epigenetic silencing of microRNA-139 in gastric cancer cells. Gastroenterology 141(6):2076–2087.e6
148. Li R-Y et al (2013) MiR-139 inhibits Mcl-1 expression and potentiates TMZ-induced apoptosis in glioma. CNS Neurosci Ther 19(7):477–483
149. Wong CC-L et al (2011) The microRNA miR-139 suppresses metastasis and progression of hepatocellular carcinoma by down-regulating Rho-kinase 2. Gastroenterology 140(1):322–331
150. Mascaux C et al (2009) Evolution of microRNA expression during human bronchial squamous carcinogenesis. Eur Respir J 33(2):352–359
151. Krishnan K, Steptoe AL, Martin HC, Wani S et al (2013) MicroRNA-182-5p targets a network of genes involved in DNA repair. RNA 19(2):230–242
152. Herbert SP, Stainier DYR (2011) Molecular control of endothelial cell behaviour during blood vessel morphogenesis. Nat Rev Mol Cell Biol 12(9):551–564
153. Suárez Y, Sessa WC (2009) MicroRNAs as novel regulators of angiogenesis. Circ Res 104(4):442–454
154. Yamakuchi M et al (2010) P53-induced microRNA-107 inhibits HIF-1 and tumor angiogenesis. Proc Natl Acad Sci U S A 107(14):6334–6339
155. Dews M et al (2006) Augmentation of tumor angiogenesis by a Myc-activated microRNA cluster. Nat Genet 38(9):1060–1065
156. Bonauer A et al (2009) MicroRNA-92a controls angiogenesis and functional recovery of ischemic tissues in mice. Science 324(5935):1710–1713
157. Krishnan K, Steptoe AL, Martin HC, Pattabiraman DR et al (2013) miR-139-5p is a regulator of metastatic pathways in breast cancer. RNA 19(12):1767–1780
158. Lal A et al (2009) miR-24-mediated downregulation of H2AX suppresses DNA repair in terminally differentiated blood cells. Nat Struct Mol Biol 16(5):492–498
159. Crosby ME et al (2009) MicroRNA regulation of DNA repair gene expression in hypoxic stress. Cancer Res 69(3):1221–1229
160. Lima RT et al (2011) MicroRNA regulation of core apoptosis pathways in cancer. Eur J Cancer 47(2):163–174
161. Warburg O (1956) On the origin of cancer cells. Science 123(3191):309–314
162. Eichner LJ et al (2010) miR-378(∗) mediates metabolic shift in breast cancer cells via the PGC-1β/ERRγ transcriptional pathway. Cell Metab 12(4):352–361
163. Sun Y et al (2012) miR-124, miR-137 and miR-340 regulate colorectal cancer growth via inhibition of the Warburg effect. Oncol Rep 28(4):1346–1352
164. Chi SW et al (2009) Argonaute HITS-CLIP decodes microRNA-mRNA interaction maps. Nature 460(7254):479–486
165. Doebele C et al (2010) Members of the microRNA-17-92 cluster exhibit a cell-intrinsic antiangiogenic function in endothelial cells. Blood 115(23):4944–4950
166. Hafner M et al (2010) Transcriptome-wide identification of RNA-binding protein and microRNA target sites by PAR-CLIP. Cell 141(1):129–141
167. Ivanovska I et al (2008) MicroRNAs in the miR-106b family regulate p21/CDKN1A and promote cell cycle progression. Mol Cell Biol 28(7):2167–2174
168. Guo H et al (2013) The regulation of Toll-like receptor 2 by miR-143 suppresses the invasion and migration of a subset of human colorectal carcinoma cells. Mol Cancer 12(1):77
169. Cho WCS, Chow ASC, Au JSK (2009) Restoration of tumour suppressor hsa-miR-145 inhibits cancer cell growth in lung adenocarcinoma patients with epidermal growth factor receptor mutation. Eur J Cancer 45(12):2197–2206
170. Myatt SS et al (2010) Definition of microRNAs that repress expression of the tumor suppressor gene FOXO1 in endometrial cancer. Cancer Res 70(1):367–377

171. Sarver AL et al (2009) Human colon cancer profiles show differential microRNA expression depending on mismatch repair status and are characteristic of undifferentiated proliferative states. BMC Cancer 9:401
172. Schaefer A et al (2010) Diagnostic and prognostic implications of microRNA profiling in prostate carcinoma. Int J Cancer 126(5):1166–1176
173. Segura MF et al (2009) Aberrant miR-182 expression promotes melanoma metastasis by repressing FOXO3 and microphthalmia-associated transcription factor. Proc Natl Acad Sci U S A 106(6):1814–1819
174. Kong W-Q et al (2012) MicroRNA-182 targets cAMP-responsive element-binding protein 1 and suppresses cell growth in human gastric adenocarcinoma. FEBS J 279(7):1252–1260
175. Sun Y et al (2010) Hsa-mir-182 suppresses lung tumorigenesis through down regulation of RGS17 expression in vitro. Biochem Biophys Res Commun 396(2):501–507
176. Zhang L et al (2011) microRNA-182 inhibits the proliferation and invasion of human lung adenocarcinoma cells through its effect on human cortical actin-associated protein. Int J Mol Med 28(3):381–388
177. Gottardo F et al (2007) Micro-RNA profiling in kidney and bladder cancers. Urol Oncol 25(5):387–392
178. Greenberg E et al (2011) Regulation of cancer aggressive features in melanoma cells by microRNAs. PLoS One 6(4):e18936
179. Hayashita Y et al (2005) A polycistronic microRNA cluster, miR-17-92, is overexpressed in human lung cancers and enhances cell proliferation. Cancer Res 65(21):9628–9632
180. Koga Y et al (2010) MicroRNA expression profiling of exfoliated colonocytes isolated from feces for colorectal cancer screening. Cancer Prev Res (Phila) 3(11):1435–1442
181. Levati L, Alvino E (2009) Altered expression of selected microRNAs in melanoma: antiproliferative and proapoptotic activity of miRNA-155. Int J Oncol 35:393–400
182. Li H et al (2011) miR-17-5p promotes human breast cancer cell migration and invasion through suppression of HBP1. Breast Cancer Res Treat 126(3):565–575
183. Luo H et al (2012) Up-regulated miR-17 promotes cell proliferation, tumour growth and cell cycle progression by targeting the RND3 tumour suppressor gene in colorectal carcinoma. Biochem J 442(2):311–321
184. Ohuchida K et al (2012) MicroRNA-10a is overexpressed in human pancreatic cancer and involved in its invasiveness partially via suppression of the HOXA1 gene. Ann Surg Oncol 19(7):2394–2402
185. Tsujiura M et al (2010) Circulating microRNAs in plasma of patients with gastric cancers. Br J Cancer 102(7):1174–1179
186. Yan H-J et al (2012) miR-17-5p inhibitor enhances chemosensitivity to gemcitabine via upregulating Bim expression in pancreatic cancer cells. Dig Dis Sci 57(12):3160–3167
187. Yang F et al (2010) miR-17-5p Promotes migration of human hepatocellular carcinoma cells through the p38 mitogen-activated protein kinase-heat shock protein 27 pathway. Hepatology 51(5):1614–1623
188. Yu J et al (2010) MicroRNA miR-17-5p is overexpressed in pancreatic cancer, associated with a poor prognosis, and involved in cancer cell proliferation and invasion. Cancer Biol Ther 10(8):748–757
189. Hossain A, Kuo MT, Saunders GF (2006) Mir-17-5p regulates breast cancer cell proliferation by inhibiting translation of AIB1 mRNA. Mol Cell Biol 26(21):8191–8201
190. Wei Q et al (2012) MiR-17-5p targets TP53INP1 and regulates cell proliferation and apoptosis of cervical cancer cells. IUBMB Life 64(8):697–704
191. He L, Hannon GJ (2004) MicroRNAs: small RNAs with a big role in gene regulation. Nat Rev Genet 5(7):522–531
192. Devi GR (2006) siRNA-based approaches in cancer therapy. Cancer Gene Ther 13(9):819–829
193. Izquierdo M (2005) Short interfering RNAs as a tool for cancer gene therapy. Cancer Gene Ther 12(3):217–227

194. Soutschek J et al (2004) Therapeutic silencing of an endogenous gene by systemic administration of modified siRNAs. Nature 432(7014):173–178
195. Watanabe T et al (2008) Endogenous siRNAs from naturally formed dsRNAs regulate transcripts in mouse oocytes. Nature 453(7194):539–543
196. Kim VN, Han J, Siomi MC (2009) Biogenesis of small RNAs in animals. Nat Rev Mol Cell Biol 10(2):126–139
197. Doench JG, Petersen CP, Sharp PA (2003) siRNAs can function as miRNAs. Genes Dev 17(4):438–442
198. Sontheimer EJ, Carthew RW (2005) Silence from within: endogenous siRNAs and miRNAs. Cell 122(1):9–12
199. Shukla R et al (2013) Endogenous retrotransposition activates oncogenic pathways in hepatocellular carcinoma. Cell 153(1):101–111
200. Brennecke J et al (2007) Discrete small RNA-generating loci as master regulators of transposon activity in Drosophila. Cell 128(6):1089–1103
201. Yan Z et al (2011) Widespread expression of piRNA-like molecules in somatic tissues. Nucleic Acids Res 39(15):6596–6607
202. Siomi MC et al (2011) PIWI-interacting small RNAs: the vanguard of genome defence. Nat Rev Mol Cell Biol 12(4):246–258
203. Cheng J et al (2011) piRNA, the new non-coding RNA, is aberrantly expressed in human cancer cells. Clin Chim Acta 412(17–18):1621–1625
204. Huang G et al (2013) Altered expression of piRNAs and their relation with clinicopathologic features of breast cancer. Clin Transl Oncol 15(7):563–568
205. Cheng J et al (2012) piR-823, a novel non-coding small RNA, demonstrates in vitro and in vivo tumor suppressive activity in human gastric cancer cells. Cancer Lett 315(1):12–17
206. Mei Y, Clark D, Mao L (2013) Novel dimensions of piRNAs in cancer. Cancer Lett 336(1): 46–52
207. Darzacq X et al (2002) Cajal body-specific small nuclear RNAs: a novel class of 2′-O-methylation and pseudouridylation guide RNAs. EMBO J 21(11):2746–2756
208. Kishore S et al (2010) The snoRNA MBII-52 (SNORD 115) is processed into smaller RNAs and regulates alternative splicing. Hum Mol Genet 19(7):1153–1164
209. Darzacq X, Kiss T (2000) Processing of intron-encoded box C/D small nucleolar RNAs lacking a 5′, 3′-terminal stem structure. Mol Cell Biol 20(13):4522–4531
210. Kiss T, Fayet-Lebaron E, Jády BE (2010) Box H/ACA small ribonucleoproteins. Mol Cell 37(5):597–606
211. Williams GT, Farzaneh F (2012) Are snoRNAs and snoRNA host genes new players in cancer? Nat Rev Cancer 12(2):84–88
212. Dong X-Y et al (2009) Implication of snoRNA U50 in human breast cancer. J Genet Genomics 36(8):447–454
213. Dong X-Y et al (2008) SnoRNA U50 is a candidate tumor-suppressor gene at 6q14.3 with a mutation associated with clinically significant prostate cancer. Hum Mol Genet 17(7): 1031–1042
214. Liao J et al (2010) Small nucleolar RNA signatures as biomarkers for non-small-cell lung cancer. Mol Cancer 9:198
215. Mei Y-P et al (2012) Small nucleolar RNA 42 acts as an oncogene in lung tumorigenesis. Oncogene 31(22):2794–2804
216. Matera AG, Terns RM, Terns MP (2007) Non-coding RNAs: lessons from the small nuclear and small nucleolar RNAs. Nat Rev Mol Cell Biol 8(3):209–220
217. Jankowska A et al (2008) Reduction of human chorionic gonadotropin beta subunit expression by modified U1 snRNA caused apoptosis in cervical cancer cells. Mol Cancer 7:26
218. Gridasova A, Henry R (2005) The p53 tumor suppressor protein represses human snRNA gene transcription by RNA polymerases II and III independently of sequence-specific DNA binding. Mol Cell Biol 25(8):3247–3260
219. Christov CP et al (2006) Functional requirement of noncoding Y RNAs for human chromosomal DNA replication. Mol Cell Biol 26(18):6993–7004

220. Chen X et al (2013) An RNA degradation machine sculpted by Ro autoantigen and noncoding RNA. Cell 153(1):166–177
221. Christov CP, Trivier E, Krude T (2008) Noncoding human Y RNAs are overexpressed in tumours and required for cell proliferation. Br J Cancer 98(5):981–988
222. Kickhoefer VA et al (2003) Identification of conserved vault RNA expression elements and a non-expressed mouse vault RNA gene. Gene 309(2):65–70
223. Van Zon A et al (2003) The vault complex. Cell Mol Life Sci 60(9):1828–1837
224. Gopinath SCB, Wadhwa R, Kumar PKR (2010) Expression of noncoding vault RNA in human malignant cells and its importance in mitoxantrone resistance. Mol Cancer Res 8(11):1536–1546
225. Kapranov P et al (2007) RNA maps reveal new RNA classes and a possible function for pervasive transcription. Science 316(June):1484–1488
226. Taft RJ et al (2009) Tiny RNAs associated with transcription start sites in animals. Nat Genet 41(5):572–578
227. Preker P, Nielsen J, Kammler S (2008) RNA exosome depletion reveals transcription upstream of active human promoters. Science 322:1851–1854
228. Seila AC et al (2008) Divergent transcription from active promoters. Science 322(5909):1849–1851
229. Taft RJ et al (2010) Nuclear-localized tiny RNAs are associated with transcription initiation and splice sites in metazoans. Nat Struct Mol Biol 17(8):1030–1034
230. Ponjavic J, Ponting CP, Lunter G (2007) Functionality or transcriptional noise? Evidence for selection within long noncoding RNAs. Genome Res 17(5):556–565
231. Cabili MN et al (2011) Integrative annotation of human large intergenic noncoding RNAs reveals global properties and specific subclasses. Genes Dev 25(18):1915–1927
232. Tano K et al (2010) MALAT-1 enhances cell motility of lung adenocarcinoma cells by influencing the expression of motility-related genes. FEBS Lett 584(22):4575–4580
233. Wang J et al (2010) CREB up-regulates long non-coding RNA, HULC expression through interaction with microRNA-372 in liver cancer. Nucleic Acids Res 38(16):5366–5383
234. Gupta RA et al (2010) Long non-coding RNA HOTAIR reprograms chromatin state to promote cancer metastasis. Nature 464(7291):1071–1076
235. Ota A et al (2004) Identification and characterization of a novel gene, C13orf25, as a target for 13q31-q32 amplification in malignant lymphoma. Cancer Res 64:3087–3095
236. Prensner JR et al (2011) Transcriptome sequencing across a prostate cancer cohort identifies PCAT-1, an unannotated lincRNA implicated in disease progression. Nat Biotechnol 29(8): 742–749
237. Nguyen VT et al (2001) 7SK small nuclear RNA binds to and inhibits the activity of CDK9/ cyclin T complexes. Nature 414(6861):322–325
238. Zhao Y, Guo Q, Chen J, Hu J, Shuwei Wang YS (2014) Role of long non-coding RNA HULC in cell proliferation, apoptosis and tumor metastasis of gastric cancer: a clinical and in vitro investigation. Oncol Rep 31(1):358–364
239. Kotake Y et al (2011) Long non-coding RNA ANRIL is required for the PRC2 recruitment to and silencing of p15(INK4B) tumor suppressor gene. Oncogene 30(16):1956–1962
240. Carpenter S et al (2013) A long noncoding RNA mediates both activation and repression of immune response genes. Science 341(6147):789–792
241. Feng J et al (1995) The RNA component of human telomerase. Science 269(5228): 1236–1241
242. Redon S, Reichenbach P, Lingner J (2010) The non-coding RNA TERRA is a natural ligand and direct inhibitor of human telomerase. Nucleic Acids Res 38(17):5797–5806
243. Rapicavoli NA et al (2013) A mammalian pseudogene lncRNA at the interface of inflammation and anti-inflammatory therapeutics. eLife 2:e00762
244. Rayet B, Gélinas C (1999) Aberrant rel/nfkb genes and activity in human cancer. Oncogene 18(49):6938–6947
245. Ji P et al (2003) MALAT-1, a novel noncoding RNA, and thymosin beta4 predict metastasis and survival in early-stage non-small cell lung cancer. Oncogene 22(39):8031–8041

246. Guo F et al (2010) Inhibition of metastasis-associated lung adenocarcinoma transcript 1 in CaSki human cervical cancer cells suppresses cell proliferation and invasion. Acta Biochim Biophys Sin 42(3):224–229
247. Emadi-Andani E et al (2014) Association of HOTAIR expression in gastric carcinoma with invasion and distant metastasis. Adv Biomed Res 3:135
248. Yuan J et al (2014) A long noncoding RNA activated by TGF-β promotes the invasion-metastasis cascade in hepatocellular carcinoma. Cancer Cell 25(5):666–681
249. Yuan S-X et al (2012) Long noncoding RNA associated with microvascular invasion in hepatocellular carcinoma promotes angiogenesis and serves as a predictor for hepatocellular carcinoma patients' poor recurrence-free survival after hepatectomy. Hepatology 56(6):2231–2241
250. Rossignol F, Vaché C, Clottes E (2002) Natural antisense transcripts of hypoxia-inducible factor 1alpha are detected in different normal and tumour human tissues. Gene 299(1–2): 135–140
251. Prensner JR et al (2014) PCAT-1, a long noncoding RNA, regulates BRCA2 and controls homologous recombination in cancer. Cancer Res 74(6):1651–1660
252. Petrovics G et al (2004) Elevated expression of PCGEM1, a prostate-specific gene with cell growth-promoting function, is associated with high-risk prostate cancer patients. Oncogene 23(2):605–611
253. Pickard MR, Mourtada-Maarabouni M, Williams GT (2013) Long non-coding RNA GAS5 regulates apoptosis in prostate cancer cell lines. Biochim Biophys Acta 1832(10):1613–1623
254. DeOcesano-Pereira C et al (2014) Long non-coding RNA INXS is a critical mediator of BCL-XS induced apoptosis. Nucleic Acids Res 42(13):8343–8355
255. Hung T et al (2011) Extensive and coordinated transcription of noncoding RNAs within cell-cycle promoters. Nat Genet 43(7):621–629
256. Kino T et al (2010) Noncoding RNA gas5 is a growth arrest- and starvation-associated repressor of the glucocorticoid receptor. Sci Signal 3(107):ra8
257. Li Z et al (2014) Long non-coding RNA UCA1 promotes glycolysis by upregulating hexokinase 2 through the mTOR-STAT3/microRNA143 pathway. Cancer Sci 105(8):951–955
258. Blank A, Dekker CA (1981) Ribonucleases of human serum, urine, cerebrospinal fluid, and leukocytes. Activity staining following electrophoresis in sodium dodecyl sulfate-polyacrylamide gels. Biochemistry 20(8):2261–2267
259. Häusler SFM et al (2010) Whole blood-derived miRNA profiles as potential new tools for ovarian cancer screening. Br J Cancer 103(5):693–700
260. Mitchell PS et al (2008) Circulating microRNAs as stable blood-based markers for cancer detection. Proc Natl Acad Sci U S A 105(30):10513–10518
261. Chen X et al (2008) Characterization of microRNAs in serum: a novel class of biomarkers for diagnosis of cancer and other diseases. Cell Res 18(10):997–1006
262. Arroyo JD et al (2011) Argonaute2 complexes carry a population of circulating microRNAs independent of vesicles in human plasma. Proc Natl Acad Sci U S A 108(12):5003–5008
263. Vickers KC et al (2011) MicroRNAs are transported in plasma and delivered to recipient cells by high-density lipoproteins. Nat Cell Biol 13(4):423–433
264. Valadi H et al (2007) Exosome-mediated transfer of mRNAs and microRNAs is a novel mechanism of genetic exchange between cells. Nat Cell Biol 9(6):654–659
265. Weber JA et al (2010) The microRNA spectrum in 12 body fluids. Clin Chem 56(11): 1733–1741
266. Xie Y et al (2010) Altered miRNA expression in sputum for diagnosis of non-small cell lung cancer. Lung Cancer 67(2):170–176
267. Link A et al (2010) Fecal MicroRNAs as novel biomarkers for colon cancer screening. Cancer Epidemiol Biomarkers Prev 19(7):1766–1774
268. Van Gils MPMQ et al (2007) The time-resolved fluorescence-based PCA3 test on urinary sediments after digital rectal examination; a Dutch multicenter validation of the diagnostic performance. Clin Cancer Res 13(3):939–943
269. Welch JS et al (2011) Use of whole-genome sequencing to diagnose a cryptic fusion oncogene. JAMA 305(15):1577–1584

270. Jones SJ et al (2010) Evolution of an adenocarcinoma in response to selection by targeted kinase inhibitors. Genome Biol 11(8):R82
271. Hudson TJ et al (2010) International network of cancer genome projects. Nature 464(7291):993–998
272. Balch CM et al (2009) Final version of 2009 AJCC melanoma staging and classification. J Clin Oncol 27(36):6199–6206
273. Berry DA et al (2005) Effect of screening and adjuvant therapy on mortality from breast cancer. N Engl J Med 353(17):1784–1792
274. Naucler P et al (2007) Human papillomavirus and Papanicolaou tests to screen for cervical cancer. N Engl J Med 357(16):1589–1597
275. Aberle D, Adams A, Berg C (2011) Reduced lung-cancer mortality with low-dose computed tomographic screening. N Engl J Med 365(5):395–409
276. Mandel JJS et al (1993) Reducing mortality from colorectal cancer by screening for fecal occult blood. N Engl J Med 328(19):1365–1371
277. Nanda K et al (2000) Accuracy of the Papanicolaou Test in screening for and follow-up of cervical cytologic abnormalities: a systematic review. Ann Intern Med 132(10):810–819
278. Jacobs I, Bast RC (1989) The CA 125 tumour-associated antigen: a review of the literature. Hum Reprod 4(1):1–12
279. Shitrit D et al (2005) Diagnostic value of CYFRA 21-1, CEA, CA 19-9, CA 15-3, and CA 125 assays in pleural effusions: analysis of 116 cases and review of the literature. Oncologist 10(7):501–507
280. Jemal A et al (2010) Global patterns of cancer incidence and mortality rates and trends. Cancer Epidemiol Biomarkers Prev 19(8):1893–1907
281. Xing L et al (2010) Early detection of squamous cell lung cancer in sputum by a panel of microRNA markers. Mod Pathol 23(8):1157–1164
282. Liu H et al (2012) Genome-wide microRNA profiles identify miR-378 as a serum biomarker for early detection of gastric cancer. Cancer Lett 316(2):196–203
283. Hessels D, Schalken JA (2009) The use of PCA3 in the diagnosis of prostate cancer. Nat Rev Urol 6(5):255–261
284. Hessels D et al (2003) DD3PCA3-based molecular urine analysis for the diagnosis of prostate cancer. Eur Urol 44(1):8–16
285. De Kok JB et al (2002) DD3(PCA3), a very sensitive and specific marker to detect prostate tumors. Cancer Res 62(9):2695–2698
286. Narod S et al (1993) Increasing incidence of breast cancer in family with BRCA1 mutation. Lancet 341(8852):1101–1102
287. Wooster R et al (1994) Localization of a breast cancer susceptibility gene, BRCA2, to chromosome 13q12-13. Science 265(5181):2088–2090
288. Easton D, Ford D, Peto J (1993) Inherited susceptibility to breast cancer. Cancer Surv 18:95–113
289. Moran A et al (2012) Risk of cancer other than breast or ovarian in individuals with BRCA1 and BRCA2 mutations. Fam Cancer 11(2):235–242
290. Lerman C, Shields AE (2004) Genetic testing for cancer susceptibility: the promise and the pitfalls. Nat Rev Cancer 4(3):235–241
291. Moskwa P et al (2011) miR-182-mediated downregulation of BRCA1 impacts DNA repair and sensitivity to PARP inhibitors. Mol Cell 41(2):210–220
292. Aaltonen K et al (2008) Familial breast cancers without mutations in BRCA1 or BRCA2 have low cyclin E and high cyclin D1 in contrast to cancers in BRCA mutation carriers. Clin Cancer Res 14(7):1976–1983
293. Pentheroudakis G, Pavlidis N (2006) Perspectives for targeted therapies in cancer of unknown primary site. Cancer Treat Rev 32(8):637–644
294. Pentheroudakis G, Golfinopoulos V, Pavlidis N (2007) Switching benchmarks in cancer of unknown primary: from autopsy to microarray. Eur J Cancer 43(14):2026–2036
295. Rosenwald S et al (2010) Validation of a microRNA-based qRT-PCR test for accurate identification of tumor tissue origin. Mod Pathol 23(6):814–823

296. Meiri E et al (2012) A second-generation microRNA-based assay for diagnosing tumor tissue origin. Oncologist 17(6):801–812
297. Viale G (2012) The current state of breast cancer classification. Ann Oncol 23 Suppl 10:x207–x210
298. Bishop JA et al (2010) Accurate classification of non-small cell lung carcinoma using a novel microRNA-based approach. Clin Cancer Res 16(2):610–619
299. Gilad S et al (2012) Classification of the four main types of lung cancer using a microRNA-based diagnostic assay. J Mol Diagn 14(5):510–517
300. Spector Y et al (2013) Development and validation of a microRNA-based diagnostic assay for classification of renal cell carcinomas. Mol Oncol 7(3):732–738
301. Benjamin H et al (2010) A diagnostic assay based on microRNA expression accurately identifies malignant pleural mesothelioma. J Mol Diagn 12(6):771–779
302. Rosenfeld N et al (2008) MicroRNAs accurately identify cancer tissue origin. Nat Biotechnol 26(4):462–469
303. Cancer T, Atlas G (2012) Comprehensive genomic characterization of squamous cell lung cancers. Nature 489(7417):519–525
304. Cancer T, Atlas G (2011) Integrated genomic analyses of ovarian carcinoma. Nature 474(7353):609–615
305. Perou CM et al (2000) Molecular portraits of human breast tumours. Nature 406(6797):747–752
306. Hudis CA (2007) Trastuzumab--mechanism of action and use in clinical practice. N Engl J Med 357(1):39–51
307. The Clinical Lung Cancer Genome Project & Network Genomic Medicine (2013) A genomics-based classification of human lung tumors. Sci Transl Med 5(209):209ra153
308. Fecher LA et al (2007) Toward a molecular classification of melanoma. J Clin Oncol 25(12):1606–1620
309. Verhaak RGW et al (2010) Integrated genomic analysis identifies clinically relevant subtypes of glioblastoma characterized by abnormalities in PDGFRA, IDH1, EGFR, and NF1. Cancer Cell 17(1):98–110
310. Blenkiron C et al (2007) MicroRNA expression profiling of human breast cancer identifies new markers of tumor subtype. Genome Biol 8(10):R214
311. Volinia S et al (2012) Breast cancer signatures for invasiveness and prognosis defined by deep sequencing of microRNA. Proc Natl Acad Sci U S A 109(8):3024–3029
312. Lowery AJ et al (2009) MicroRNA signatures predict oestrogen receptor, progesterone receptor and HER2/neu receptor status in breast cancer. Breast Cancer Res 11(3):R27
313. Cascione L et al (2013) Integrated microRNA and mRNA signatures associated with survival in triple negative breast cancer. PLoS One 8(2):e55910
314. Spahn M et al (2010) Expression of microRNA-221 is progressively reduced in aggressive prostate cancer and metastasis and predicts clinical recurrence. Int J Cancer 127(2):394–403
315. Patnaik SK et al (2010) Evaluation of microRNA expression profiles that may predict recurrence of localized stage I non-small cell lung cancer after surgical resection. Cancer Res 70(1):36–45
316. Gupta GP, Massagué J (2006) Cancer metastasis: building a framework. Cell 127(4): 679–695
317. Leary RJ et al (2010) Development of personalized tumor biomarkers using massively parallel sequencing. Sci Transl Med 2(20):20ra14
318. McBride DJ et al (2010) Use of cancer-specific genomic rearrangements to quantify disease burden in plasma from patients with solid tumors. Genes Chromosomes Cancer 49(11):1062–1069
319. Hauptman N, Glavac D (2013) MicroRNAs and long non-coding RNAs: prospects in diagnostics and therapy of cancer. Radiol Oncol 47(4):311–318
320. Ling H, Fabbri M, Calin GA (2013) MicroRNAs and other non-coding RNAs as targets for anticancer drug development. Nat Rev Drug Discov 12(11):847–865

321. Nana-Sinkam SP, Croce CM (2013) Clinical applications for microRNAs in cancer. Clin Pharmacol Ther 93(1):98–104
322. Begley CG, Ellis LM (2012) Raise standards for preclinical cancer research. Nature 483:531–533
323. Prinz F, Schlange T, Asadullah K (2011) Believe it or not: how much can we rely on published data on potential drug targets? Nat Rev Drug Discov 10(9):712
324. Bustin SA et al (2009) The MIQE guidelines: minimum information for publication of quantitative real-time PCR experiments. Clin Chem 55(4):611–622
325. Mortazavi A et al (2008) Mapping and quantifying mammalian transcriptomes by RNA-Seq. Nat Methods 5(7):621–628
326. Su Z et al (2014) A comprehensive assessment of RNA-seq accuracy, reproducibility and information content by the Sequencing Quality Control Consortium. Nat Biotechnol 32(9)
327. Baechler EC et al (2004) Expression levels for many genes in human peripheral blood cells are highly sensitive to ex vivo incubation. Genes Immun 5(5):347–353
328. Waddington C (1942) Canalization of development and the inheritance of acquired characters. Nature 150(3811):563–565
329. Gygi SP et al (1999) Correlation between protein and mRNA abundance in yeast correlation between protein and mRNA abundance in yeast. Mol Cell Biol 19(3):1720–1730
330. Hornstein E, Shomron N (2006) Canalization of development by microRNAs. Nat Genet 38(Suppl):S20–S24
331. Li X et al (2009) A microRNA imparts robustness against environmental fluctuation during development. Cell 137(2):273–282
332. Kim Y-K et al (2012) Short structured RNAs with low GC content are selectively lost during extraction from a small number of cells. Mol Cell 46(6):893–895
333. Bossuyt PM et al (2004) Towards complete and accurate reporting of studies of diagnostic accuracy: the STARD initiative. Fam Pract 21(1):4–10
334. Ochodo EA, Bossuyt PM (2013) Reporting the accuracy of diagnostic tests: the STARD initiative 10 years on. Clin Chem 59(6):917–919
335. McShane LM et al (2005) REporting recommendations for tumor MARKer prognostic studies (REMARK). Nat Clin Pract Oncol 2(8):416–422

Proteomics Methods

Keith Ashman, Greg Rice, and Murray Mitchell

Abbreviations

2DPAGE	Two dimensional polyacrylamide gel electrophoresis
CSF	Cerebrospinal fluid
CVF	Cervico vaginal fluid
DIGE	Fluorescent difference gel electrophoresis
ELISA	Enzyme linked immunosorbent assay
IHC	Immunohistochemistry
MALDI-MS	Matrix assisted laser desorption ionization mass spectrometry
PCR	Polymerase chain reaction
RIA	Radioimmunoassay
SDS-PAGE	Sodium dodecyl sulphate polyacrylamide gel electrophoresis
WB	Western Blotting

K. Ashman (✉) • G. Rice (✉) • M. Mitchell
University of Queensland Centre for Clinical Research,
Building 71/918, Royal Brisbane & Women's Hospital Campus,
Herston, QLD 4029, Australia
e-mail: k.ashman@uq.edu.au; g.rice@uq.edu.au; murray.mitchell@uq.edu.au

S.R. Lakhani, S.B. Fox (eds.), *Molecular Pathology in Cancer Research*,
DOI 10.1007/978-1-4939-6643-1_10

Introduction

The key to treating patients is early and accurate diagnosis. Molecular pathology provides diagnostic information by measuring the presence and concentration of distinct molecular species (DNA, metabolites, proteins and lipids), so that changes in levels between health and disease can be used to guide diagnosis and therapy. Biological systems are dynamic. They contain many molecular species that interact with each other, resulting in complex series of physico-chemical, spatial and temporal changes. Since proteins carry out the majority of the biochemical reactions in cells, as well as performing many key signalling functions and forming structural elements, it follows that measuring the concentration and state of proteins both in their biochemical and temporal context will yield means of distinguishing health from disease. This is often called biomarker analysis [1, 2]. The term proteome was coined by Wilkins [3] and led to the field of protein biochemistry being dubbed proteomics [4]. The proteome was originally defined as the proteins present in a cell or organism at any one time. The proteome is clearly dynamic, and unlike the genome varies highly over time and biological location (e.g. Plasma vs. Serum vs. Tissue vs. Urine vs. CSF vs. CVF vs. Saliva, etc.). Another important consideration in analysing proteins especially in a higher eukaryote such as man is the critical role played by post-translational modifications in cell signalling [5] which not only includes such changes as phosphorylation, glycosylation, ubiquitination and methylation but also activation through proteolytic cleavage, e.g., the conversion of prothrombin to thrombin or the excision of the c-peptide to yield active insulin. This all implies the proteome is information rich but poses a significant analytical challenge. As there is no protein equivalent of the DNA polymerase chain reaction, PCR, various sample preparation and analytical strategies are used to identify and quantify proteins and peptides. The following pages describe the methods currently in use.

Sample Preparation

Robust and reproducible sample preparation is an essential component in any analytical technique. There are numerous possible sources of samples for proteomics analysis: serum vs. plasma, urine, fresh frozen vs. paraffin embedded tissue, etc. Something that is often underappreciated is the experimental design and statistical interpretation of clinical samples. Experimental design and interpretation are closely linked. In proteomic profiling experiments, a set of reference disease specimens and a balanced number of controls may be sufficient since only a hypothesis of potential biomarker candidates will be built and subsequently more stringent validation will occur in a second verification step. Figure 1 shows a typical assay development workflow using mass spectrometry as the analytical technique. The key point is, as the assay is moved

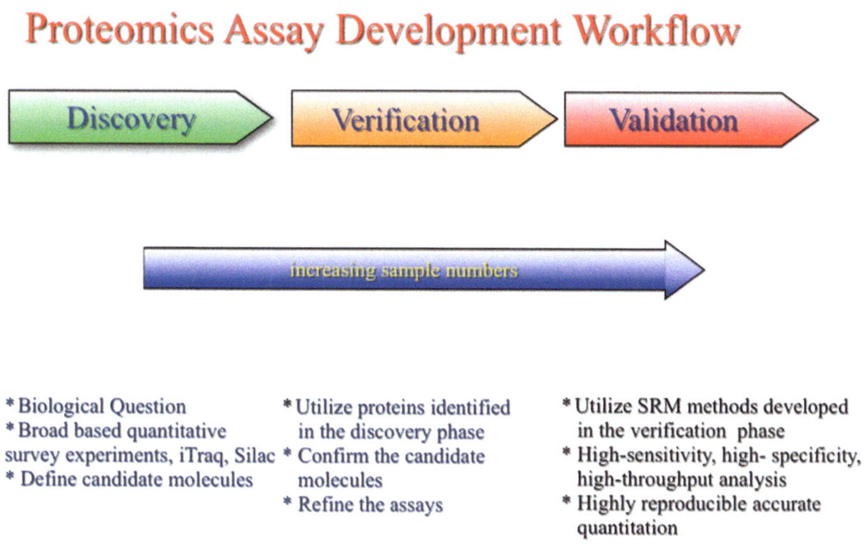

Fig. 1 Mass spectrometry-based assay development

towards clinical use the number of samples must be increased to ensure both the robustness of the method and assessing individual variation.

To make meaningful clinical conclusions from follow-up verification studies, it is essential to design the experimental workflow around a suitably sized controlled collection of clinical specimens [6, 7]. This starts with the selection of a sufficient number of representative patients and suitable case controls to achieve statistical significance in terms of differential biomarker abundance. Without proper design, the collected information might well be useless to link a certain molecular pattern to a given disease state. Follow-up reports should provide statistical statements about the significance of each finding at the individual patient level.

The most frequently analysed clinical sample is blood, since its collection is rapid and non-invasive. From the proteomics standpoint, however, there are a number of issues to consider when analysing the protein content of blood or more specifically serum and plasma [8]. The dynamic range of concentration of individual proteins in blood covers 10 orders of magnitude, with albumin at 34–54 mg/mL and cytokines such as IL1-beta 0.16 ± 0.17 pg/mL (http://www.copewithcytokines.de) [9, 10] This poses significant analytical problems since the linear dynamic range of the current analytical methods is only 4–5 orders of magnitude. Hence, to measure the low concentration analytes either an enrichment step, e.g. an antibody and or a signal amplification, e.g. ELISA is required. The collection protocol can have a significant affect on the analysis, the timing, method of sampling and even the type of tube can affect the result [11].

Table 1 Methods of protein analysis

Protein property		Method	
Size	Size exclusion chromatography	SDS PAGE	Mass spectrometry
Charge	Ion exchange chromatography	Isoelectric focussing	Capillary electrophoresis
Hydrophobicity	Reversed phase HPLC	Hydrophobic interaction	Hydrophilic interaction
Affinity	Antibodies, lectins	Binding proteins, e.g. protein A	Metal chelation
Activity	Enzyme substrate conversion	Reaction inhibition	
Structure	Mass spectrometry	Nuclear magnetic resonance	X-Ray crystallography

Proteome Analysis Methods

Two Dimensional Polyacrylamide Gel Electrophoresis (2DPAGE)

2D-PAGE is a form of polyacrylamide gel electrophoresis in which proteins are separated in two dimensions oriented at right angles to each other [12–14]. It makes use of two distinct physical properties of proteins, charge and mass (see Table 1). The first dimension, isoelectric focusing, separates proteins on the basis of their net charge while the second dimension, SDS-PAGE, further separates the proteins according to their mass. Small changes in charge and mass can easily be detected by this method, because it is rare that two different proteins will resolve to the same place in both dimensions. 2D-PAGE in combination with mass spectrometry to identify proteins was used for the first large scale comparative proteomics experiments [15, 16]. The introduction of fluorescent difference gel electrophoresis (DIGE) technology [17] and the use of an internal standard [18] made it possible to quantify differentially expressed proteins in a series of samples. Software has been developed to image the gels and quantify the proteins based on the strength of the fluorescent signal or if another staining technique is used (e.g. Coomassie blue, Sypro Ruby, silver), the intensity of the staining [19]. The protein(s) in the spots are identified by excision and in gel digestion with Trypsin then analysing the resulting peptides by mass spectrometry [20, 21].

Immuno-Blotting or Western Blotting

The technique of protein blotting or Western blotting (WB) [22, 23] separates proteins by polyacrylamide gel electrophoresis, either on one dimensional sodium dodecyl sulphate polyacrylamide gels (SDS-PAGE) or 2D-PAGE, then subsequently transfers them to an adsorbent membrane support under the influence of an electric current. After transfer the membrane is blocked, with a detergent or protein solution to prevent further proteins binding non-specifically to the membrane. The membrane is then incubated with an antibody against a specific target protein. A second incubation with a secondary reporter antibody, coupled to an enzyme that generates a detectable signal, is then used to determine if the primary antibody has bound to its target. The developed membrane can then be analysed by densitometry to obtain semi-quantitative data. It is important to control for sample loading between samples by measuring in the same way the levels of a protein common to all samples, e.g., actin or tubulin [24].

Radioimmunoassay

RIA was the first type of immunoassay to be developed [25] for the measurement of insulin in serum. Immunoassays now play a prominent role in the clinical laboratory analysis for analytes such as proteins, hormones, drugs and nucleic acids [26]. To perform a radioimmunoassay, a known quantity of an antigen is made radioactive, frequently by labelling it with gamma-radioactive isotopes of iodine attached to tyrosine. The radiolabeled antigen is then mixed with a known amount of antibody for that antigen, and as a result forms a labelled complex. Then, a sample of serum from a patient containing an unknown quantity of that same antigen is added. This causes the unlabelled (or 'cold') antigen from the serum to compete with the radiolabeled antigen ('hot') for antibody binding sites. As the concentration of 'cold' antigen is increased, more of it binds to the antibody, displacing the radiolabeled variant, and reducing the ratio of antibody-bound radiolabeled antigen to free radiolabeled antigen. The bound antigens are then separated from the unbound ones, and the radioactivity of the free antigen remaining in the supernatant is measured using a gamma counter. Using known standards, a binding curve can then be generated which allows the amount of antigen in the patient's serum to be derived.

Enzyme Linked Immunosorbent Assay

ELISAs are 96 or 384 well polystyrene plate-based immunoassays designed for detecting and quantifying substances such as peptides, proteins, antibodies and hormones. In its simplest form an antigen is immobilized on a solid surface and then complexed with an antibody that is linked to an enzyme. Detection is accomplished by assessing the conjugated enzyme activity via incubation with a substrate to produce a detectable product. A second common type of ELISA, referred to as a sandwich assay (see Fig. 2), first coats the plate well with a primary antibody, which can capture an antigen from solution. Non-bound material is washed away and a second antibody, which recognizes another epitope on the captured antigen and has been coupled to an enzyme is added. Excess antibody is washed away before adding

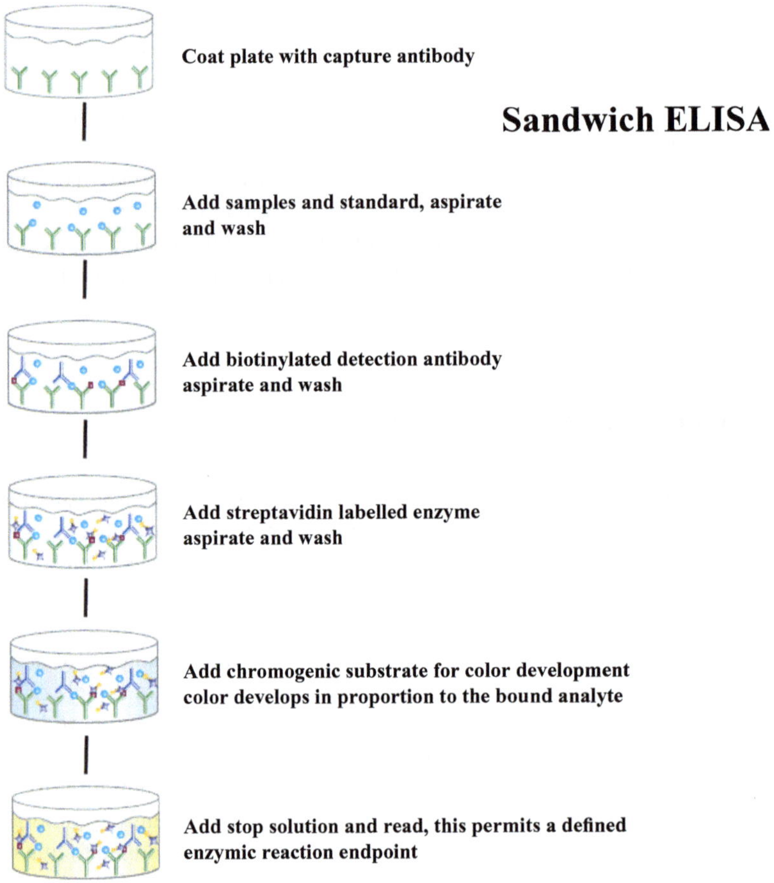

Fig. 2 An enzyme linked immunoabsorbent assay workflow

substrate to produce a detectable product. The advantage of this form of ELISA is that it contains both a capture and signal amplification step, but it does require two antibodies that recognize two different non-competing epitopes on the analyte of interest.

Array Technologies

Protein Microarrays

A protein microarray, also known as a protein chip, is a solid surface (typically glass) on which thousands of different proteins (e.g. antigens, antibodies, enzymes, substrates, etc.) are immobilized in discrete spatial locations, forming a high-density protein dot matrix. Depending on their application, protein microarrays can be classified into two types: analytical and functional protein microarrays. Analytical protein microarrays are usually composed of well-characterized biomolecules with specific binding activities, such as antibodies, to analyse the components of complex biological samples (e.g. serum and cell lysates) or to determine whether a sample contains a specific protein of interest. They have been used for protein expression profiling, biomarker identification, cell surface marker/glycosylation profiling, clinical diagnosis and environmental/food safety analysis. Functional protein microarrays are constructed by printing a large number of individually purified proteins, and are mainly used to comprehensively query the biochemical properties and activities of the immobilized proteins. In principle, it is feasible to print arrays comprised of virtually all the annotated proteins of a given organism, effectively comprising a whole proteome microarray [27–29]. Another type of polypeptide array is based on synthetic peptides that are synthesized in situ on a cellulose membrane. The peptides can then be used to probe for specific interactions [30, 31].

Antibody Arrays

This type of array can be considered a special type of protein array. Since antibodies share the same general structure they offer some advantages in terms of defined support coupling chemistries. Depending on whether the antibodies are immobilized on a planar or spherical surface, antibody arrays have been classified into planar and suspension/bead formats, respectively [32]. Planar antibody arrays represent a versatile platform, with many potential clinical applications. The main planar label-based formats comprise one-antibody and sandwich assays. Multiplex protein suspension arrays have a number of advantages over current analyte quantification technologies, including measurement of many biomarkers (theoretically, up to 100 different analytes) in a single sample; wider operational dynamic range and increased sensitivity and specificity derived from multivariate modelling of combinations of biomarker analytes. This system utilizes a sandwich ELISA-like protocol, in which capture antibodies are coupled to spectrally distinct beads. Biotinylated sandwich antibody and streptavidin-phycoerythrin fluorophores are used as a reporter complex. Bead identity and analyte-specific fluorescence are assessed by passing the beads through a flow cytometer [33].

Both these formats may be described as micro-ELISA assays offering the important advantages of multiplexing and requiring smaller amounts of sample.

Immunohistochemistry

Immunohistochemistry (IHC) uses anatomical, immunological and biochemical techniques to identify discrete tissue components by the interaction of target antigens with specific antibodies tagged with a visible label. IHC makes it possible to visualize the distribution of specific cellular components within cells and in their tissue spatial context. The intensity of the labelling provides some quantitative

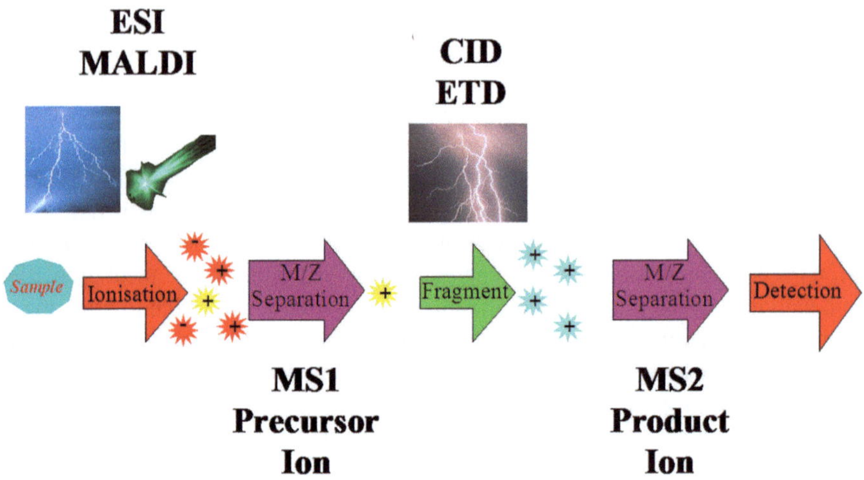

Fig. 3 A generic mass spectrometer. *ESI* electrospray ionization, *MALDI* matrix assisted laser desorption ionization

information. While there are multiple approaches and permutations in IHC methodology, all of the steps involved are separated into two groups: sample preparation and labelling. The principle of IHC has been known since the 1930s, but it was not until 1942 that the first IHC study was reported [34]. The study used FITC-labelled antibodies to identify *Pneumococcal* antigens in infected tissue. Since then, improvements have been made in protein conjugation, tissue fixation methods, detection labels and microscopy, making immunohistochemistry a routine and essential tool in diagnostic and research laboratories. A detailed discussion is beyond the scope of this article but the reader is referred to [35–37].

Mass Spectrometry

Mass spectrometry is already an important clinical tool, widely used for the measurement of drug metabolites, steroids, for the detection on inborn metabolic disease in newborn infants [38]. Recently, it has been applied to bacterial typing and identification [39]. Advances in speed, resolution and sensitivity as well as the ability to perform multiplexed assays will likely see this technology find even wider application in clinical diagnosis [40]. Mass spectrometric measurements are carried out in the gas phase on ionized analytes. The development of the soft ionization techniques electrospray ionization (ESI) [41–43] and Matrix Assisted Laser Desorption Ionization (MALDI) [43, 44] that allow proteins and peptides to be readily transferred to the gas phase has transformed mass spectrometry into the mainstay of proteome analysis. A mass spectrometer consists of an ion source, to transfer the analytes into the gas phase, a mass analyzer that measures the mass-to-charge ratio (*m*/*z*) of the ionized analytes and a detector that registers the number of ions at each *m*/*z* value (Fig. 3). ESI ionizes the analytes out of a solution and is

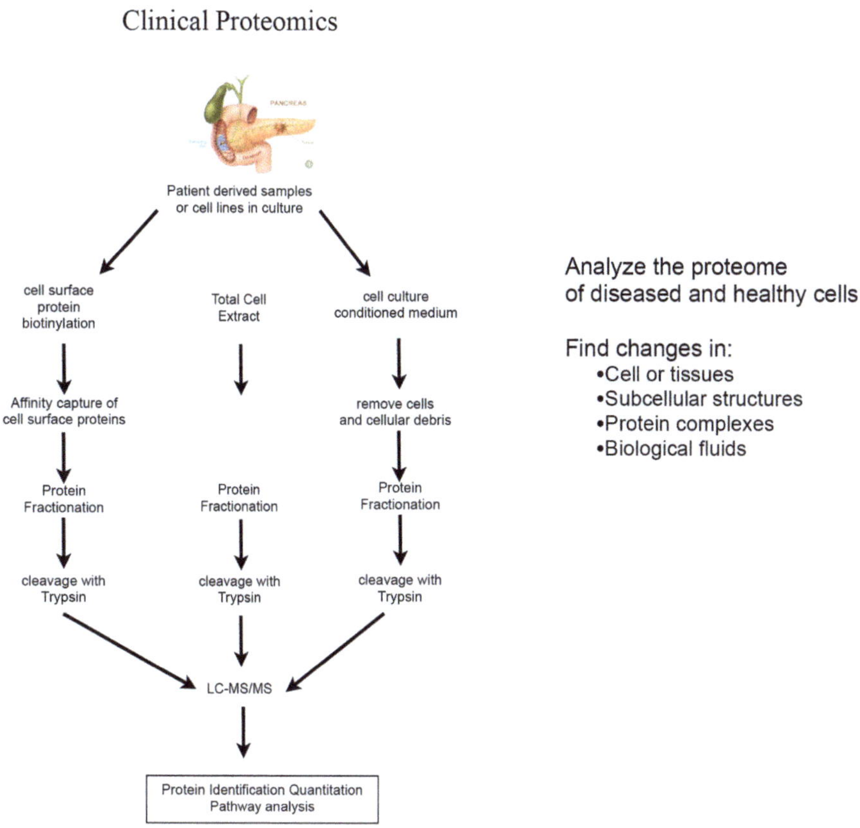

Fig. 4 Mass spectrometry in clinical proteomics

therefore readily coupled to liquid-based (for example, chromatographic and electrophoretic) separation tools. MALDI sublimates and ionizes the samples out of a dry, crystalline matrix via laser pulses. MALDI-MS is normally used to analyse relatively simple peptide mixtures, whereas integrated liquid-chromatography ESI-MS systems (LC-MS) are preferred for the analysis of complex samples.

Mass spectrometry-based proteomics strategies (Fig. 4) can be divided into targeted and non-targeted (shotgun) approaches. The non-targeted approach aims to identify as many protein species as possible in a biological matrix such as a tissue, cell lysate or plasma [45]. It is generally a combination of separation methods, see Table 1, designed to separate proteins and peptides to the maximum achievable degree. The holy grail is to be able to characterize entire proteomes completely, which is currently not attainable. It is also important to remember that the proteome is highly dynamic and especially in higher eukaryotes such as humans post-translational modification of a protein(s) can have a profound effect on the function. It is therefore necessary not only to detect the presence of a protein but also to know its quantity and molecular state. The non-targeted strategy is applied at the discovery stage in the biomarker development process (Fig. 1). Stable isotope labelling is being applied extensively in this scenario as it allows the quantitative comparison of large sets of proteins (iTraq [46] TMT [47] SILAC [48–51]) (Fig. 4).

Targeted proteomics in contrast aims to obtain quantitative information from defined sets of proteins that have been identified as important players in biological process either in a non-targeted proteome analysis or by such methods as gene expression analysis on DNA or RNA microarrays. The field of quantitative proteomics is enthusiastically embracing selected reaction monitoring (SRM) MS methods for targeted quantitative proteomic analysis [52]. The alternate term multiple reaction monitoring (MRM) will soon be superseded by IUPAC. Since SRM makes use of the analysis of a selected product ion (termed Q3) from a specific precursor ion (termed Q1), this provides excellent selectivity, enabling high S/N and very sensitive detection of the analyte from 'noisy' backgrounds. These features are well suited for proteomic applications allowing for the detection of target peptides in highly complex biological mixtures (i.e. proteolytically cleaved entire cell lysates, human plasma, etc.). The underlying principle of SRM in proteomic applications is that the selected set of precursor and product ions contain sufficient information to represent the target peptide of interest, and thereby its protein of origin. Libraries of SRM-based data are being actively established, e.g., SRM protein atlas (http://www.srmatlas.org/ [53]). Quantitation is achieved by including in the assay a stable heavy isotope labelled version of the target peptide(s) as an internal standard(s) and applying the method of isotope dilution [54]. A variation on the general principal of isotope dilution termed SISCAPA [55, 56] uses anti-peptide antibodies to capture peptides of interest, primarily from serum, again with a spiked in synthetic isotopically labelled heavy internal standard peptide to improve the selectivity and sensitivity of the method [57, 58].

Phosphoproteomics

Phosphorylation of tyrosine, serine and threonine amino acid residues is a reversible and dynamic protein post-translational modification that plays important roles in the regulation of the cellular signalling pathways, which control many biological processes, including cell growth, differentiation, invasion, metastasis and apoptosis. Abnormal protein phosphorylation is known to cause or be a consequence of many diseases, including cancer [5, 59, 60]. The deregulation of protein kinase activity with its resulting change in protein phosphorylation states has been implicated in the onset of tumour formation and cancer progression [61]. The phospho-status of a protein is dynamically controlled by protein kinases and counteracting phosphatases. Therefore, monitoring of kinase and phosphatase activities, identification of specific phosphorylation sites and assessment of their functional significance are of crucial importance to understand development and homeostasis of diseases such as cancer, where the cell proliferative control mechanisms are severely compromised. Recent advances especially in the area in mass spectrometry-based phosphoproteomics have opened new possibilities to reach an unprecedented depth and a proteome-wide understanding of phosphorylation processes [62, 63].

Tissue Imaging Mass Spectrometry

Analysis of clinical tissues using matrix assisted laser desorption ionization (MALDI) imaging mass spectrometry (IMS) is a powerful way to assess spatial expression of molecules linked with histopathology and associated clinical information [64]. MALDI-IMS can be utilized as a biomarker discovery tool as it facilitates a pathology-directed, unbiased approach to identifying the cellular origins and relative concentrations of biomarker candidates across an entire tissue section. The MALDI technique is well suited to this application since it permits the ionization of diverse biomolecules, including peptides, proteins, oligonucleotides, sugars and lipids [65]. Once a biomarker of interest has been identified, properties related to its localization, abundance, regulation and function can be assessed across multiple tissues to better understand disease progression at the molecular level.

Mass Cytometry (CyTOF) Single Cell Analysis

The recent development of 'mass cytometer' technology [66–68] offers an exciting new approach to single cell analysis. It is premised on the use of elements, or stable isotopes, as tags instead of fluorophores, with measurement of the tags using an Inductively Coupled Plasma Mass Spectrometer (ICP-MS). The advantage lies in the large number of available elements and stable isotopes (potentially greater than 100), the high resolution of the mass spectrometer between detection channels, and the large dynamic range (linearity) of detection of the ICP-MS. These benefits, and others, result in the ability to perform multi-parameter assays of high order (up to 100) in single cells without the need for mathematical correction of overlap, and with large dynamic range both for a given target biomarker and between different biomarkers (Fig. 5).

Bacterial Identification by Mass Spectrometry

The general workflow of microorganism profiling by Matrix Assisted Laser Desorption Mass Spectrometry (MALDI) is straightforward (Fig. 6). Starting from a single colony or other biological material sample deposition followed by the addition of MALDI matrix is performed within a few minutes. After sample drying and loading into the instrument spectra are rapidly acquired. The instrument used is a time of flight mass spectrometer operated in linear mode (TOFMS) and since MALDI produces predominantly singly charged ions the mass peaks in a typical spectrum require less interpretation than those generated in an electrospray instrument. The high reproducibility of the methodology is based on the measurement of constantly expressed high-abundant proteins, e.g., ribosomal proteins [69]. The observed mass range of spectra is between 2000 and 20,000 Da. Analysis in this range contains very few measurable metabolites. The data are transferred to dedicated software package for rapid identification and/or

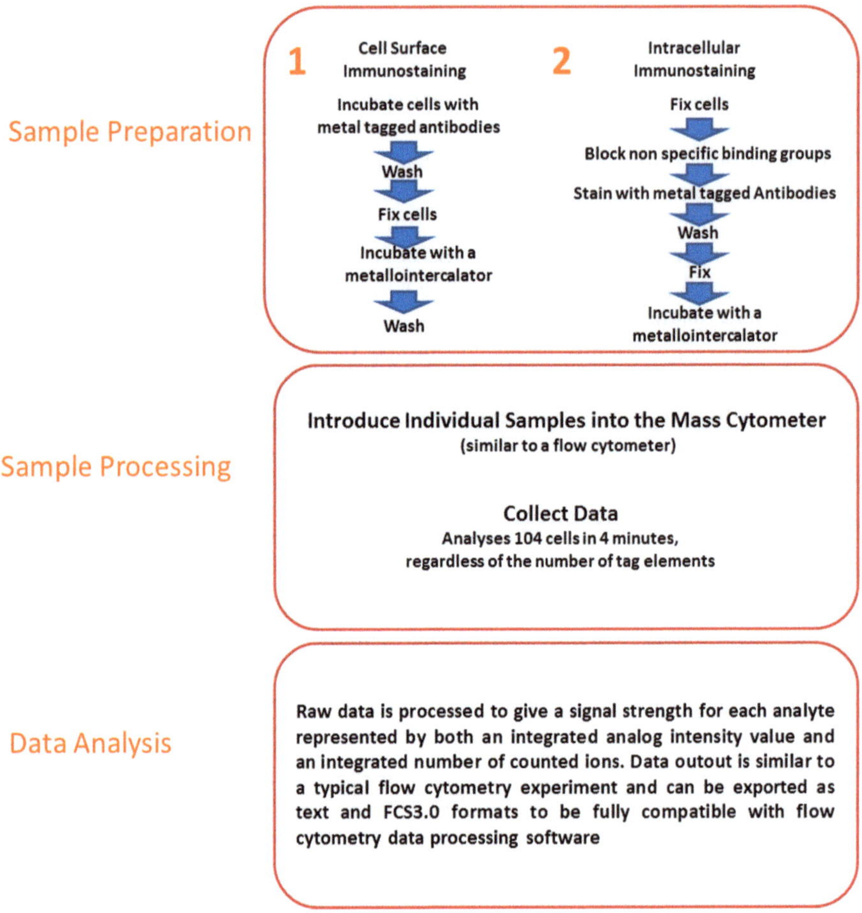

Fig. 5 The CyTof workflow

classification by comparing the set of masses measured with a library of spectra. In contrast to PCR or metabolic pathway analysis, MALDI-TOF MS needs no initial assessment such as gram staining, oxidase test of unknown samples or choice of PCR primers. Each class of sample material is treated in the same manner. The technology is an excellent alternative to classical microbiological identification and classification techniques [39, 70, 71].

Single vs. Multiple Biomarkers

Multiplexed measurement is logical for biological discovery with proteins because they constitutively function within networks, pathways, complexes and families [72, 73]. The consequence of this is that measuring multiple biomarkers will provide

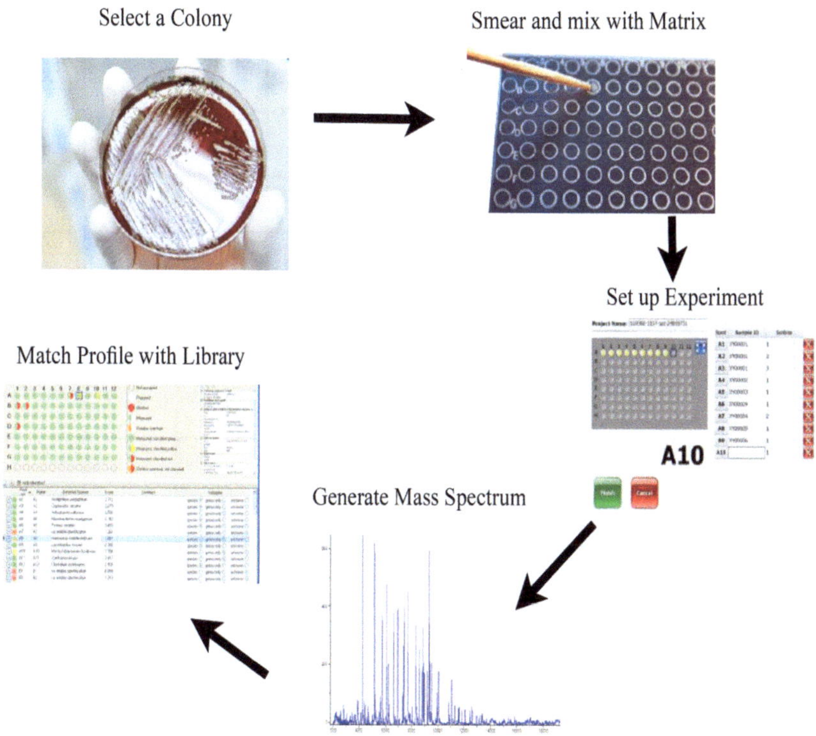

Fig. 6 Bacterial identification by MALDI mass spectrometry

better predictive and therefore diagnostic capabilities. There is a growing consensus that panels of markers may be able to supply the specificity and sensitivity that individual markers lack. For example, a panel combining four known biomarkers (leptin, prolactin, osteopontin, insulin-like growth factor II), none of which used alone could distinguish patients from the controls, achieved a sensitivity and specificity of 95 % for the diagnosis of ovarian cancer [74]. In another example, multiple plasma biomarkers were measured at 11 weeks of gestation in women who experienced normal pregnancy outcomes ($n=14$) and women who developed gestational diabetes ($n=14$). Of the biomarkers considered, receiver operator characteristic curves (ROC) for three biomarkers (adiponectin, insulin and blood glucose) are presented together with an ROC based on the predicted posterior probability values (ppv) generated by a classification model that combined information from all three biomarkers (Fig. 7). The model out performed individual biomarkers based upon the area under the ROC (model = 0.94; adiponectin = 0.867; insulin = 0.872 and glucose = 0.827). This simple example demonstrates the benefit of a multimarker approach for improving diagnostic efficiency [75].

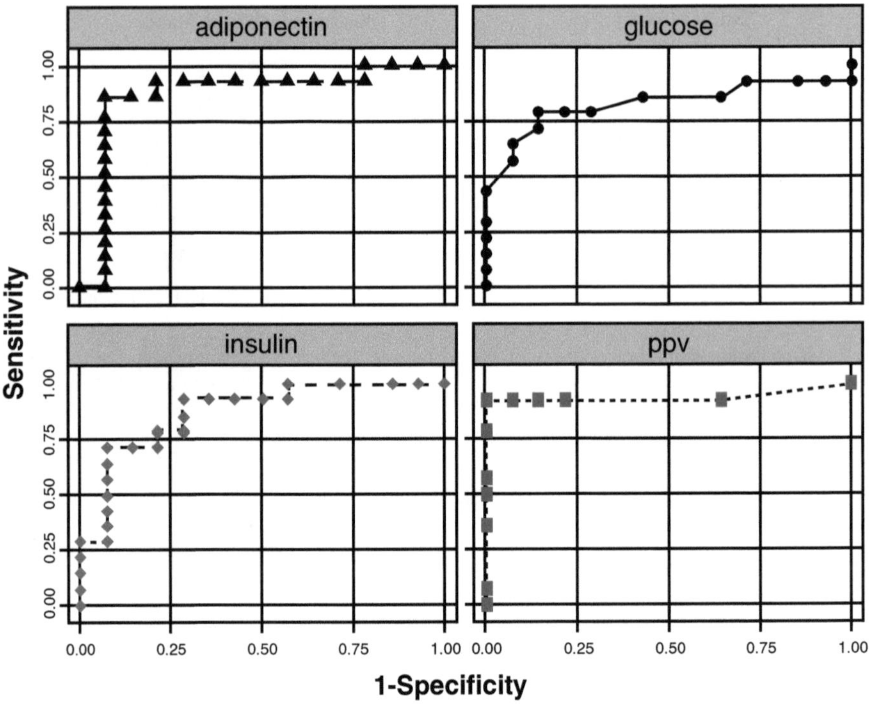

Fig. 7 A comparison of ROC curves of the performance of individual biomarkers (adiponectin, glucose and insulin) and a combined model (ppv) to correctly classify women who subsequently developed gestational diabetes

Conclusion

Can current proteomic technologies deliver clinically useful biomarkers? The opinion of these authors is yes, but it will require more development of the sample preparation, data acquisition and analysis steps. Precise and accurate absolute quantification of proteins represents a challenging task, impaired by multiple potential sources of error. These errors, however, can be minimized to a satisfactory level, if sample preparation, measurement and data analysis are adjusted to the respective sample type under investigation, and if each step of the workflow is conducted thoroughly and reproducibly. In order to demonstrate the clinical utility of any diagnostic test it is necessary to show that:

(a) the analyte can be reliably and consistently measured, i.e. it requires a robust and well-controlled protocol;
(b) the test has a combined sensitivity and specificity that will with high probability segregate disease from health and
(c) the use of the test will improve the clinical outcome of patients by targeting interventions, i.e. well documented follow-up.

To implement appropriate experimental designs, it is important to have in mind the purpose of any assay. Studies can be performed to ascertain their utility for screening (who is sick?), risk assessment (who may get sick?), what is the correct therapy (personalized medicine), assessment of clinical outcome (is the treatment working). Laboratories have to be resourced not only to discover biomarkers but also to verify their clinical utility. Analysing a few samples to compare their differences with no reasonable attempt at verification or validation should no longer be pursued. It is essential that the analytical laboratory scientist and clinical researchers work closely with practicing physicians, so that their combined skills are brought to bear. Finding clinically useful biomarkers is not going to be easy, so we should stop treating it like it is.

Acknowledgements KA acknowledges and thanks the Rotary Club of Williamstown, Victoria, Australia for partial salary support through the RoCan program

References

1. Blonder J, Issaq HJ, Veenstra TD (2011) Proteomic biomarker discovery: it's more than just mass spectrometry. Electrophoresis 32(13):1541–1548. doi:10.1002/elps.201000585
2. Issaq HJ, Waybright TJ, Veenstra TD (2011) Cancer biomarker discovery: opportunities and pitfalls in analytical methods. Electrophoresis 32(9):967–975. doi:10.1002/elps.201000588
3. Wasinger VC, Cordwell SJ, Cerpa-Poljak A, Yan JX, Gooley AA, Wilkins MR, Duncan MW, Harris R, Williams KL, Humphery-Smith I (1995) Progress with gene-product mapping of the Mollicutes: Mycoplasma genitalium. Electrophoresis 16(7):1090–1094
4. Blackstock WP, Weir MP (1999) Proteomics: quantitative and physical mapping of cellular proteins. Trends Biotechnol 17(3):121–127
5. Deribe YL, Pawson T, Dikic I (2010) Post-translational modifications in signal integration. Nat Struct Mol Biol 17(6):666–672. doi:10.1038/nsmb.1842
6. Rifai N, Gillette MA, Carr SA (2006) Protein biomarker discovery and validation: the long and uncertain path to clinical utility. Nat Biotechnol 24(8):971–983. doi:10.1038/nbt1235
7. Whiteley G (2008) Bringing diagnostic technologies to the clinical laboratory: rigor, regulation, and reality. Proteomics Clin Appl 2(10-11):1378–1385. doi:10.1002/prca.200780170
8. Liu X, Valentine SJ, Plasencia MD, Trimpin S, Naylor S, Clemmer DE (2007) Mapping the human plasma proteome by SCX-LC-IMS-MS. J Am Soc Mass Spectrom 18(7):1249–1264. doi:10.1016/j.jasms.2007.04.012
9. Anderson NL, Polanski M, Pieper R, Gatlin T, Tirumalai RS, Conrads TP, Veenstra TD, Adkins JN, Pounds JG, Fagan R, Lobley A (2004) The human plasma proteome: a nonredundant list developed by combination of four separate sources. Mol Cell Proteomics 3(4):311–326. doi:10.1074/mcp.M300127-MCP200
10. Polanski M, Anderson NL (2007) A list of candidate cancer biomarkers for targeted proteomics. Biomark Insights 1:1–48
11. Randall SA, McKay MJ, Molloy MP (2010) Evaluation of blood collection tubes using selected reaction monitoring MS: implications for proteomic biomarker studies. Proteomics 10(10):2050–2056. doi:10.1002/pmic.200900517
12. Gorg A, Drews O, Luck C, Weiland F, Weiss W (2009) 2-DE with IPGs. Electrophoresis 30(Suppl 1):S122–S132. doi:10.1002/elps.200900051
13. Kaltschmidt E, Wittmann HG (1969) Ribosomal proteins: VI. Preparative polyacrylamide del electrophoresis as applied to the isolation of ribosomal proteins. Anal Biochem 30(1): 132–141

14. O'Farrell PH (1975) High resolution two-dimensional electrophoresis of proteins. J Biol Chem 250(10):4007–4021
15. Perrot M, Sagliocco F, Mini T, Monribot C, Schneider U, Shevchenko A, Mann M, Jeno P, Boucherie H (1999) Two-dimensional gel protein database of Saccharomyces cerevisiae (update 1999). Electrophoresis 20(11):2280–2298. doi:10.1002/(SICI)1522-2683(19990801)20:11<2280::AID-ELPS2280>3.0.CO;2-Q
16. Shevchenko A, Jensen ON, Podtelejnikov AV, Sagliocco F, Wilm M, Vorm O, Mortensen P, Boucherie H, Mann M (1996) Linking genome and proteome by mass spectrometry: large-scale identification of yeast proteins from two dimensional gels. Proc Natl Acad Sci U S A 93(25):14440–14445
17. Unlu M, Morgan ME, Minden JS (1997) Difference gel electrophoresis: a single gel method for detecting changes in protein extracts. Electrophoresis 18(11):2071–2077. doi:10.1002/elps.1150181133
18. Alban A, David SO, Bjorkesten L, Andersson C, Sloge E, Lewis S, Currie I (2003) A novel experimental design for comparative two-dimensional gel analysis: two-dimensional difference gel electrophoresis incorporating a pooled internal standard. Proteomics 3(1):36–44. doi:10.1002/pmic.200390006
19. Gauci VJ, Wright EP, Coorssen JR (2011) Quantitative proteomics: assessing the spectrum of in-gel protein detection methods. J Chem Biol 4(1):3–29. doi:10.1007/s12154-010-0043-5
20. Sabido E, Selevsek N, Aebersold R (2011) Mass spectrometry-based proteomics for systems biology. Curr Opin Biotechnol. doi:10.1016/j.copbio.2011.11.014
21. Shevchenko A, Tomas H, Havlis J, Olsen JV, Mann M (2006) In-gel digestion for mass spectrometric characterization of proteins and proteomes. Nat Protoc 1(6):2856–2860. doi:10.1038/Nprot.2006.468
22. LeGendre N (1990) Immobilon-P transfer membrane: applications and utility in protein biochemical analysis. Biotechniques 9(6 Suppl):788–805
23. Towbin H, Staehelin T, Gordon J (1979) Electrophoretic transfer of proteins from polyacrylamide gels to nitrocellulose sheets: procedure and some applications. Proc Natl Acad Sci U S A 76(9):4350–4354
24. Kurien BT, Dorri Y, Dillon S, Dsouza A, Scofield RH (2011) An overview of Western blotting for determining antibody specificities for immunohistochemistry. Methods Mol Biol 717:55–67. doi:10.1007/978-1-61779-024-9_3
25. Yalow RS, Berson SA (1959) Assay of plasma insulin in human subjects by immunological methods. Nature 184(4699):1648–1649
26. Wu AHB (2006) A selected history and future of immunoassay development and applications in clinical chemistry. Clin Chim Acta 369(2):119–124. doi:10.1016/J.Cca.2006.02.045
27. Hu SH, Xie Z, Qian J, Blackshaw S, Zhu H (2011) Functional protein microarray technology. Wires Syst Biol Med 3(3):255–268. doi:10.1002/Wsbm.118
28. Joos T, Bachmann J (2009) Protein microarrays: potentials and limitations. Front Biosci 14:4376–4385
29. Yu X, Schneiderhan-Marra N, Hsu HY, Bachmann J, Joos TO (2009) Protein microarrays: effective tools for the study of inflammatory diseases. Methods Mol Biol 577:199–214. doi:10.1007/978-1-60761-232-2_15
30. Katz C, Levy-Beladev L, Rotem-Bamberger S, Rito T, Rudiger SG, Friedler A (2011) Studying protein-protein interactions using peptide arrays. Chem Soc Rev 40(5):2131–2145. doi:10.1039/c0cs00029a
31. Winkler DF, Hilpert K, Brandt O, Hancock RE (2009) Synthesis of peptide arrays using SPOT-technology and the CelluSpots-method. Methods Mol Biol 570:157–174. doi:10.1007/978-1-60327-394-7_5
32. Sanchez-Carbayo M (2011) Antibody microarrays as tools for biomarker discovery. Methods Mol Biol 785:159–182. doi:10.1007/978-1-61779-286-1_11
33. Rice GE, Georgiou HM, Ahmed N, Shi G, Kruppa G (2006) Translational proteomics: developing a predictive capacity – a review. Placenta 27 Suppl A:S76–S86. doi:10.1016/j.placenta.2005.11.003

34. Coons AH, Creech HJ, Jones RN, Berliner E (1942) The demonstration of pneumococcal antigen in tissues by the use of fluorescent antibody. J Immunol 45(3):159–170
35. Abbondanzo SL (1999) Paraffin immunohistochemistry as an adjunct to hematopathology. Ann Diagn Pathol 3(5):318–327. doi:10.1053/ADPA00300318
36. Shi SR, Shi Y, Taylor CR (2011) Antigen retrieval immunohistochemistry: review and future prospects in research and diagnosis over two decades. J Histochem Cytochem 59(1):13–32. doi:10.1369/jhc.2010.957191
37. Taylor CR (2011) New revised Clinical and Laboratory Standards Institute Guidelines for Immunohistochemistry and Immunocytochemistry. Appl Immunohistochem Mol Morphol 19(4):289–290. doi:10.1097/PAI.0b013e31821b505b
38. Shushan B (2010) A review of clinical diagnostic applications of liquid chromatography-tandem mass spectrometry. Mass Spectrom Rev 29(6):930–44. doi:10.1002/mas.20295
39. Stevenson LG, Drake SK, Murray PR (2010) Rapid identification of bacteria in positive blood culture broths by matrix-assisted laser desorption ionization-time of flight mass spectrometry. J Clin Microbiol 48(2):444–447. doi:10.1128/JCM.01541-09
40. Parker CE, Pearson TW, Anderson NL, Borchers CH (2010) Mass-spectrometry-based clinical proteomics—a review and prospective. Analyst 135(8):1830–1838. doi:10.1039/c0an00105h
41. Fenn JB (2003) Electrospray wings for molecular elephants (Nobel lecture). Angew Chem Int Ed Engl 42(33):3871–3894. doi:10.1002/anie.200300605
42. Fenn JB, Mann M, Meng CK, Wong SF, Whitehouse CM (1989) Electrospray ionization for mass spectrometry of large biomolecules. Science 246(4926):64–71
43. Hillenkamp F, Karas M (1990) Mass spectrometry of peptides and proteins by matrix-assisted ultraviolet laser desorption/ionization. Methods Enzymol 193:280–295
44. Karas M, Hillenkamp F (1988) Laser desorption ionization of proteins with molecular masses exceeding 10,000 daltons. Anal Chem 60(20):2299–2301
45. Jungblut PR, Schluter H (2011) Towards the analysis of protein species: an overview about strategies and methods. Amino Acids 41(2):219–222. doi:10.1007/s00726-010-0828-4
46. Hunt T, Huang Y, Ross P, Pillai S, Purkayastha S, Pappin D (2004) Protein expression analysis and biomarker identification and quantification using multiplexed isobaric tagging technology – iTRAQ reagents. Mol Cell Proteomics 3(10):S286
47. Dayon L, Hainard A, Licker V, Turck N, Kuhn K, Hochstrasser DF, Burkhard PR, Sanchez JC (2008) Relative quantification of proteins in human cerebrospinal fluids by MS/MS using 6-plex isobaric tags. Anal Chem 80(8):2921–2931. doi:10.1021/ac702422x
48. Ashman K, Ruppen Canas MI, Luque-Garcia JL, Garcia Martinez F (2011) Stable isotopic labeling for proteomics. In: Sample preparation in biological mass spectrometry. doi:10.1007/978-94-007-0828-0_27
49. Coombs KM (2011) Quantitative proteomics of complex mixtures. Expert Rev Proteomics 8(5):659–677. doi:10.1586/epr.11.55
50. Monetti M, Nagaraj N, Sharma K, Mann M (2011) Large-scale phosphosite quantification in tissues by a spike-in SILAC method. Nat Methods 8(8):U655–U674. doi:10.1038/Nmeth.1647
51. Ong SE, Blagoev B, Kratchmarova I, Kristensen DB, Steen H, Pandey A, Mann M (2002) Stable isotope labeling by amino acids in cell culture, SILAC, as a simple and accurate approach to expression proteomics. Mol Cell Proteomics 1(5):376–386. doi:10.1074/Mcp.M200025-Mcp200
52. Rauh M (2012) LC-MS/MS for protein and peptide quantification in clinical chemistry. J Chromatogr B Analyt Technol Biomed Life Sci 883–884:59–67. doi:10.1016/j.jchromb.2011.09.030
53. Farrah T, Deutsch EW, Kreisberg R, Sun Z, Campbell DS, Mendoza L, Kusebauch U, Brusniak MY, Huttenhain R, Schiess R, Selevsek N, Aebersold R, Moritz RL (2012) PASSEL: The PeptideAtlas SRM Experiment Library. Proteomics. doi:10.1002/pmic.201100515
54. Ciccimaro E, Blair IA (2010) Stable-isotope dilution LC-MS for quantitative biomarker analysis. Bioanalysis 2(2):311–341. doi:10.4155/bio.09.185
55. Anderson NL, Anderson NG, Haines LR, Hardie DB, Olafson RW, Pearson TW (2004) Mass spectrometric quantitation of peptides and proteins using Stable Isotope Standards and Capture by Anti-Peptide Antibodies (SISCAPA). J Proteome Res 3(2):235–244

56. Whiteaker JR, Zhao L, Anderson L, Paulovich AG (2009) An automated and multiplexed method for high throughput peptide immunoaffinity enrichment and multiple reaction monitoring mass spectrometry-based quantification of protein biomarkers. Mol Cell Proteomics. doi:10.1074/mcp.M900254-MCP200. pii: M900254-MCP200
57. Addona TA, Abbatiello SE, Schilling B, Skates SJ, Mani DR, Bunk DM, Spiegelman CH, Zimmerman LJ, Ham AJL, Keshishian H, Hall SC, Allen S, Blackman RK, Borchers CH, Buck C, Cardasis HL, Cusack MP, Dodder NG, Gibson BW, Held JM, Hiltke T, Jackson A, Johansen EB, Kinsinger CR, Li J, Mesri M, Neubert TA, Niles RK, Pulsipher TC, Ransohoff D, Rodriguez H, Rudnick PA, Smith D, Tabb DL, Tegeler TJ, Variyath AM, Vega-Montoto LJ, Wahlander A, Waldemarson S, Wang M, Whiteaker JR, Zhao L, Anderson NL, Fisher SJ, Liebler DC, Paulovich AG, Regnier FE, Tempst P, Carr SA (2009) Multi-site assessment of the precision and reproducibility of multiple reaction monitoring-based measurements of proteins in plasma (vol 27, pg 633, 2009). Nat Biotechnol 27(9):864. doi:10.1038/Nbt0909-864b
58. Anderson NL, Jackson A, Smith D, Hardie D, Borchers C, Pearson TW (2009) SISCAPA peptide enrichment on magnetic beads using an in-line bead trap device. Mol Cell Proteomics 8(5):995–1005. doi:10.1074/Mcp.M800446-Mcp200
59. Dissmeyer N, Schnittger A (2011) The age of protein kinases. Methods Mol Biol 779:7–52. doi:10.1007/978-1-61779-264-9_2
60. Pawson T, Kofler M (2009) Kinome signaling through regulated protein-protein interactions in normal and cancer cells. Curr Opin Cell Biol 21(2):147–153. doi:10.1016/j.ceb.2009.02.005
61. Malumbres M, Barbacid M (2009) Cell cycle, CDKs and cancer: a changing paradigm. Nat Rev Cancer 9(3):153–166. doi:10.1038/nrc2602
62. Iwai LK, Benoist C, Mathis D, White FM (2010) Quantitative phosphoproteomic analysis of T cell receptor signaling in diabetes prone and resistant mice. J Proteome Res 9(6):3135–3145. doi:10.1021/pr100035b
63. Thingholm TE, Jensen ON, Larsen MR (2009) Analytical strategies for phosphoproteomics. Proteomics 9(6):1451–1468. doi:10.1002/pmic.200800454
64. Cazares LH, Troyer DA, Wang B, Drake RR, Semmes OJ (2011) MALDI tissue imaging: from biomarker discovery to clinical applications. Anal Bioanal Chem 401(1):17–27. doi:10.1007/s00216-011-5003-6
65. Vestal ML (2009) Modern MALDI time-of-flight mass spectrometry. J Mass Spectrom 44(3):303–317. doi:10.1002/jms.1537
66. Bandura DR, Baranov VI, Ornatsky OI, Antonov A, Kinach R, Lou X, Pavlov S, Vorobiev S, Dick JE, Tanner SD (2009) Mass cytometry: technique for real time single cell multitarget immunoassay based on inductively coupled plasma time-of-flight mass spectrometry. Anal Chem 81(16):6813–6822. doi:10.1021/ac901049w
67. Bendall SC, Simonds EF, Qiu P, el Amir AD, Krutzik PO, Finck R, Bruggner RV, Melamed R, Trejo A, Ornatsky OI, Balderas RS, Plevritis SK, Sachs K, Pe'er D, Tanner SD, Nolan GP (2011) Single-cell mass cytometry of differential immune and drug responses across a human hematopoietic continuum. Science 332(6030):687–696. doi:10.1126/science.1198704
68. Janes MR, Rommel C (2011) Next-generation flow cytometry. Nat Biotechnol 29(7):602–604. doi:10.1038/nbt.1919
69. Ryzhov V, Fenselau C (2001) Characterization of the protein subset desorbed by MALDI from whole bacterial cells. Anal Chem 73(4):746–50
70. Fenselau C, Demirev PA (2001) Characterization of intact microorganisms by MALDI mass spectrometry. Mass Spectrom Rev 20(4):157–171. doi:10.1002/mas.10004
71. Holland RD, Wilkes JG, Rafii F, Sutherland JB, Persons CC, Voorhees KJ, Lay JO Jr (1996) Rapid identification of intact whole bacteria based on spectral patterns using matrix-assisted laser desorption/ionization with time-of-flight mass spectrometry. Rapid Commun Mass Spectrom 10(10):1227–1232. doi:10.1002/(SICI)1097-0231(19960731)10:10<1227::AID-RCM659>3.0.CO;2-6
72. Kathiresan S, Gona P, Larson MG, Vita JA, Mitchell GF, Tofler GH, Levy D, Newton-Cheh C, Wang TJ, Benjamin EJ, Vasan RS (2006) Cross-sectional relations of multiple biomarkers from distinct biological pathways to brachial artery endothelial function. Circulation 113(7):938–945. doi:10.1161/CIRCULATIONAHA.105.580233

73. Wang TJ, Gona P, Larson MG, Tofler GH, Levy D, Newton-Cheh C, Jacques PF, Rifai N, Selhub J, Robins SJ, Benjamin EJ, D'Agostino RB, Vasan RS (2006) Multiple biomarkers for the prediction of first major cardiovascular events and death. N Engl J Med 355(25):2631–2639. doi:10.1056/NEJMoa055373
74. Mor G, Visintin I, Lai Y, Zhao H, Schwartz P, Rutherford T, Yue L, Bray-Ward P, Ward DC (2005) Serum protein markers for early detection of ovarian cancer. Proc Natl Acad Sci U S A 102(21):7677–7682. doi:10.1073/pnas.0502178102
75. Georgiou HM, Lappas M, Georgiou GM, Marita A, Bryant VJ, Hiscock R, Permezel M, Khalil Z, Rice GE (2008) Screening for biomarkers predictive of gestational diabetes mellitus. Acta Diabetol 45(3):157–165. doi:10.1007/s00592-008-0037-8

The Clinical Application of Proteomics

Keith Ashman, Murray Mitchell, and Gregory Rice

Abbreviations

CRC	Colorectal cancer
CSF	Cerebrospinal fluid
CVF	Cervico vaginal fluid
DRE	Digital rectal exam
ELISA	Enzyme Linked Immunosorbent Assay
FDA	Federal Drug Administration
HUPO	Human Proteome Organisation
PSA	Prostate-specific antigen
ROC	Receiver operator characteristic curves
SOP	Standard Operating Procedure

Introduction

The key to the successful treatment of patients is early and accurate diagnosis. Molecular pathology provides diagnostic information by measuring the presence and concentration of distinct molecular species (DNA, metabolites, proteins and lipids), so that changes in levels between health and disease can be used to guide diagnosis and therapy. Since proteins carry out the majority of the biochemical reactions in cells, as well as performing many key signalling and structural functions, it follows that measuring the concentration

K. Ashman (✉) • M. Mitchell • G. Rice (✉)
University of Queensland Centre for Clinical Research,
Building 71/918, Royal Brisbane & Women's Hospital Campus,
Herston, QLD 4029, Australia
e-mail: k.ashman@uq.edu.au; murray.mitchell@uq.edu.au; g.rice@uq.edu.au

S.R. Lakhani, S.B. Fox (eds.), *Molecular Pathology in Cancer Research*,
DOI 10.1007/978-1-4939-6643-1_11

and state of proteins both in their biochemical and temporal context will yield means of distinguishing health from disease. This is often called biomarker analysis [1, 2]. The term proteome was coined by Wilkins and co-workers [3] and led to the field of protein biochemistry being dubbed proteomics [4]. The proteome was originally defined as the proteins present in a cell or organism at any one time but has come to be used in a wider context, which covers all aspects of protein measurement. The proteome is clearly dynamic, and unlike the genome varies highly and rapidly over time and biological location (e.g. plasma vs. serum vs. tissue vs. urine vs. CSF vs. CVF vs. saliva, etc.). Another important consideration in analysing proteins especially in a higher eukaryote such as man is the critical role played by post-translational modifications in cell signalling [5] which not only includes such changes as phosphorylation, glycosylation, ubiquitination and methylation but also activation through proteolytic cleavage, e.g. the conversion of prothrombin to thrombin or the excision of the c-peptide to yield active insulin. This all implies that the proteome is information rich but poses a significant analytical challenge. Despite considerable progress in protein analytical techniques, there are still only a small number of validated protein assays in use in the clinic. This relates strongly to the problem of measuring molecules that vary highly in their concentration, structure as well as spatial and temporal distribution. The study of cancer biomarker proteins began in 1847 with the discovery by Bence-Jones [6] of what turned out, more than 100 years later, to be a tumour-produced free antibody light chain 'Bence Jones protein' in the urine of a multiple myeloma patient [7] where it was present in large quantities and could be revealed by simple heat denaturation. 140 years later this protein was demonstrated to be present also in the serum [8], and in 1998 a routine immunodiagnostic test was approved by the FDA. Hormones produced by tumours were also detected early on. Adrenocorticotropic hormone (ACTH), calcitonin and chorionic gonadotropin (hCG), for example, are elevated in specific cancer types, though not with the tumour specificity of Bence-Jones proteins. Polanski and Anderson have pointed out that despite large numbers of proteins being identified as potential cancer biomarkers, little has been done to systematically validate that potential and they generated a list of proteins that could or should be further investigated [9]. The following pages describe some examples of protein analyses that are already used in a clinical diagnostic context.

The Analysis of Proteins

Whatever the source of the molecules to be measured, robust and reproducible sample preparation is the key in any analytical technique. There are numerous possible sources of samples for proteomics analysis: serum vs. plasma, urine, fresh frozen vs. paraffin embedded tissue, etc. Something that is often under appreciated is the experimental design and statistical interpretation of studies and results using clinical samples. Table 1 shows the process that leads to the development of a new clinical assay. It is important to understand that as an assay is progresses towards routine clinical use the number of samples tested must be increased to ensure it is robust, as well as defining the range of concentrations that are typically present in a patient population.

Table 1 Stages of development of a clinical biomarker test

Phase	Process	Sample selection	Sample size	Time frame	Outcome
Phase I—Preclinical discovery	Hypothesis-driven identification of protein candidates	Model systems of disease	10s	6–12 months	List of potential biomarker candidates
Phase II—Preclinical verification	Development of robust assays and testing	Proof of concept: disease and control subjects	10–50s	6–12 months	Detect disease in plasma, assess technology
Phase III—Preclinical validation	Defined protein signature for the study objective	Selection of samples with known disease outcome	100–500 s	6–12 months	Define criteria for clinical evaluation
Phase IV—Clinical evaluation	Trail the assay in the clinical context and determine the accuracy	Prospective collection from intended population	500–1000s	>24 months	Sensitivity and specificity of the diagnostic test
Phase V—Disease control	Assess the impact of the test on the population	Randomly from target population	Many 1000s	Many years	Effect on disease management

Requirements: a clear study question and clinical need; definition of a clinical objective for the study, including predefined hypotheses, e.g., assessment of the stage of cancer malignancy, effectiveness of treatment

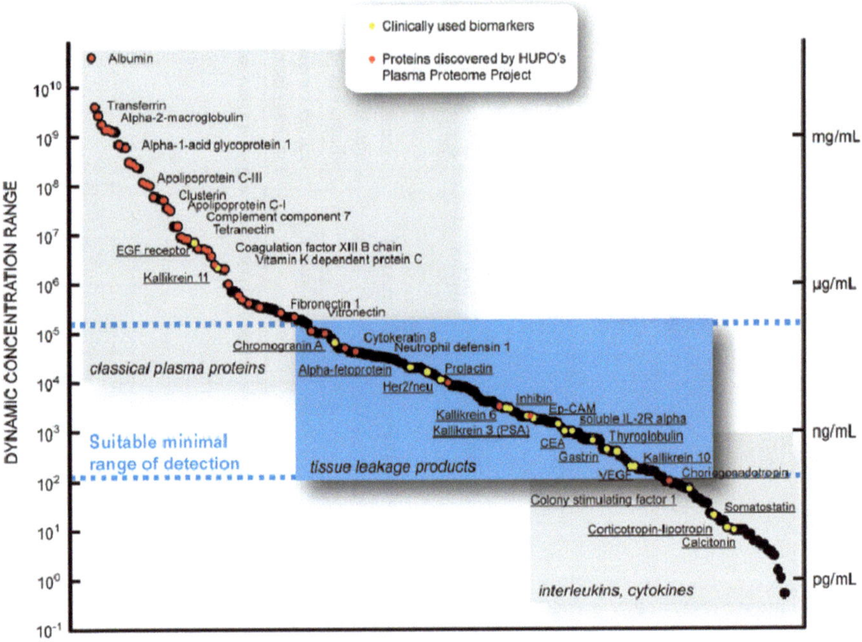

Fig. 1 Dynamic plasma protein concentration range and the three main plasma protein categories are shown as reported by Anderson and Anderson [13]. *Red dots* indicate proteins identified by the HUPO plasma proteome project (PPP) [14] and *yellow dots* represent currently used biomarkers in the clinic. A suitable minimal range of detection for biomarker targeting in plasma is shown with *dotted lines*. Adapted from Schiess et al. [15]

To make meaningful clinical conclusions from follow-up verification studies, it is essential to design the experimental workflow around a suitably sized collection of clinical specimens [10, 11]. This starts with the selection of a sufficient number of representative patients and suitable case controls to achieve statistical significance in terms of differential biomarker abundance. Without proper design, control of standard operating procedures (SOP) and data integrity, the information may be difficult to translate into clinical practice and/or not relevant to the actual clinical cohort. Follow-up reports should provide statistical statements about the significance of each finding on the individual patient level.

The most frequently analysed clinical sample is blood, since its collection is rapid and non-invasive. From the proteomics standpoint, however, there are a number of issues to consider when analysing the protein content of blood or more specifically serum and plasma [12]. The dynamic range of concentration of individual proteins in blood covers 10 orders of magnitude (see Fig. 1), with albumin at 34–54 mg/mL and cytokines such as IL1-beta 0.16 ± 0.17 pg/mL (http://www.copewithcytokines.de) [9, 16]. This poses significant analytical problems since the linear dynamic range of the current analytical methods is only 4–5 orders of magnitude. Hence, to measure the low concentration analytes either an enrichment step, e.g. an antibody and or a signal amplification, e.g. ELISA (see chapter on proteomic methods) is required. The collection protocol can also have a significant affect on the analysis, the timing, method of sampling and even the type of tube can also affect the result [17].

Some examples that are already in the clinical laboratory use include analysis of genetic variants of the protein transthyretin and detection of carbohydrate-deficient transferrin [18–20]. These are applications of targeted proteomics, i.e. analysis of structural changes in specific proteins. The following pages discuss several examples of protein analysis in a clinical context, while Table 2 gives an overview of the proteins that are currently used in cancer diagnosis/prognosis.

Example 1: Prostate-Specific Antigen

Prostate-specific antigen (PSA) is a 33-kd protein produced by the cells of the prostate gland. It is a single-chain glycoprotein of 237 amino acid residues, 4 carbohydrate side chains and multiple disulfide bonds. It is homologous with the proteases of the kallikrein family. PSA may be referred to as human glandular kallikrein (hK)-3 to distinguish it from hK-2, another prostate cancer marker with which it shares 80 % homology. A third kallikrein, hK-1, is found mainly in pancreatic and renal tissue but shows 73 and 84 % homology with PSA. Because of the similarities between these kallikreins, concern exists that both polyclonal and monoclonal assays may have cross-reactivity, which could affect PSA measurements. It has been demonstrated that very few monoclonal anti-PSA immunoglobulin Gs (IgGs) cross-react with hK-2 [58, 59]. Epitopes have been identified that are unique to PSA without possessing cross-reactivity to hK-2. This has led to the development of ultrasensitive immunoassays that are specific for PSA and hK-2, as well as assays that are fully cross-reactive with both proteins. PSA is a neutral serine protease with biochemical attributes that are similar to the proteases involved in blood clotting. The role of proteases in the coagulation process has been studied extensively and applies to all serine proteases, including PSA. PSA splits the seminal vesicle proteins seminogelin I and II, resulting in liquefaction of the seminal coagulum. The complete gene encoding PSA has been sequenced and localised to chromosome 19.

The PSA test measures the concentration of PSA in a man's blood. The results are usually reported as nanograms of PSA per millilitre (ng/mL) of blood. The blood level of PSA is often elevated in men with prostate cancer, and the PSA test was originally approved by the FDA in 1986 to monitor the progression of prostate cancer in men who had already been diagnosed with the disease. In 1994, the FDA approved the use of the PSA test in conjunction with a digital rectal exam (DRE) to test asymptomatic men for prostate cancer. Men who report prostate symptoms often undergo PSA testing (along with a DRE) to help doctors determine the nature of the problem.

In addition to prostate cancer, a number of benign (not cancerous) conditions can cause a man's PSA level to rise. The most frequent benign prostate conditions that cause an elevation in PSA level are prostatitis (inflammation of the prostate) and benign prostatic hyperplasia (BPH) (enlargement of the prostate). There is no evidence that prostatitis or BPH leads to prostate cancer, but it is possible for a man to have one or both of these conditions and to develop prostate cancer as well [60].

Table 2 Proteins measured in cancer pathology

Tumour marker	Cancer type	Tissue or fluid analysed	How used	References
ALK gene rearrangements (tissue marker). A gene that makes a protein called anaplastic lymphoma kinase (ALK), which may be involved in cell growth	Non-small cell lung cancer and anaplastic large cell lymphoma	Tumour	To help determine treatment and prognosis	[21]
Alpha-fetoprotein (AFP). A protein normally produced by a foetus. AFP levels are usually undetectable in the blood of healthy adult men or women (who are not pregnant). An elevated level of AFP suggests the presence of either a primary liver cancer or germ cell tumour	Germ cell cancers of ovaries and testes (non-seminomatous, particularly embryonal and yolk sac, testicular cancers). Some primary liver cancers (hepatocellular)	Blood serum	To help diagnose liver cancer and follow response to treatment; to assess stage, prognosis and response to treatment of germ cell tumours	[22, 23]
Bence-Jones Proteins is a monoclonal globulin protein or immunoglobulin light chain with a molecular weight of 22–24 kDa. The Bence-Jones protein was described by the English physician Henry Bence Jones in 1847 and published in 1848	Multiple myeloma Waldenstrom's macroglobulinemia chronic lymphocytic leukaemia	Urine, blood	Diagnostic of multiple myeloma in the context of end-organ manifestations such as renal failure, lytic (or 'punched out') bone lesions, anaemia or large numbers of plasma cells in the bone marrow of patients. Bence-Jones proteins are present in 2/3 of multiple myeloma cases	[6, 7]
Beta-2-microglobulin (B2M). A small protein normally found on the surface of many cells, including lymphocytes, and in small amounts in the blood and urine. An increased amount in the blood or urine may be a sign of certain diseases, including some types of cancer, such as multiple myeloma or lymphoma	Multiple myeloma, chronic lymphocytic leukaemia and some lymphomas	Blood, urine or cerebrospinal fluid	To determine prognosis and follow response to treatment	[24]

Beta-human chorionic gonadotropin (Beta-hCG). A hormone found in the blood and urine during pregnancy	Choriocarcinoma and testicular cancer	Urine or blood	To assess stage, prognosis and response to treatment	[25–27]
BCR-ABL fusion gene. A gene formed when pieces of chromosomes 9 and 22 break off and trade places. The ABL gene from chromosome 9 joins to the BCR gene on chromosome 22 to form the BCR-ABL fusion gene	Chronic myeloid leukaemia	Blood and/or bone marrow	To confirm diagnosis and monitor disease status	[28, 29]
BRAF mutation V600E. A specific mutation (change) in the BRAF gene, which makes a protein that is involved in sending signals in cells and in cell growth. This BRAF gene mutation may be found in some types of cancer, including melanoma and colorectal cancer	Cutaneous melanoma and colorectal cancer	Tumour	To predict response to targeted therapies	[30]
CA15-3/CA27.29 (Carbohydrate antigen 15-3, 27,29)	Breast cancer	Blood	To assess whether treatment is working or disease has recurred	[31]
CA19-9 (Carbohydrate antigen 19-9)	Pancreatic cancer, gallbladder cancer, bile duct cancer and gastric cancer	Blood	To assess whether treatment is working	[31]
CA-125 may be found in high amounts in the blood of patients with certain types of cancer, including ovarian cancer. CA-125 levels may also help monitor how well cancer treatments are working or if cancer has come back. Also called cancer antigen 125	Ovarian cancer	Blood	To help in diagnosis, assessment of response to treatment and evaluation of recurrence	[32]

(continued)

Table 2 (continued)

Tumour marker	Cancer type	Tissue or fluid analysed	How used	References
Calcitonin. A hormone formed by the C cells of the thyroid gland. It helps maintain a healthy level of calcium in the blood. When the calcium level is too high, calcitonin lowers it	Medullary thyroid cancer	Blood	To aid in diagnosis, check whether treatment is working and assess recurrence	[33]
Carcinoembryonic antigen (CEA) may be found in the blood of people who have colon cancer, other types of cancer or diseases or who smoke tobacco. Levels may help keep track of how well cancer treatments are working or if cancer has come back	Colorectal cancer and breast cancer	Blood	To check whether colorectal cancer has spread; to look for breast cancer recurrence and assess response to treatment	[34, 35]
CD20 A protein found on B cells (a type of white blood cell). It may be found in higher than normal amounts in patients with certain types of B-cell lymphomas and leukemias. Also called CD20 antigen	Non-Hodgkin lymphoma	Blood	To determine whether treatment with a targeted therapy is appropriate	[36, 37]
Chromogranin A. A protein found inside neuroendocrine cells, which release chromogranin A and certain hormones into the blood. Also called CgA	Neuroendocrine tumours	Blood	To help in diagnosis, assessment of treatment response and evaluation of recurrence	[38, 39]
Chromosomes 3, 7, 17 and *9p21*. Homozygous loss of band 9p21, the site for the tumour suppressor gene *P16*, is a known early genetic event in the development of papillary carcinoma and urothelial carcinoma in situ (CIS). Increased chromosomal instability and aneuploidy have been implicated in tumour progression	Bladder cancer	Urine	To help in monitoring for tumour recurrence	[40]

Cytokeratin fragments 21-1. A type of protein found on epithelial cells, which line the inside and outside surfaces of the body. Cytokeratins help form the tissues of the hair, nails and the outer layer of the skin	Lung cancer	Blood	To help in monitoring for recurrence	[41, 42]
Epidermal growth factor receptor, EGFR mutation analysis. The protein found on the surface of some cells and to which epidermal growth factor binds, causing the cells to divide. It is found at abnormally high levels on the surface of many types of cancer cells, so these cells may divide excessively in the presence of epidermal growth factor. Also called ErbB1 and HER1	Non-small cell lung cancer	Tumour	To help determine treatment and prognosis	[43]
Oestrogen receptor (ER)/Progesterone receptor (PR). Proteins found inside the cells of the female reproductive tissue, some other types of tissue and some cancer cells. The hormone oestrogen will bind to the receptors inside and may cause the cells to grow	Breast cancer	Tumour	To determine whether treatment with hormonal therapy (such as tamoxifen) is appropriate	[44]
Fibrin/fibrinogen (also called Factor Ia) is a fibrous, non-globular protein involved in the clotting of blood. It is formed from fibrinogen by the protease thrombin. It can be released by the proteolytic activity associated with invasive tumour cells	Bladder cancer	Urine	To monitor progression and response to treatment	[45]

(continued)

Table 2 (continued)

Tumour marker	Cancer type	Tissue or fluid analysed	How used	References
Human epididymis protein 4 (HE4) is the product of the *WFDC2* (*HE4*) gene that is overexpressed in patients with ovarian carcinoma	Ovarian cancer	Blood	To assess disease progression and monitor for recurrence	[46]
Human EGF receptor 2 (HER2/neu). A protein involved in normal cell growth. It is found on some types of cancer cells, including breast and ovarian. Cancer cells removed from the body may be tested for the presence of HER2/neu to help decide the best type of treatment. HER2/neu is a type of receptor tyrosine kinase. Also called c-erbB-2 and human epidermal growth factor receptor 2	Breast cancer, gastric cancer and esophageal cancer	Tumour	To determine whether treatment with trastuzumab is appropriate	[47]
KIT Mast/stem cell growth factor receptor (SCFR), also known as proto-oncogene c-Kit or tyrosine-protein kinase Kit or CD117, is a protein that in humans is encoded by the KIT gene. CD117 is a proto-oncogene, meaning that overexpression or mutations of this protein can lead to cancer	Gastrointestinal stromal tumour and mucosal melanoma	Tumour	To help in diagnosing and determining treatment	[48]
KRAS mutation analysis. The *KRAS* gene encodes the KRAS protein that regulates 2 such signalling pathways: PI3K/PTEN/AKT and RAF/MEK/ERK. These pathways are targets of anti-cancer drugs that are currently in development. Drugs targeting EGFR, which controls these pathways upstream from KRAS, are already available	Colorectal cancer, pancreatic and non-small cell lung cancer	Tumour	To determine whether treatment with a particular type of targeted therapy is appropriate	[49]

Lactate dehydrogenase (LDH). One of a group of proteins found in the blood and other body tissues and involved in energy production in cells. An increased amount of lactate dehydrogenase in the blood may be a sign of tissue damage and some types of cancer or other diseases	Germ cell tumours	Blood	To assess stage, prognosis and response to treatment	[50]
Nuclear matrix protein 22 (NMP22). Nuclear matrix proteins (NMPs) make up the internal structural framework of the nucleus and are associated with functions such as DNA replication and RNA synthesis	Bladder cancer	Urine	To monitor response to treatment	[51]
Prostate-specific antigen (PSA). A protein made by the prostate gland and found in the blood. Prostate-specific antigen blood levels may be higher than normal in men who have prostate cancer, benign prostatic hyperplasia (BPH) or infection or inflammation of the prostate gland	Prostate cancer	Blood	To help in diagnosis, assess response to treatment and look for recurrence	[52]
Thyroglobulin. The form that thyroid hormone takes when stored in the cells of the thyroid. If the thyroid has been removed, thyroglobulin should not show up on a blood test. Doctors measure thyroglobulin level in blood to detect thyroid cancer cells that remain in the body after treatment	Thyroid cancer	Tumour	To evaluate response to treatment and look for recurrence	[53, 54]

(continued)

Table 2 (continued)

Tumour marker	Cancer type	Tissue or fluid analysed	How used	References
Urokinase plasminogen activator (uPA) and plasminogen activator inhibitor (PAI-1). A protein that is made in the kidney and found in the urine	Breast cancer	Tumour	To determine aggressiveness of cancer and guide treatment	[55]
5-Protein signature (Ova1) is an FDA-approved test for the evaluation of an ovarian mass prior to surgery. This new test combines the results of five immunoassays (CA-125 II, transthyretin [prealbumin], apolipoprotein A1, b2-microglobulin and transferrin). Using a unique proprietary algorithm to produce a single numerical score, OVA1 indicates a woman's likelihood of malignancy	Ovarian cancer	Blood	To pre-operatively assess pelvic mass for suspected ovarian cancer	[56]
21-Gene signature (Oncotype DX). Oncotype DX analyses a panel of 21 genes within a tumour to determine a Recurrence Score. The Recurrence Score is a number between 0 and 100 that corresponds to a specific likelihood of breast cancer recurrence within 10 years of the initial diagnosis	Breast cancer	Tumour	To evaluate risk of recurrence	[57]
70-Gene signature (MammaPrint). MammaPrint is carried out on tumour tissue that is fixed in formalin or fresh, which is put in an mRNA preservative immediately after the operation has taken place. The expression of 70 specific genes is measured six times	Breast cancer	Tumour	To evaluate risk of recurrence	[57]

Example 2: Carcinoembryonic Antigen

Carcinoembryonic antigen (CEA) is a glycoprotein involved in cell adhesion. It is normally produced during fetal development, but the production of CEA stops before birth. It, therefore, is not usually present in the blood of healthy adults, although it is detectable in heavy smokers. CEA is a glycosyl phosphatidyl inositol (GPI)-cell surface anchored glycoprotein whose specialised sialofucosylated glycoforms serve as functional colon carcinoma L-selectin and E-selectin ligands, which may be critical to the metastatic dissemination of colon carcinoma cells. The CEA protein consists of 668 amino acids, and has a configuration that is similar to that of other members of the immunoglobulin gene superfamily. The protein extends out from the cell membrane into the extracellular space, and is anchored through a hydrophobic C-terminal region. Most of the final molecular weight of CEA is provided by N-linked glycosylation.

CEA is most useful to monitor treatment of cancer patients. It is used for patients who have had surgery, to measure response to therapy and to monitor whether the disease has recurred. A blood test for CEA in this circumstance is used as a tumour marker, i.e. an indicator of whether the cancer is present or not. CEA is used as a marker for bowel cancer in particular, but may be measured where other forms of cancer are present. It has been found helpful in monitoring some patients with cancer of the rectum, lung, breast, liver, pancreas, stomach and ovary. Not all cancers produce CEA, and a level within the given reference range does not guarantee that cancer (even the kinds known to produce CEA) is not present, therefore the CEA test is not used for screening the general population. There are test kits available that use antibodies to detect CEA. The test device is similar to a pregnancy test (see Fig. 2), but uses serum/plasma as the sample, not urine.

Example 3: KRAS

The use of companion-test molecular assays for oncologic treatment decisions is becoming increasingly important. These tests are used to determine whether patients are eligible to receive a targeted therapy. In the case of *KRAS* mutation testing for colorectal cancer (CRC), the intent is to avoid unnecessary toxicity and monetary costs for patients who are not likely to respond to anti-EGFR therapies by screening them before initiating therapy. The cost of cetuximab and panitumumab therapy has been estimated to be approximately $100,000/patient/year [61]. Mutations in the *RAS* gene family (*HRAS, KRAS* and *NRAS*) have been observed in a variety of cancers. They are activating mutations that result in continual signal transduction, stimulating downstream signalling pathways involved in cell growth, proliferation, invasion and metastasis. The *KRAS* gene encodes the KRAS protein that regulates 2 such signalling pathways: PI3K/PTEN/AKT and RAF/MEK/ERK. These pathways are targets of anti-cancer drugs that are currently in development. Drugs targeting EGFR, which controls these pathways upstream from KRAS, are already available.

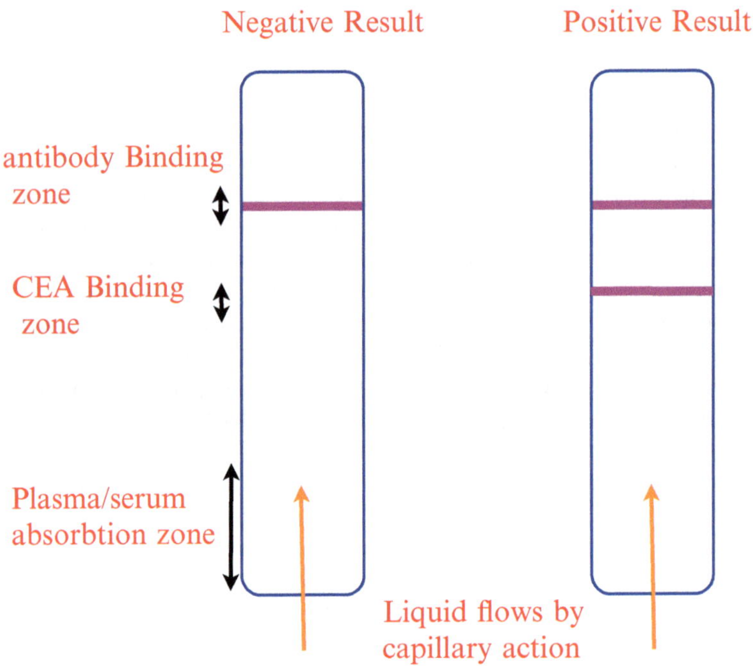

Fig. 2 Testing for CEA with a commercial testing kit

When bound to its ligand, EGFR stimulates tyrosine kinase activity, leading to activation of KRAS and the signalling pathways.

Current therapies targeting EGFR are used to treat colorectal cancer (CRC) and non-small cell lung cancer (NSCLC) and employ either (1) monoclonal antibodies (e.g. cetuximab and panitumumab) that prevent ligand binding and EGFR activation or (2) tyrosine kinase inhibitors (e.g. erlotinib) that prevent activation of the signalling pathways. If, however, the signalling pathways are activated independent of EGFR, as happens when the *KRAS* gene is mutated, these drugs are rendered ineffective.

KRAS mutations frequently found in neoplasms include those at exon 1 (codons 12 and 13) and exon 2 (codon 61) Table 3. Mutations in *KRAS* codons 12 and 13 have been associated with lack of response to EGFR-targeted therapies in both CRC and NSCLC patients (Table 4) [62].

Mass Spectrometry to Detect KRAS Point Mutations at the Protein Level

Mass spectrometry is already an important clinical tool, widely used for the measurement of drug metabolites, steroids, for the detection on inborn metabolic disease in newborn infants [63]. It is also beginning to show promise in diagnostic

Table 3 Number and type of mutations, affected codons and corresponding altered amino acids in exon 2, codon 12 and 13 of the KRAS gene detected in 1018 metastatic colorectal cancers [62]

Codon	Type of point mutation	Number of point mutations (% of all tumours)
12	c.35G4A (p.G12D)	144 (14.1 %)
	c.35G4T (p.G12V)	87 (8.5 %)
	c.34G4T (p.G12C)	32 (3.1 %)
	c.34G4A (p.G12S)	26 (2.6 %)
	c.35G4C (p.G12A)	24 (2.4 %)
	c.34G4C (p.G12R)	5 (0.5 %)
	c.34G4T, c.35G4T (p.G12F)	2 (0.2 %)
	c.34G4A, c.35G4T (p.G12I)	1 (0.1 %)
13	c.38G4A (p.G13D) 75	(7.3 %)
	c.37G4T (p.G13C) 3	(0.3 %)
	c.37G4C (p.G13R) 1	(0.1 %)
	Wild type	618 (60.7 %)

Table 4 *RAS* mutation frequency

Cancer	Frequency (%)
Adenocarcinoma	
Lung	30
Colon	43
Pancreas	80
Thyroid cancer	
Follicular	53
Undifferentiated	60
Myeloid disorders	
MDS (myelodysplastic syndromes)	27
AML (Acute myeloid leukaemia)	27

proteomics [64–66]. Targeted proteomics aims to obtain quantitative information from defined sets of proteins that have been identified as important players in a biological process. The field of quantitative proteomics is enthusiastically embracing selected reaction monitoring (SRM) MS methods for targeted quantitative proteomic analysis [67].

The underlying principle of SRM in proteomic applications is that the selected set of precursor and product ions contain sufficient information to represent the target peptide of interest, and thereby its protein of origin. Libraries of SRM based data are being actively established, e.g. SRM protein atlas (http://www.srmatlas.org/ [68]). Quantitation is achieved by including a stable heavy isotope labelled version of the target peptide(s) as an internal standard in the assay and applying the method of isotope dilution [69]. A variation on the general principle of isotope dilution, termed SISCAPA [70, 71], uses anti-peptide antibodies to capture peptides of interest, primarily from serum, again with a spiked in synthetic isotopically labelled

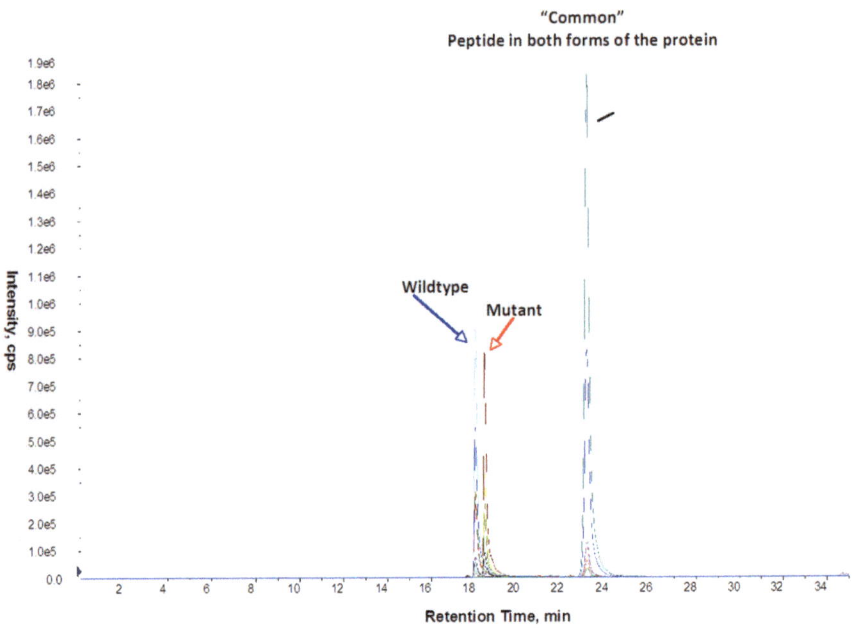

Fig. 3 Detection of mutant KRAS by targeted mass spectrometry

heavy internal standard peptide, to improve the selectivity and sensitivity of the method [64, 72]. Recently this technique has been applied to develop a quantitative proteomics assay for wild type and mutant K12D KRAS. The method achieved low attomole detection sensitivity, and required only a few milligrams of tumour tissue [65]. The result of such an assay is shown in Fig. 3.

Single vs. Multiple Biomarkers

OVA5

Multiplexed measurement is logical for biological discovery with proteins because they constitutively function within networks, pathways, complexes and families [73, 74]. The consequence of this is that measuring multiple biomarkers will provide better predictive and therefore diagnostic capabilities. There is a growing consensus that panels of markers may be able to supply the specificity and sensitivity that individual markers lack. For example, a panel combining four known biomarkers (leptin, prolactin, osteopontin and insulin-like growth factor II), none of which used alone could distinguish patients from the controls, achieved a sensitivity and specificity of 95 % for the diagnosis of ovarian cancer [75].

In another example, multiple plasma biomarkers were measured at 11 weeks of gestation in women who experienced normal pregnancy outcomes ($n=14$) and women who developed gestational diabetes ($n=14$). Of the biomarkers considered, receiver operator characteristic curves (ROC) for three biomarkers (adiponectin, insulin and blood glucose) are presented together with an ROC based on the predicted posterior probability values (ppv) generated by a classification model that combined information from all three biomarkers. The model out performed individual biomarkers based upon the area under the ROC (model=0.94; adiponectin) [76].

OVA1

In September 2009, the OVA1 test (submitted to FDA by Vermillion and later acquired by Quest Diagnostics) was the first de novo 510(k)-cleared IVDMIA that was protein based (21 CFR 866.6050, ovarian adnexal mass assessment score test system). The test was launched in March 2010 by Quest Diagnostics Inc. The test combines into a single score the results of five protein biomarkers that change due to the presence of ovarian cancer. It is indicated for women who are older than 18 years, who have an ovarian adnexal mass present for which surgery is planned, and have not yet been referred to an oncologist. The OVA1 test is an aid to further assess the likelihood that malignancy is present when the physician's independent clinical and radiological evaluation does not indicate malignancy. The five serum biomarkers that comprise the test are prealbumin, apolipoprotein A-1, transferrin, beta-2-microglobulin and CA125, and they are measured using standard immunoassays. The results are analysed with software to produce a single result ranging from 1 to 10 to classify the likelihood that a woman's pelvic mass is cancerous or benign. The OVA1 test was developed in collaboration with academic medical centres testing more than 2500 clinical samples [77]. Extensive analytical studies were done to evaluate repeatability and reproducibility of the OVA1 test result and each of the five component proteins. These studies supported the 510(k). In early 2007 Vermillion began a multicenter prospective clinical trial to demonstrate the clinical performance of the OVA1 test. Clinical specimens were collected at 27 sites, and test performance was determined based on 516 evaluable subjects who underwent surgery to remove a documented ovarian tumour and for whom a pathology result was available. After surgery, the specimen was examined by a surgical pathologist using routine procedures. The ability of physicians to predict malignancy without the OVA1 test was compared with the ability of physicians and the OVA1 test via dual assessment to predict malignancy. With dual assessment, 80 % of cancers missed by clinician impression alone were detected, and the sensitivity and negative predictive value were each more than 90 %.

Conclusion

Do current proteomic technologies deliver clinically useful results? The opinion of these authors is yes, and the examples described here support that conclusion. However, the potential of techniques such as mass spectrometry to expand the range of proteins analysed has yet to be realised. The required sensitivity and the ability to multiplex assays have been demonstrated, but it will require more development of the sample preparation, data acquisition and analysis steps. Precise and accurate absolute quantification of proteins represents a challenging task, impaired by multiple potential sources of error. These errors, however, can be minimised to a satisfactory level, if sample preparation, measurement and data analysis are adjusted to the respective sample type under investigation, and if each step of the workflow is conducted thoroughly and reproducibly. As pointed out in the previous chapter, to demonstrate the clinical utility of any diagnostic test it is essential to show that:

(a) the analyte can be reliably and consistently measured, i.e., it requires a robust and well-controlled protocol;
(b) the test has a combined sensitivity and specificity that will permit the clear and consistant diagnosis of a disease state from a healthy state, leading to the correct treatment regime.
(c) the use of the test will improve the clinical outcome of patients by targeting precise effective interventions, and providing well-documented follow-up.

To implement appropriate assay designs, it is important to have in mind the purpose of any assay. Studies can be performed to ascertain their utility for screening (who is sick?), risk assessment (who may get sick?), what is the correct therapy (personalised medicine) and assessment of clinical outcome (is the treatment working). The achievement of these goals requires strong multi-disciplinary teams that combine the respective skills and knowledge of clinicians, pathologists and researchers. Analysing a few samples to compare their differences with no reasonable attempt at verification or validation will not result in new diagnostic tests that can be used to make a more personalised medicine a reality. Finding clinically useful molecular biomarkers is not easy, but it is essential if diagnosis is to improve, and it will only happen if clinicians, analytical scientists and patients work together.

Acknowledgments Keith Ashman acknowledges and thanks the Rotary Club of Williamstown, Victoria, Australia for partial salary support through the RoCan program. Gregory Rice was in receipt of an NHMRC Principal Research Fellowship.

References

1. Blonder J, Issaq HJ, Veenstra TD (2011) Proteomic biomarker discovery: it's more than just mass spectrometry. Electrophoresis 32(13):1541–1548. doi:10.1002/elps.201000585
2. Issaq HJ, Waybright TJ, Veenstra TD (2011) Cancer biomarker discovery: opportunities and pitfalls in analytical methods. Electrophoresis 32(9):967–975. doi:10.1002/elps.201000588

3. Wasinger VC, Cordwell SJ, Cerpa-Poljak A, Yan JX, Gooley AA, Wilkins MR, Duncan MW, Harris R, Williams KL, Humphery-Smith I (1995) Progress with gene-product mapping of the Mollicutes: Mycoplasma genitalium. Electrophoresis 16(7):1090–1094
4. Blackstock WP, Weir MP (1999) Proteomics: quantitative and physical mapping of cellular proteins. Trends Biotechnol 17(3):121–127
5. Deribe YL, Pawson T, Dikic I (2010) Post-translational modifications in signal integration. Nat Struct Mol Biol 17(6):666–672. doi:10.1038/nsmb.1842
6. Bence-Jones H (1848) On a new substance occurring in the urine of a patient with mollities ossium. Philos Trans R Soc Lond 138:55–62
7. Kyle RA (1994) Multiple myeloma: how did it begin? Mayo Clin Proc 69(7):680–683
8. Sinclair D, Dagg JH, Smith JG, Stott DI (1986) The incidence and possible relevance of Bence-Jones protein in the sera of patients with multiple myeloma. Br J Haematol 62(4):689–694
9. Polanski M, Anderson NL (2007) A list of candidate cancer biomarkers for targeted proteomics. Biomark Insights 1:1–48
10. Rifai N, Gillette MA, Carr SA (2006) Protein biomarker discovery and validation: the long and uncertain path to clinical utility. Nat Biotechnol 24(8):971–983. doi:10.1038/nbt1235
11. Whiteley G (2008) Bringing diagnostic technologies to the clinical laboratory: rigor, regulation, and reality. Proteomics Clin Appl 2(10-11):1378–1385. doi:10.1002/prca.200780170
12. Liu X, Valentine SJ, Plasencia MD, Trimpin S, Naylor S, Clemmer DE (2007) Mapping the human plasma proteome by SCX-LC-IMS-MS. J Am Soc Mass Spectrom 18(7):1249–1264. doi:10.1016/j.jasms.2007.04.012
13. Anderson NL, Anderson NG (2002) The human plasma proteome: history, character, and diagnostic prospects. Mol Cell Proteomics 1(11):845–867
14. States DJ, Omenn GS, Blackwell TW, Fermin D, Eng J, Speicher DW, Hanash SM (2006) Challenges in deriving high-confidence protein identifications from data gathered by a HUPO plasma proteome collaborative study. Nat Biotechnol 24(3):333–338. doi:10.1038/nbt1183
15. Schiess R, Wollscheid B, Aebersold R (2009) Targeted proteomic strategy for clinical biomarker discovery. Mol Oncol 3(1):33–44. doi:10.1016/j.molonc.2008.12.001
16. Anderson NL, Polanski M, Pieper R, Gatlin T, Tirumalai RS, Conrads TP, Veenstra TD, Adkins JN, Pounds JG, Fagan R, Lobley A (2004) The human plasma proteome: a nonredundant list developed by combination of four separate sources. Mol Cell Proteomics 3(4):311–326. doi:10.1074/mcp.M300127-MCP200
17. Randall SA, McKay MJ, Molloy MP (2010) Evaluation of blood collection tubes using selected reaction monitoring MS: implications for proteomic biomarker studies. Proteomics 10(10):2050–2056. doi:10.1002/pmic.200900517
18. Bergen HR 3rd, Zeldenrust SR, Butz ML, Snow DS, Dyck PJ, Klein CJ, O'Brien JF, Thibodeau SN, Muddiman DC (2004) Identification of transthyretin variants by sequential proteomic and genomic analysis. Clin Chem 50(9):1544–1552. doi:10.1373/clinchem.2004.033266
19. Bergen HR, Lacey JM, O'Brien JF, Naylor S (2001) Online single-step analysis of blood proteins: the transferrin story. Anal Biochem 296(1):122–129. doi:10.1006/abio.2001.5232
20. Nepomuceno AI, Mason CJ, Muddiman DC, Bergen HR 3rd, Zeldenrust SR (2004) Detection of genetic variants of transthyretin by liquid chromatography-dual electrospray ionization fourier-transform ion-cyclotron-resonance mass spectrometry. Clin Chem 50(9):1535–1543. doi:10.1373/clinchem.2004.033274
21. Paik JH, Choe G, Kim H, Choe JY, Lee HJ, Lee CT, Lee JS, Jheon S, Chung JH (2011) Screening of anaplastic lymphoma kinase rearrangement by immunohistochemistry in non-small cell lung cancer: correlation with fluorescence in situ hybridization. J Thorac Oncol 6(3):466–472. doi:10.1097/JTO.0b013e31820b82e8
22. Bertino G, Neri S, Bruno CM, Ardiri AM, Calvagno GS, Malaguarnera M, Toro A, Clementi S, Bertino N, Di Carlo I (2011) Diagnostic and prognostic value of alpha-fetoprotein, des-gamma-carboxy prothrombin and squamous cell carcinoma antigen immunoglobulin M complexes in hepatocellular carcinoma. Minerva Med 102(5):363–371
23. Sherman M (2011) Current status of alpha-fetoprotein testing. Gastroenterol Hepatol (N Y) 7(2):113–114

24. Fonseca R, San Miguel J (2007) Prognostic factors and staging in multiple myeloma. Hematol Oncol Clin North Am 21(6):1115–1140. doi:10.1016/j.hoc.2007.08.010, ix
25. Cole LA (2012) hCG, the wonder of today's science. Reprod Biol Endocrinol 10:24. doi:10.1186/1477-7827-10-24
26. Stenman UH (2004) Standardization of assays for human chorionic gonadotropin. Clin Chem 50(5):798–800. doi:10.1373/clinchem.2003.031013
27. Stenman UH, Alfthan H, Hotakainen K (2004) Human chorionic gonadotropin in cancer. Clin Biochem 37(7):549–561. doi:10.1016/j.clinbiochem.2004.05.008
28. Dekking E, van der Velden VH, Bottcher S, Bruggemann M, Sonneveld E, Koning-Goedheer A, Boeckx N, Lucio P, Sedek L, Szczepanski T, Kalina T, Kovac M, Evans P, Hoogeveen PG, Flores-Montero J, Orfao A, Comans-Bitter WM, Staal FJ, van Dongen JJ (2010) Detection of fusion genes at the protein level in leukemia patients via the flow cytometric immunobead assay. Best Pract Res Clin Haematol 23(3):333–345. doi:10.1016/j.beha.2010.09.010
29. Foroni L, Wilson G, Gerrard G, Mason J, Grimwade D, White HE, de Castro DG, Austin S, Awan A, Burt E, Clench T, Farruggia J, Hancock J, Irvine AE, Kizilors A, Langabeer S, Milner BJ, Nickless G, Schuh A, Sproul A, Wang L, Wickham C, Cross NC (2011) Guidelines for the measurement of BCR-ABL1 transcripts in chronic myeloid leukaemia. Br J Haematol 153(2):179–190. doi:10.1111/j.1365-2141.2011.08603.x
30. Safaee Ardekani G, Jafarnejad SM, Tan L, Saeedi A, Li G (2012) The prognostic value of BRAF mutation in colorectal cancer and melanoma: a systematic review and meta-analysis. PLoS One 7(10), e47054. doi:10.1371/journal.pone.0047054
31. Cabrera-Abreu JC, Smellie WS, Bowley R, Shaw N (2012) Best practice in primary care pathology: review 13. J Clin Pathol 65(2):97–100. doi:10.1136/jclinpath-2011-200292
32. Moore LE, Pfeiffer RM, Zhang Z, Lu KH, Fung ET, Bast RC Jr (2012) Proteomic biomarkers in combination with CA 125 for detection of epithelial ovarian cancer using prediagnostic serum samples from the Prostate, Lung, Colorectal, and Ovarian (PLCO) Cancer Screening Trial. Cancer 118(1):91–100. doi:10.1002/cncr.26241
33. Ahmed SR, Ball DW (2011) Clinical review: incidentally discovered medullary thyroid cancer: diagnostic strategies and treatment. J Clin Endocrinol Metab 96(5):1237–1245. doi:10.1210/jc.2010-2359
34. Bacolod MD, Barany F (2011) Molecular profiling of colon tumors: the search for clinically relevant biomarkers of progression, prognosis, therapeutics, and predisposition. Ann Surg Oncol 18(13):3694–3700. doi:10.1245/s10434-011-1615-5
35. Grunnet M, Sorensen JB (2012) Carcinoembryonic antigen (CEA) as tumor marker in lung cancer. Lung Cancer 76(2):138–143. doi:10.1016/j.lungcan.2011.11.012
36. Freedman A (2012) Follicular lymphoma: 2012 update on diagnosis and management. Am J Hematol 87(10):988–995. doi:10.1002/ajh.23313
37. Jacobson CA, Freedman AS (2012) Early stage follicular lymphoma, current management and controversies. Curr Opin Oncol 24(5):475–479. doi:10.1097/CCO.0b013e328356898b
38. Ramachandran R, Bech P, Murphy KG, Dhillo WS, Meeran KM, Chapman RS, Caplin M, Ghatei MA, Bloom SR, Martin NM (2012) Improved diagnostic accuracy for neuroendocrine neoplasms using two chromogranin A assays. Clin Endocrinol (Oxf) 76(6):831–836. doi:10.1111/j.1365-2265.2011.04319.x
39. Syversen U, Ramstad H, Gamme K, Qvigstad G, Falkmer S, Waldum HL (2004) Clinical significance of elevated serum chromogranin A levels. Scand J Gastroenterol 39(10):969–973. doi:10.1080/00365520410003362
40. Caraway NP, Khanna A, Fernandez RL, Payne L, Bassett RL Jr, Zhang HZ, Kamat A, Katz RL (2010) Fluorescence in situ hybridization for detecting urothelial carcinoma: a clinicopathologic study. Cancer Cytopathol 118(5):259–268. doi:10.1002/cncy.20099
41. De Petris L, Branden E, Herrmann R, Sanchez BC, Koyi H, Linderholm B, Lewensohn R, Linder S, Lehtio J (2011) Diagnostic and prognostic role of plasma levels of two forms of cytokeratin 18 in patients with non-small-cell lung cancer. Eur J Cancer 47(1):131–137. doi:10.1016/j.ejca.2010.08.006
42. Holdenrieder S, Stieber P, Liska V, Treska V, Topolcan O, Dreslerova J, Matejka VM, Finek J, Holubec L (2012) Cytokeratin serum biomarkers in patients with colorectal cancer. Anticancer Res 32(5):1971–1976

43. Cheng L, Alexander RE, Maclennan GT, Cummings OW, Montironi R, Lopez-Beltran A, Cramer HM, Davidson DD, Zhang S (2012) Molecular pathology of lung cancer: key to personalized medicine. Mod Pathol 25(3):347–369. doi:10.1038/modpathol.2011.215
44. Lim E, Metzger-Filho O, Winer EP (2012) The natural history of hormone receptor-positive breast cancer. Oncology (Williston Park) 26(8):688–694, 696
45. Jeong S, Park Y, Cho Y, Kim YR, Kim HS (2012) Diagnostic values of urine CYFRA21-1, NMP22, UBC, and FDP for the detection of bladder cancer. Clin Chim Acta 414C:93–100. doi:10.1016/j.cca.2012.08.018
46. Montagnana M, Danese E, Giudici S, Franchi M, Guidi GC, Plebani M, Lippi G (2011) HE4 in ovarian cancer: from discovery to clinical application. Adv Clin Chem 55:1–20
47. Molina R, Escudero JM, Munoz M, Auge JM, Filella X (2012) Circulating levels of HER-2/neu oncoprotein in breast cancer. Clin Chem Lab Med 50(1):5–21. doi:10.1515/cclm.2011.822
48. Maleddu A, Pantaleo MA, Nannini M, Biasco G (2011) The role of mutational analysis of KIT and PDGFRA in gastrointestinal stromal tumors in a clinical setting. J Transl Med 9:75. doi:10.1186/1479-5876-9-75
49. Heideman DA, Lurkin I, Doeleman M, Smit EF, Verheul HM, Meijer GA, Snijders PJ, Thunnissen E, Zwarthoff EC (2012) KRAS and BRAF mutation analysis in routine molecular diagnostics: comparison of three testing methods on formalin-fixed, paraffin-embedded tumor-derived DNA. J Mol Diagn 14(3):247–255. doi:10.1016/j.jmoldx.2012.01.011
50. Venkitaraman R, Johnson B, Huddart RA, Parker CC, Horwich A, Dearnaley DP (2007) The utility of lactate dehydrogenase in the follow-up of testicular germ cell tumours. BJU Int 100(1):30–32. doi:10.1111/j.1464-410X.2007.06905.x
51. Shariat SF, Savage C, Chromecki TF, Sun M, Scherr DS, Lee RK, Lughezzani G, Remzi M, Marberger MJ, Karakiewicz PI, Vickers AJ (2011) Assessing the clinical benefit of nuclear matrix protein 22 in the surveillance of patients with nonmuscle-invasive bladder cancer and negative cytology: a decision-curve analysis. Cancer 117(13):2892–2897. doi:10.1002/cncr.25903
52. Lumen N, Fonteyne V, De Meerleer G, De Visschere P, Ost P, Oosterlinck W, Villeirs G (2012) Screening and early diagnosis of prostate cancer: an update. Acta Clin Belg 67(4):270–275
53. Cox AE, LeBeau SO (2011) Diagnosis and treatment of differentiated thyroid carcinoma. Radiol Clin North Am 49(3):453–462. doi:10.1016/j.rcl.2011.02.006, vi
54. Spencer CA (2011) Clinical review: clinical utility of thyroglobulin antibody (TgAb) measurements for patients with differentiated thyroid cancers (DTC). J Clin Endocrinol Metab 96(12):3615–3627. doi:10.1210/jc.2011-1740
55. Iwaki T, Urano T, Umemura K (2012) PAI-1, progress in understanding the clinical problem and its aetiology. Br J Haematol 157(3):291–298. doi:10.1111/j.1365-2141.2012.09074.x
56. Fung ET (2010) A recipe for proteomics diagnostic test development: the OVA1 test, from biomarker discovery to FDA clearance. Clin Chem 56(2):327–329. doi:10.1373/clinchem.2009.140855
57. Retel VP, Joore MA, van Harten WH (2012) Head-to-head comparison of the 70-gene signature versus the 21-gene assay: cost-effectiveness and the effect of compliance. Breast Cancer Res Treat 131(2):627–636. doi:10.1007/s10549-011-1769-7
58. Lovgren J, Piironen T, Overmo C, Dowell B, Karp M, Pettersson K, Lilja H, Lundwall A (1995) Production of recombinant PSA and HK2 and analysis of their immunologic cross-reactivity. Biochem Biophys Res Commun 213(3):888–895
59. Pettersson K, Piironen T, Seppala M, Liukkonen L, Christensson A, Matikainen MT, Suonpaa M, Lovgren T, Lilja H (1995) Free and complexed prostate-specific antigen (PSA): in vitro stability, epitope map, and development of immunofluorometric assays for specific and sensitive detection of free PSA and PSA-alpha 1-antichymotrypsin complex. Clin Chem 41(10):1480–1488
60. Schroder FH, Hugosson J, Roobol MJ, Tammela TL, Ciatto S, Nelen V, Kwiatkowski M, Lujan M, Lilja H, Zappa M, Denis LJ, Recker F, Paez A, Maattanen L, Bangma CH, Aus G, Carlsson S, Villers A, Rebillard X, van der Kwast T, Kujala PM, Blijenberg BG, Stenman UH, Huber A, Taari K, Hakama M, Moss SM, de Koning HJ, Auvinen A (2012) Prostate-cancer mortality at 11 years of follow-up. N Engl J Med 366(11):981–990. doi:10.1056/NEJMoa1113135
61. Tigue CC, Fitzner KA, Alkhatib M, Schmid E, Bennett CL (2007) The value of innovation: the economics of targeted drugs for cancer. Target Oncol 2(2):113–119. doi:10.1007/S11523-007-0043-8

62. Neumann J, Zeindl-Eberhart E, Kirchner T, Jung A (2009) Frequency and type of KRAS mutations in routine diagnostic analysis of metastatic colorectal cancer. Pathol Res Pract 205(12):858–862. doi:10.1016/J.Prp.2009.07.010
63. Shushan B (2010) A review of clinical diagnostic applications of liquid chromatography-tandem mass spectrometry. Mass Spectrom Rev 29(6):930–944. doi:10.1002/mas.20295
64. Addona TA, Abbatiello SE, Schilling B, Skates SJ, Mani DR, Bunk DM, Spiegelman CH, Zimmerman LJ, Ham AJL, Keshishian H, Hall SC, Allen S, Blackman RK, Borchers CH, Buck C, Cardasis HL, Cusack MP, Dodder NG, Gibson BW, Held JM, Hiltke T, Jackson A, Johansen EB, Kinsinger CR, Li J, Mesri M, Neubert TA, Niles RK, Pulsipher TC, Ransohoff D, Rodriguez H, Rudnick PA, Smith D, Tabb DL, Tegeler TJ, Variyath AM, Vega-Montoto LJ, Wahlander A, Waldemarson S, Wang M, Whiteaker JR, Zhao L, Anderson NL, Fisher SJ, Liebler DC, Paulovich AG, Regnier FE, Tempst P, Carr SA (2009) Multi-site assessment of the precision and reproducibility of multiple reaction monitoring-based measurements of proteins in plasma (vol 27, pg 633, 2009). Nat Biotechnol 27(9):864. doi:10.1038/Nbt0909-864b
65. Ruppen-Canas I, Lopez-Casas PP, Garcia F, Ximenez-Embun P, Munoz M, Morelli MP, Real FX, Serna A, Hidalgo M, Ashman K (2012) An improved quantitative mass spectrometry analysis of tumor specific mutant proteins at high sensitivity. Proteomics 12(9):1319–1327. doi:10.1002/pmic.201100611
66. Wang Q, Chaerkady R, Wu J, Hwang HJ, Papadopoulos N, Kopelovich L, Maitra A, Matthaei H, Eshleman JR, Hruban RH, Kinzler KW, Pandey A, Vogelstein B (2011) Mutant proteins as cancer-specific biomarkers. Proc Natl Acad Sci U S A 108(6):2444–2449. doi:10.1073/pnas.1019203108
67. Rauh M (2012) LC-MS/MS for protein and peptide quantification in clinical chemistry. J Chromatogr B Analyt Technol Biomed Life Sci 883–884:59–67. doi:10.1016/j.jchromb.2011.09.030
68. Farrah T, Deutsch EW, Kreisberg R, Sun Z, Campbell DS, Mendoza L, Kusebauch U, Brusniak MY, Huttenhain R, Schiess R, Selevsek N, Aebersold R, Moritz RL (2012) PASSEL: The PeptideAtlas SRM Experiment Library. Proteomics. doi:10.1002/pmic.201100515
69. Ciccimaro E, Blair IA (2010) Stable-isotope dilution LC-MS for quantitative biomarker analysis. Bioanalysis 2(2):311–341. doi:10.4155/bio.09.185
70. Anderson NL, Anderson NG, Haines LR, Hardie DB, Olafson RW, Pearson TW (2004) Mass spectrometric quantitation of peptides and proteins using Stable Isotope Standards and Capture by Anti-Peptide Antibodies (SISCAPA). J Proteome Res 3(2):235–244
71. Whiteaker JR, Zhao L, Anderson L, Paulovich AG (2009) An automated and multiplexed method for high throughput peptide immunoaffinity enrichment and multiple reaction monitoring mass spectrometry-based quantification of protein biomarkers. Mol Cell Proteomics. doi:10.1074/mcp.M900254-MCP200. pii: M900254-MCP200
72. Anderson NL, Jackson A, Smith D, Hardie D, Borchers C, Pearson TW (2009) SISCAPA peptide enrichment on magnetic beads using an in-line bead trap device. Mol Cell Proteomics 8(5):995–1005. doi:10.1074/Mcp.M800446-Mcp200
73. Kathiresan S, Gona P, Larson MG, Vita JA, Mitchell GF, Tofler GH, Levy D, Newton-Cheh C, Wang TJ, Benjamin EJ, Vasan RS (2006) Cross-sectional relations of multiple biomarkers from distinct biological pathways to brachial artery endothelial function. Circulation 113(7):938–945. doi:10.1161/CIRCULATIONAHA.105.580233
74. Wang TJ, Gona P, Larson MG, Tofler GH, Levy D, Newton-Cheh C, Jacques PF, Rifai N, Selhub J, Robins SJ, Benjamin EJ, D'Agostino RB, Vasan RS (2006) Multiple biomarkers for the prediction of first major cardiovascular events and death. N Engl J Med 355(25):2631–2639. doi:10.1056/NEJMoa055373
75. Mor G, Visintin I, Lai Y, Zhao H, Schwartz P, Rutherford T, Yue L, Bray-Ward P, Ward DC (2005) Serum protein markers for early detection of ovarian cancer. Proc Natl Acad Sci U S A 102(21):7677–7682. doi:10.1073/pnas.0502178102
76. Mitchell MD, Rice G (2102) Early pregnancy screening for complications of pregnancy: proteomic profiling approaches. In: Zheng J (ed) Recent advances in research on the human placenta. Intech
77. Brian M. Nolen and Anna E. Lokshin (2013) Biomarker Testing for Ovarian Cancer: Clinical Utility of Multiplex Assays. Mol Diagn Ther. Jun; 17(3): 139–146

Analysis of DNA Methylation in Clinical Samples: Methods and Applications

Alexander Dobrovic

Methods for Analysis of DNA Methylation: Introduction and Overview

DNA methylation usually occurs at the cytosine of a CpG dinucleotide. This chapter will focus on DNA methylation detection methods relevant to analysis of the clinical cancer specimens that would be found in a pathology laboratory. Certain key applications that are in use in cancer diagnostics will also be considered. However, much of the methodological considerations will also be applicable to the pre-clinical research laboratory with cell line or xenograft models of cancer.

As DNA methylation is a covalent modification of cytosine in a CpG dinucleotide context, its consequent stability makes it an attractive target for diagnostics. However, assessing DNA methylation has not yet become commonplace in the diagnostic molecular pathology laboratory. Before tests involving methylation analysis become part of routine practice, extensive validation is necessary as has been recently reviewed [1, 2]. Consensus is difficult as major discrepancies can occur between studies due to different methodologies and different regions being analysed. For this reason, meta-analyses evaluating DNA methylation biomarkers that do not take these discrepancies into account should be treated with caution.

Methylation analysis is more complex than mutational analysis. Although the methylation of individual CpG dinucleotides can have a profound effect (for example, on transcription factor binding), the methylation density of multiple

---

A. Dobrovic (✉)
Translational Genomics and Epigenomics Laboratory, Olivia Newton-John Cancer Research Institute, Heidelberg, VIC, Australia

School of Cancer Medicine, La Trobe University, Bundoora, VIC, Australia

Department of Pathology, University of Melbourne, Parkville, VIC, Australia
e-mail: alex.dobrovic@onjcri.org.au

S.R. Lakhani, S.B. Fox (eds.), *Molecular Pathology in Cancer Research*,
DOI 10.1007/978-1-4939-6643-1_12

adjacent CpG dinucleotides in a region is usually more important. Thus, analysis first requires the recognition of a key region. This is often not a trivial task, especially when methylation needs to be related to transcriptional inactivation. As methylation needs to be considered in terms of the combinatorial modifications of multiple adjacent CpG dinucleotides in a region, different PCR-based methodologies can yield varying results [3] and an appropriate methodology must be chosen [1, 4].

Methodologies for DNA methylation analysis can be broadly divided into two categories. Locus-specific methodologies analyse single regions. Global methodologies are often more complex, require bioinformatic processing in order to analyse the information, and thus usually entail significantly longer turn around times for both methodology and analysis, making them currently less clinically applicable.

Currently, the principal question likely to be asked in the specialist pathology laboratory is whether a specific gene or region is methylated in a specific pathological situation; for example, is the *MLH1* gene promoter methylated in a colorectal cancer specimen showing MLH1 negative immunohistochemistry? The locus-specific methodologies can be divided into those that interrogate single sites for DNA methylation and those that examine a region for DNA methylation. The former are of limited use unless multiplexed (thereby becoming a methodology that examines a region) as methylation should be considered over multiple CpGs in a region.

Each of the various methods used to study locus-specific methylation has its characteristic advantages and disadvantages, particularly when dealing with complex methylation patterns (reviewed in [3]). The methods chosen should be dictated by the nature of the sample that needs to be analysed and the information required.

Many of the methodologies developed for methylation analysis either were not widely adopted or have fallen out of general use. Some of these can be found in older reviews of methylation methodology (e.g. [5]).

Much of this chapter will focus on methodologies that are most likely to be utilised by diagnostic laboratories. These will be mainly polymerase chain reaction (PCR) based. There will be an inevitable bias to those methodologies that the author of this chapter has had direct experience with. The reader is thus encouraged to read some of the other excellent recent reviews that are available (e.g. [6]) as a complement to this chapter.

5-Methylcytosine can undergo active demethylation mediated by the TET (ten-eleven translocation) family of enzymes via 5-hydroxymethylcytosine (5hmC) to 5-formylcytosine and 5-carboxylcytosine, which are then removed by the base excision repair enzyme thymine DNA glycosylase (reviewed in [7]). The significance of these intermediates in cellular processes is still poorly understood. 5hmC is depleted in many tumours indicating a key role for this epigenetic modification [8]. It should be noted that most non-global methodologies do not differentiate 5-methylcytosine from its breakdown intermediates. It is unclear what the implication of this will be for diagnostics.

Bisulfite Modification

DNA methylation information is lost once the DNA is amplified in vitro. Thus, to retain methylation information during PCR, the DNA must be modified. Sodium bisulfite treatment, which deaminates cytosine to uracil, is the method of choice for modification.

5-methylcytosine but not cytosine is resistant to deamination during sodium bisulfite treatment [9]. Thus, following PCR of bisulfite-modified DNA, the remaining cytosines are a direct readout of the methylated cytosines in the original DNA template [10, 11]. Accordingly, methylation can be determined by comparison to the original sequence.

The key parameters for effective bisulfite modification have been described [12, 13]. It is highly advisable to use one of the kits that are available as the consequences of incomplete bisulfite conversion can be serious for accurate interpretation of methylation information, particularly when using methylation-specific PCR (see below). Incomplete bisulfite conversion can be readily observed by sequencing as some of the non-CpG cytosines will remain as cytosines. In practice, there is a low rate (1 % or less) of non-conversion.

PCR-Based Methods for the Detection of DNA Methylation

Importantly, the amount of tissue available for molecular studies in the molecular pathology laboratory is often limited and the DNA from formalin-fixed paraffin-embedded (FFPE) tissues is frequently highly fragmented (reviewed in [14]). All of the PCR-based techniques can potentially use DNA made from FFPE tissues. Perhaps the most important adaptation is the use of short amplicons to compensate for the fragmentation. In most methods, bisulfite-modified DNA is amplified with strand-specific primers framing the region of interest, which undergoes sequencing or other analysis.

Analysis of tumour material is further complicated, both because specimens can contain substantial amounts of normal cells with normal methylation patterns, and because there can be intra-tumoural heterogeneity of methylation patterns.

Sanger Sequencing and Pyrosequencing

The bisulfite-modification-based genomic sequencing methodology [10, 15] revolutionised methylation analysis. The PCR product is directly sequenced providing a readout of the mean methylation of each CpG in the sequence. This type of direct sequencing methylation analysis is still widely used.

The gold standard for analysis of methylation is clonal sequencing as it reveals the complexity of epialleles present in a sample. For tumour samples, sequence information from many clones is desirable, as a considerable amount of the sequences may be derived from normal tissue, even when the tumour content is enriched. Individual PCR products are cloned prior to sequencing of multiple clones to determine the methylation of each cytosine in individual DNA molecules [10]. An early example of clonal sequencing revealed the complexity of epiallelic methylation patterns in the promoter region of the retinoblastoma gene.

Originally, clonal sequencing was performed by cloning PCR products into plasmid vectors and sequencing multiple individual clones. Limiting dilution prior to PCR amplification proved to be a much more cost-effective path to clonal sequencing [16, 17]. In this digital approach, epialleles are separated prior to PCR amplification. Separation eliminates PCR bias and enables accurate quantification. Samples for sequencing can be chosen based on their melting pattern (see below) leading to significant cost reductions [16, 17]. Nevertheless, Sanger sequencing of individual epialleles remains an expensive proposition.

Pyrosequencing was introduced as an alternative to Sanger sequencing for direct sequencing [18–20]. Pyrosequencing has now established itself as a leading sequencing methodology for the analysis of clinical samples [4]. Pyrosequencing provides accurate sequencing immediately adjacent to the sequencing primer, allowing sequencing of PCR products with small inserts. Moreover, it enables quantitative assessment of methylation at each CpG site with sensitivities approaching 5–10 % [21].

Deep sequencing of methylation is now becoming increasingly common, either at individual loci [22] or at multiple multiplexed sites [23]. Massively parallel sequencing of amplicons or captured sequences is likely to become the principal tool for methylation analysis in the clinic. Due to the complexity of the data that can be obtained, synoptic algorithms must be developed for its visualisation [24].

Methylation-Specific PCR

A variety of readily performed methylation detection methods that did not involve sequencing were developed for the cost-effective analysis of multiple samples. Methylation-specific PCR (MSP) was the first of these [25] and is still widely used to interrogate whether a given region is methylated. The primers specifically amplify methylated DNA. MSP is based on the principle that PCR primers with a mismatched 3′ end will not extend under stringent conditions and thus the 3′ ends of both primers are designed to overlay CpGs in the bisulfite-modified DNA. This usually involves the placing of one or more CpG Cs at the 3′ base or near to the 3′ end. If a band is seen on a gel after PCR, it is concluded that the sample is methylated. Use of second pair of primers specific for unmethylated sequences was reported in the original publication [25], and is often used but rarely useful as there is always a degree of normal cell content in tumour samples. MSP is the most sensitive non-quantitative technique available and can detect very low levels of methylation.

MSP is a useful rapid method to screen for methylation when the regions are heavily methylated. It is less suited when regions show highly variable methylation, as has been reported for many genes including the promoter CpG island of *CDKN2A* (p15) [26, 27] as exact matches to the primer sequences occur less frequently and methylation can be underestimated.

An important limitation of the MSP approach is its susceptibility to false positives. Considerable overestimates for the methylation frequency of particular genes have been reported by some laboratories when MSP is used (reviewed in [28, 29]). Incomplete bisulfite conversion can lead to false positives. Alternatively, amplification can occur across the 3′ mismatch(es), especially if the annealing temperature is too low. Thus, appropriate negative controls such as a cell line with known methylation status are critical. Raising the annealing temperature and the use of a hot-start methodology can minimise this problem.

The sensitivity of MSP can also lead to (false) positives because of the amplification of rare populations of methylated sequences either in the tumour or in the normal somatic tissue [29]. For some genes, peripheral blood is not suitable as a negative control as low levels of methylation are present [30]. If the tumour is heterogeneous, with only a small proportion of methylated cells, it would not be correct to call the tumour methylated for that particular gene. It is difficult to determine whether the signal arises from the predominant proportion of cells or from a small subpopulation thereof.

The above limitations arise because standard MSP is a non-quantitative methodology. For this reason, it is recommended that a quantitative MSP assay using real-time PCR is always used. Real-time PCR analysis also requires no further manipulation after the PCR step, which allows high-throughput analysis and eliminates problems resulting from possible cross-contamination by PCR products when they are removed from their amplification tubes or wells.

Quantitative Methylation-Specific PCR

There are several adaptations of MSP using real-time PCR for methylation analysis. MSP can be carried out with the TaqMan methodology, which uses a fluorescent probe to monitor amplification [31], an approach that has become known as MethyLight [32]. Bisulfite-modified DNA is amplified using MSP primers and monitored by a TaqMan probe containing a 5′ fluorescent reporter and a 3′ quencher which binds inside the amplified region. The 5′ to 3′ nuclease activity of *Taq* DNA polymerase cleaves the hybridised probe, separating the fluorescent reporter from the quencher each time an amplification occurs. The intensity of the fluorescent signal is proportional to the amount of PCR product, allowing quantification of the PCR reaction.

MethyLight is capable of detecting methylated alleles in the presence of a 10,000-fold excess of unmethylated alleles [32]. However, it should be noted that

10 ng of bisulfite-modified DNA contains at best 6600 allelic templates and often much less dependent on DNA fragmentation both before and post-bisulfite treatment [12]. Methylation of candidate genes needs to be scored relative to standard curves constructed from dilutions of DNA that is methylated for the gene in question. These dilutions can be made in DNA prepared from a cell line or normal DNA that is correspondingly unmethylated.

Most commonly used MethyLight probes contain several CpG sites in their sequence. According to the probe design and the PCR conditions used, the TaqMan probe may bind only if all the probe target CpG sites are also methylated or may bind when only some of the sites are methylated. This may result in underestimating of the methylation levels when the methylation is heterogeneous.

A simpler real-time approach is to convert an established MSP protocol to a quantitative protocol using a dye such as SYBR Green which fluoresces when it intercalates into double-stranded DNA [33]. It is critical that the MSP reaction is highly specific, as intercalating dye-based methods cannot readily distinguish between specific and non-specific reactions in the way that probe-based assays can.

One solution to the problem of specificity in dye-based qMSP is to follow the qPCR step with a high resolution melting (HRM) analysis as in the SMART-MSP protocol [34]. The same intercalating dye enables both the qPCR and the HRM analysis. The assay is typically performed with no CpGs between the primers though other designs are possible. Non-specific amplification is identified by an atypical melting pattern.

DNA Melting Analysis

Several methylation typing methods are based directly on melting analysis [35, 36]. Melting monitors the fluorescence of a double-stranded DNA-binding dye as the double-stranded PCR product is slowly denatured by increasing temperature. The analysis is done in the same tube or well as the PCR, which allows high throughput and reduces potential problems resulting from PCR product contamination.

These methods screen a number of CpG sites occurring close together in the same PCR amplicon. DNA denatures in discrete segments called melting domains as the concentration of a denaturant or the temperature increases. The melting of a domain is determined by its sequence. After bisulfite modification, unmethylated sequences or partially methylated sequences are less cytosine rich and more thymine rich than methylated sequences and consequently melt at a lower temperature. Therefore, amplicons will have melting temperatures according to their degree of methylation.

Melting methodologies became more widely adopted after the introduction of high-resolution melting capable machines. Methylation-sensitive high-resolution melting (MS-HRM) [35] is now widely used. In many cases, MS-HRM can achieve sensitivities as low as 0.1–1 % methylated sequences. This latter sensitivity is useful when it is desirable to assess low-level methylation such as with constitutional mosaic methylation [37] or for the detection of circulating tumour DNA [38].

Use of short PCR amplicons is desirable not only to gain maximum information from fragmented DNA but also to reduce the complexity of the melting patterns by reducing the amount of melting domains. When multiple amplicons with small difference in methylation are present, heteroduplexes will form in the later stages of PCR, where already formed complementary strands increasingly out-compete primers for duplex formation. The resultant complex melting profiles allow visualisation of the heterogeneity of methylation but confound accurate quantification of methylated epialleles [17].

MS-HRM is particularly powerful when teamed with Pyrosequencing. Melting analysis replaces the quality control using agarose gel analysis normally used prior to the Pyrosequencing step and also provides important qualitative information about the degree of epiallelic heterogeneity in its own right [39].

An important adaptation of MS-HRM is the use of limiting dilution so that amplification can occur from individual templates (digital MS-HRM). This can either be directly analysed to count methylated and unmethylated templates [16] or used to analyse more complex methylation patterns in which individual reactions can be directly sequenced [17]. MS-HRM can thus be used effectively as the first step for genomic sequencing studies of methylation as both a quality measure and for cost reduction. PCR amplicons that are clearly unmethylated do not need to be sequenced [17].

Primer Design

Primer design is a critical aspect of PCR-based analysis of bisulfite-modified DNA. Primers can be designed to analyse any region but are often designed so that the amplified region overlaps the transcriptional start site and consists mainly of sequence prior to the start site.

Primer design is often challenging. Several websites for designing methylation primers have become available but do not in our opinion replace careful manual design. An important consequence of bisulfite modification is that the two DNA strands are no longer complementary. Accordingly, forward and reverse primers are designed using either the sense or antisense strand after modification. Primers are generally designed to amplify from the modified sense strand, but when it is difficult to design primers, the modified antisense strand can be considered. We always check the primers using the Macintosh freeware program Amplify to identify potential primer dimers as well as non-specific primer binding within the region of interest.

Primers are either designed to amplify both methylated and unmethylated sequences (methylation independent PCR (MIP) primers) or only methylated sequences (methylation-specific PCR (MSP) primers). For most methodologies, MIP primers are used.

Careful primer choice is important to minimise the amplification of unmodified or incompletely modified DNA. This is particularly important for MSP-based methods. Primers are chosen to have one or more Ts derived from a non-CpG cytosine at or as near as possible to the 3′ end of each primer.

For MSP, regions that contain frequent CpG sites are chosen and often the 3′ end of the primers is positioned over several CpG dinucleotides. For maximum specificity, it is desirable to have the 3′ end corresponding to a cytosine of a CpG dinucleotide, but this is not always possible. MSP is best suited for screening genes where detailed methylation patterns are already known and the 3′ ends can be placed at the most frequently methylated sites.

Quite different requirements are necessary for MIP primers that amplify all converted sequences regardless of methylation status prior to analysis [5, 40]. MIP primers are positioned to avoid CpG sites or including as few as possible and placing them as far as possible to the 5′ end of the primer. For HRM, we try to retain one cytosine of a CpG as close as possible to the 5′ end of the primer which can subsequently be used to steer amplification in favour of methylated sequences. Because methylated sequences can have an amplification disadvantage, manipulation of annealing temperature can either be used to minimise PCR bias or to favour amplification of rare methylated sequences.

It is preferable to use primers with high melting temperatures to help with specificity, particularly because the DNA sequence is less complex after bisulfite modification due to the depletion of cytosines. The last five 3′ nucleotides preferably should contain two or three (but not more) G's (C's on the anti-sense strand) to stabilise the 3′ end of the primer. The Web program OligoCalc: Oligonucleotide Properties Calculator is used to check the melting temperatures (http://www.basic.northwestern.edu/biotools/oligocalc.html). The most reliable results are when the salt adjusted Tm is set at 5 °C above the intended annealing temperature.

The use of nesting is not recommended because of the increased potential for PCR contamination. There is no need for nesting when the PCR conditions are sufficiently sensitive to amplify a single template.

Bisulfite Independent Methods

Instead of using bisulfite modification, some methodologies exploit the properties of restriction enzymes with a CpG in their recognition sequence that only cleaves DNA if the cytosine is unmethylated. The first methodologies to analyse methylation relied on Southern blotting analysis using these restriction enzymes (e.g. [41]). These were followed by PCR-based methods in which PCR amplification was used to determine if the restriction enzyme cuts the pre-amplified sequence [42]. Incomplete enzyme digestion can be a problem. This approach has now been refined in the MS-RE PCR protocol to allow analysis of multiple sites from a small starting amount of DNA [43].

Methylation-specific multiplex ligation-dependent probe amplification (MS-MLPA) is an adaption of multiplex ligation-dependent probe amplification that also enables methylation analysis of multiple CpG sites [44]. Each assay can analyse up to 40 CpG sites. The PCR products are separated on an automated DNA sequencer or equivalent. MS-MLPA requires little prior expertise in methylation analysis to run the assay, especially as no bisulfite modification needs to be performed, and assays can be purchased directly from the commercial supplier.

Global DNA Methylation Analysis

It would be clearly advantageous to analyse cancer specimens (and matched normal tissues) for the methylation of dozens if not hundreds or thousands of CpG islands at the same time. Such methods can be economical with the often limited amounts of tissues available from cancer biopsies. As always, methods that can deal with highly fragmented DNA are desirable.

Earlier important methodologies to attempt global methylation analysis like restriction landmark genomic sequencing [45] and differential methylation hybridisation [46] are now superseded. As with single locus assays, some methods require bisulfite modification while others utilise methylation-sensitive restriction enzymes. Other methods use the capability of some DNA-binding proteins to bind specifically to methylated sequences. Currently, the most promising approaches are array based or sequencing based.

Deep methylome sequencing, which is based on multiple parallel sequencing, would be the ideal technology for use in methylation analysis due to its complete genomic coverage at single base resolution and reproducibility (reviewed in [47]). However, cost of both the reagents and the bioinformatic analysis restrict this approach to research applications.

Infinium 450 K Arrays

The platform that currently dominates global methylation analysis is the Infinium 450 K array (Illumina HumanMethylation450 BeadChip) which analyses more than 485,000 CpG sites across the genome. Although, this is a fraction of the approximately 30,000,000 CpGs in the genome, the CpGs are chosen such that they cover 99 % of RefSeq genes, 96 % of CpG islands and the CpG shore flanking regions, as well as microRNA promoter regions and CpNpG methylated cytosines previously identified in stem cells. Recently, an 850 K array was introduced.

Like most array-based approaches, Infinium 450 K arrays are oligonucleotide based. Bisulfite-modified DNA samples are amplified, made single-stranded, and hybridised to the arrayed oligonucleotides. The proportional methylation of each

CpG site, referred to as the beta value, is determined by one of either of two chemistries [48]. In the Infinium type I chemistry (also used in the older Infinium HumanMethylation27 BeadChip array), unmethylated and methylated CpGs are measured by different probes using the same colour channel. The majority of CpG sites are assayed by the Infinium type II chemistry where unmethylated (red channel) and methylated (green channel) cytosines are quantified by a single probe that undergoes primer extension with differently labeled nucleotides (basically an adaptation of the methylation-sensitive single nucleotide primer extension [49]). As the Infinium 27 K array was developed first, proportionately more CpGs measured by type I map to CpG islands.

The Infinium II probes have been reported to be less sensitive for the detection of both high and low methylation values and to display greater variance between replicates [48, 50, 51]. It is thus safer to consider anything with a beta value of less than 0.2 as unmethylated [48]. In addition, despite the claimed single CpG resolution, probes also often contain other CpG sites which can affect hybridisation depending on their methylation status. Because of this, the beta values may not directly correspond to the actual methylation of the CpG dinucleotide being examined.

Despite these reservations, the Infinium 450 K arrays have had conspicuous success as a discovery tool. Importantly, DNA from most archival FFPE tissues can be used with 450 k beadchips [52]. In many cases, results can be improved by using a proprietary reagent supplied by Illumina.

In addition, epigenome-wide association studies (EWAS) are being used to determine the role of DNA methylation in complex disease states including cancer predisposition. Infinium 450 K arrays have become the platform of choice for these studies. The most widely used tissue for EWAS studies is peripheral blood. However, in contrast to genomic GWAS studies where genotype is constant across somatic cells, differential DNA methylation across cell types and changes that occur over time add considerable complexity that may confound the analyses [53].

Clinical Applications of DNA Methylation

Quality Control Issues

As with all PCR-based tests, utmost care must be taken to ensure that there is no carry-over of PCR amplicons into the PCR setup areas. Thus, negative extraction controls are advisable. This is particularly important to avoid false positives when minimal residual disease is being determined. The appropriate methylated and unmethylated controls should also be used. In addition, a sample of unmodified DNA should be run to ensure that no amplification is observed under the conditions being used, particularly when an assay is being developed.

DNA Methylation as a Biomarker for Minimal Residual Cancer

DNA methylation biomarkers are being investigated for the early detection of cancer or for monitoring response to treatment. Tumours show characteristic profiles of methylated biomarkers that are specific for given types and subtypes of cancer [54, 55]. Remarkably, similar profiles may also be seen for the same type of cancer across species. Thus, there is considerable potential for using methylation of panels of recurrently methylated genes as a source of tumour-type specific biomarkers. It is, however, critical to verify that the biomarkers are generally not methylated to any significant degree in normal tissue, particularly the blood [30].

Currently, the best example of a tumour-specific methylation biomarker is the *GSTP1* gene, which is methylated in 95 % of prostate cancers but not in benign prostatic hyperplasia [56]. In patients with prostate cancer, *GSTP1* promoter hypermethylation was detected in 72 % of plasma or serum samples, 50 % of ejaculates, and 36 % of urine samples after prostate massage [57]. For most other cancers, such very high-frequency methylation biomarkers have not yet been identified. On the other hand, with the right choice of a limited number of high-frequency methylated genes, greater than 90 % coverage can be readily achieved. This contrasts with mutations. Although there are high-frequency, single-site mutations known for some cancers it is virtually impossible to construct a panel of such mutations to cover more than 90 % of patients even in cancers such as colorectal cancer and melanoma that do have high-frequency recurrent mutations.

The detection of methylated sequences in the circulating free DNA of cancer patients is increasingly being recognised as a non-invasive monitoring method for cancer. In cancer patients, the circulating free DNA in the plasma has a varying amount of tumour-derived DNA (ctDNA). The ctDNA derives from apoptosis or necrosis of tumour cells. Detection of ctDNA requires using a DNA biomarker that is found in the cancer cells and not in the normal cells. Such DNA alterations have been detected in derived the plasma or serum of most patients even some with small or even in situ lesions [58]. Plasma DNA concentration in breast cancer patients can reach several hundred ng/mL [59], though after treatment levels can be in the normal range of less than ten ng/ml.

The source of normal DNA in plasma is principally from hematopoietic cells [60]. Thus, it is likely that DNA from normal leukocytes will affect biomarker specificity if methylated. SMART-MSP was used to evaluate the methylation levels in normal peripheral blood mononuclear cells for a panel of genes that are commonly methylated in breast cancer (*APC*, *BRCA1*, *CDH1*, *CDKN2A*, *DAPK1*, *GSTP1*, *HIC1*, *RARB*, *RASSF1A*, and *TWIST1*) [30]. Some of the biomarkers were found to be methylated at low levels in peripheral blood mononuclear cells, whereas methylation was undetectable in others.

RASSF1A is one of the biomarkers that is unmethylated in peripheral blood mononuclear cells [30]. *RASSF1A* is methylated in more than 80 % of breast cancers as well as ductal carcinoma in situ [61]. *RASSF1A* was recently studied in consecutive serum samples from patients with locally advanced breast cancer during neoadjuvant chemotherapy [38]. In four patients who achieved complete pathological response, *RASSF1A* methylation became undetectable early during therapy. In contrast, in the patients that had partial or minimal pathological response, serum *RASSF1A* methylation persisted for longer or throughout the treatment.

The utility of methylation as an early detection tumour biomarker has been suggested by numerous studies though few of these studies have been validated let alone passed into clinical use. Methylation of the *p16 (CDKN2A)* tumour suppressor gene, which was shown to be an early change in lung cancer, was detected in the sputum of 3/7 patients with squamous cell carcinoma compared to 5/26 high-risk cancer-free individuals [62]. In ductal lavage specimens, methylated alleles of the cyclin D2 (*CCND2*) and *RARβ2* genes were detected in fluid from patients with endoscopically detected carcinomas and ductal carcinoma in situ, but rarely in fluid from healthy ducts [63]. However, it is the *GSTP1* methylation biomarker in prostate cancer that is likely to have the greatest utility in the still difficult area of early detection.

DNA Methylation as a Predictive Biomarker

DNA Methylation of DNA Repair Genes

Loss of gene expression due to DNA methylation leads to a wide variety of potentially targetable lesions in cancer. DNA repair genes are a good example of this. Many therapies targeting cancer cells are effective because they cause DNA damage that the cancer cells are deficient in repairing. Whereas *de novo* silencing of DNA repair genes can lead to accelerated carcinogenesis by increasing the mutation rate, loss of DNA repair capacity can also be the Achilles' heel for the tumour during chemotherapy and or radiotherapy. Knowledge of the affected pathways can thus lead to rational choice of therapy.

A compelling example is methylation of the O^6 methylguanine DNA methyltransferase (*MGMT*) gene, which removes small alkyl groups from the O^6 position of guanine. *MGMT* is methylated in a variety of cancers. In gliomas, *MGMT* promoter methylation is associated with response to chemotherapy with alkylating agents due to their inability to repair alkylation damage [64, 65]. Thus, loss of activity of MGMT, which may initially favour tumourigenesis, is now responsible for the tumour's exquisite sensitivity to alkylating agents. Whereas attention has been focused on glioma, *MGMT* methylation should also indicate sensitivity to alkylating agents in other cancers such as colorectal cancer where alkylating agents are rarely used and melanoma where temozolomide is a therapeutic option [66, 67].

More recently, there has been interest in *BRCA1* methylation as a predictive biomarker for response to platinum compounds or PARP inhibitors in breast and ovarian cancer. Whereas the TCGA ovarian cancer study indicated that based on survival, *BRCA1* methylated tumours clustered with wild type tumours rather than *BRCA1* mutated tumours [68], further work needs to be done in this because pre-clinical data indicates that *BRCA1* methylated tumours are responsive to platinum compounds or PARP inhibitors [69, 70].

DNA Methylation and the Evaluation of Lynch Syndrome

Mutations in one or more of the genes involved in mismatch repair underlie hereditary non-polyposis colorectal cancer (HNPCC) and other tumour types, such as endometrial cancer, belonging to Lynch syndrome. Deciding which patients to screen for mutations in the mismatch repair genes, and which of the genes to test, is still an issue in the identification of new pedigrees with Lynch syndrome.

Tumour tissue from an affected member of a pedigree is thus first screened by immunohistochemistry for the MLH1, PMS2, MSH2, and MSH6 enzymes. When immunohistochemistry is negative for MLH1 and/or PMS2 (which form a heterodimer), *MLH1* mutation testing is indicated. However, the *MLH1* locus undergoes frequent methylation in colorectal tumours [71], particularly those tumours with the CpG island methylator (CIMP) phenotype [72]. Tumours are considered CIMP positive if the majority of a panel of promoter region CpG islands including that of *MLH1* are methylated.

Both *MLH1*-methylated tumours and tumours arising in patients with a germline *MLH1* mutation have the same phenotype: microsatellite instability and negative immunohistochemistry for MLH1 and PMS2. *MLH1* methylation in tumours of this phenotype is associated with a very low likelihood of there being a *MLH1* germline mutation [73]. Most CIMP tumours also have *BRAF* V600E mutations and testing for this mutation has been extensively utilised to identify patients that will not be tested for *MLH1* mutations [74]. However, it makes more sense to test directly for *MLH1* methylation when immunohistochemistry is negative for MLH1. Moreover, *MLH1* methylated endometrial cancers do not have BRAF mutations so *MLH1* methylation is the only option in this instance [75].

The Future

Barriers to the implementation of methylation-based biomarkers have been discussed in recent reviews [1, 2]. Nevertheless epigenetic alterations in tumours (or their gene expression consequences) will be increasingly considered for decision-making in precision medicine. Following the increase of interest in non-invasive monitoring of tumours, there will be increasing use of tumour-specific DNA methylation biomarkers for liquid biopsies [76].

Methylation data will ideally be integrated with mutational and gene expression to enable precision medicine. It is likely that methylation analysis of a panel of key genes will become a part of profiling for predictive medicine. Considerable validation work will need to be done before this becomes a reality. Appropriate diagnostic platforms, probably based on multiple parallel sequencing, need to be developed to enable DNA methylation detection across multiple regions. Methodologies used will need to be compatible with degraded DNA. Quantitative approaches need to replace non-quantitative methodologies.

Acknowledgments I acknowledge my colleagues in DNA methylation, both those whom I have cited and those whose contributions I have omitted due to lack of space. I would like to thank Thomas Mikeska and Basant Ebaid for helpful comments on several drafts. Funding underlying this work is from the National Breast Cancer Foundation of Australia (CG-10-04, CG-12-07, PS-15-048), Cancer Australia (ID 1009892), the Cancer Council of Victoria, and the Victorian Cancer Agency (TRP13025, TRP13026). The Olivia Newton-John Cancer Research Institute is supported by the Victorian Government Operational and Infrastructure Support Grant.

References

1. Mikeska T, Bock C, Do H, Dobrovic A (2012) DNA methylation biomarkers in cancer: progress towards clinical implementation. Expert Rev Mol Diagn 12:473–487
2. Mikeska T, Craig JM (2014) DNA methylation biomarkers: cancer and beyond. Genes (Basel) 5:821–864
3. Mikeska T, Candiloro I, Dobrovic A (2010) The methodological implications of heterogeneous DNA methylation for the use of methylation as a biomarker. Epigenomics 2:561–573
4. BLUEPRINT Consortium (2016) Quantitative comparison of DNA methylation assays for biomarker development and clinical applications. Nat Biotechnol 34:726–737
5. Dobrovic A (2005) Methods for analysis of DNA methylation. In Coleman WB, Tsongalis GJ (eds) Molecular diagnostics for the clinical laboratorian. 2nd Edn. The Humana Press Inc, Totowa, pp 149–160
6. Kristensen LS, Treppendahl MB, Grønbæk K (2013) Analysis of epigenetic modifications of DNA in human cells. Curr Protoc Hum Genet 77:20.2.1–20.2.22
7. Kohli RM, Zhang Y (2013) TET enzymes. TDG and the dynamics of DNA demethylation. Nature 502:472–479.
8. Haffner MC, Chaux A, Meeker AK et al (2011) Global 5-hydroxymethylcytosine content is significantly reduced in tissue stem/progenitor cell compartments and in human cancers. Oncotarget 2:627–637
9. Hayatsu H, Wataya Y, Kai K (1970) The addition of sodium bisulfite to uracil and cytosine. J Am Chem Soc 92:724–726
10. Frommer M, Mcdonald LE, Millar DS et al (1992) A genomic sequencing protocol that yields a positive display of 5-methylcytosine residues in individual DNA strands. Proc Natl Acad Sci U S A 89:1827–1831
11. Wang RY, Gehrke CW, Ehrlich M (1980) Comparison of bisulfite modification of 5-methyldeoxycytidine and deoxycytidine residues. Nucleic Acids Res 8:4777–4790
12. Ehrich M, Zoll S, Sur S, van den Boom D (2007) A new method for accurate assessment of DNA quality after bisulfite treatment. Nucleic Acids Res 35, e29
13. Grunau C, Clark SJ, Rosenthal A (2001) Bisulfite genomic sequencing: systematic investigation of critical experimental parameters. Nucleic Acids Res 29, e65
14. Do H, Dobrovic A (2015) Sequence artifacts in DNA from formalin-fixed tissues: causes and strategies for minimization. Clin Chem 61:64–71
15. Clark SJ, Harrison J, Paul CL, Frommer M (1994) High sensitivity mapping of methylated cytosines. Nucleic Acids Res 22:2990–2997

16. Snell C, Krypuy M, Wong EM et al (2008) BRCA1 promoter methylation in peripheral blood DNA of mutation negative familial breast cancer patients with a BRCA1 tumour phenotype. Breast Cancer Res 10:R12
17. Candiloro IL, Mikeska T, Hokland P, Dobrovic A (2008) Rapid analysis of heterogeneously methylated DNA using digital methylation-sensitive high resolution melting: application to the CDKN2B (p15) gene. Epigenetics Chromatin 1:7
18. Colella S, Shen L, Baggerly KA et al (2003) Sensitive and quantitative universal pyrosequencing methylation analysis of CpG sites. Biotechniques 35:146–150
19. Tost J, Dunker J, Glynne Gut I (2003) Analysis and quantification of multiple methylation variable positions in CpG islands by pyrosequencing. Biotechniques 35:152–156
20. Uhlmann K, Brinckmann A, Toliat MR et al (2002) Evaluation of a potential epigenetic biomarker by quantitative methyl-single nucleotide polymorphism analysis. Electrophoresis 23:4072–4079
21. Mikeska T, Felsberg J, Hewitt CA, Dobrovic A (2011) Analysing DNA methylation using bisulfite pyrosequencing. In Epigenetics protocols II. Humana Press, Totowa; Methods Mol Biol. 791:33–53
22. Varley KE, Mutch DG, Edmonston TB et al (2009) Intra-tumor heterogeneity of MLH1 promoter methylation revealed by deep single molecule bisulfite sequencing. Nucleic Acids Res 37:4603–4612
23. Korbie D, Lin E, Wall D (2015) Multiplex bisulfite PCR resequencing of clinical FFPE DNA. Clin Epigenet 7:28
24. Wong NC, Pope BJ, Candiloro IL et al (2016) MethPat: a tool for the analysis and visualisation of complex methylation patterns obtained by massively parallel sequencing. BMC Bioinformatics 17:98
25. Herman JG, Graff JR, Myohanen S et al (1996) Methylation-specific PCR: a novel PCR assay for methylation status of CpG islands. Proc Natl Acad Sci U S A 93:9821–9826
26. Aggerholm A, Guldberg P, Hokland M, Hokland P (1999) Extensive intra- and inter-individual heterogeneity of p15INK4B methylation in acute myeloid leukemia. Cancer Res 59:436–441
27. Dodge JE, List AF, Futscher BW (1998) Selective variegated methylation of the p15 CpG island in acute myeloid leukemia. Int J Cancer 78:561–567
28. Do H, Wong N, Murone C et al (2014) A critical re-assessment of DNA repair gene promoter methylation in non-small cell lung carcinoma. Sci Rep 4:4186
29. Lim AM, Candiloro ILM, Wong N et al (2014) Quantitative methodology is critical for assessing DNA methylation and impacts on correlation with patient outcome. Clin Epigenet 6:22
30. Kristensen LS, Raynor M, Candiloro IL, Dobrovic A (2012) Methylation profiling of normal individuals reveals mosaic promoter methylation of cancer associated genes. Oncotarget 3:450–461
31. Lo YMD, Wong IHN, Zhang J et al (1999) Quantitative analysis of aberrant p16 methylation using real-time quantitative methylation-specific polymerase chain reaction. Cancer Res 59:3899–3903
32. Eads CA, Danenberg KD, Kawakami K et al (2000) MethyLight: a high-throughput assay to measure DNA methylation. Nucleic Acids Res 28, e32
33. Chan MW, Chu ES, To KF, Leung WK (2004) Quantitative detection of methylated SOCS-1, a tumor suppressor gene, by a modified protocol of quantitative real time methylation- specific PCR using SYBR green and its use in early gastric cancer detection. Biotechnol Lett 26:1289–1293
34. Kristensen LS, Mikeska T, Krypuy M, Dobrovic A (2008) Sensitive melting analysis after real time-methylation sensitive PCR (SMART-MSP): high-throughput and probe-free quantitative DNA methylation detection. Nucleic Acids Res 36, e42
35. Wojdacz TK, Dobrovic A (2007) Methylation-sensitive high resolution melting (MS-HRM): a new approach for sensitive and high-throughput assessment of methylation. Nucleic Acids Res 35, e41
36. Worm J, Aggerholm A, Guldberg P (2001) In-tube DNA methylation profiling by fluorescence melting curve analysis. Clin Chem 47:1183–1189
37. Wong IH, Lo YM, Johnson J (2001) Epigenetic tumor markers in plasma and serum: biology and applications to molecular diagnosis and disease monitoring. Ann N Y Acad Sci 945:36–50
38. Avraham A, Uhlmann R, Shperber A et al (2012) Serum DNA methylation for monitoring response to neoadjuvant chemotherapy in breast cancer patients. Int J Cancer 131:E1166–E1172

39. Candiloro IL, Mikeska T, Dobrovic A (2011) Assessing combined methylation-sensitive high resolution melting and pyrosequencing for the analysis of heterogeneous DNA methylation. Epigenetics 6:500–507
40. Wojdacz TK, Hansen LL, Dobrovic A (2008) A new approach to primer design for the control of PCR bias in methylation studies. BMC Res Notes 1:54
41. Dobrovic A, Simpfendorfer D (1997) Methylation of the *BRCA1* gene in sporadic breast cancer. Cancer Res 57:3347–3350
42. Singer-Sam J, Grant M, LeBon JM et al (1990) Use of a HpaII-polymerase chain reaction assay to study DNA methylation in the Pgk-1 CpG island of mouse embryos at the time of X-chromosome inactivation. Mol Cell Biol 10:4987
43. Melnikov AA, Gartenhaus RB, Levenson AS et al (2005) MSRE-PCR for analysis of gene-specific DNA methylation. Nucleic Acids Res 33, e93
44. Nygren AO, Ameziane N, Duarte HM et al (2005) Methylation-specific MLPA (MS-MLPA): simultaneous detection of CpG methylation and copy number changes of up to 40 sequences. Nucleic Acids Res 33, e128
45. Costello JF, Plass C, Cavenee WK (2002) Restriction landmark genome scanning. Methods Mol Biol 200:53–70
46. Huang TH, Perry MR, Laux DE (1999) Methylation profiling of CpG islands in human breast cancer cells. Hum Mol Genet 8:459–470
47. Shull AY, Noonepalle SK, Lee EJ, Choi JH, Shi H (2015) Sequencing the cancer methylome. Methods Mol Biol 1238:627–651
48. Bibikova M, Barnes B, Tsan C et al (2011) High density DNA methylation array with single CpG site resolution. Genomics 98:288–295
49. Gonzalgo ML, Jones PA (1997) Rapid quantitation of methylation differences at specific sites using methylation-sensitive single nucleotide primer extension (Ms-SNuPE). Nucleic Acids Res 25:2529–2531
50. Price ME, Cotton AM, Lam LL et al (2013) Additional annotation enhances potential for biologically-relevant analysis of the Illumina Infinium HumanMethylation450 BeadChip array. Epigenetics Chromatin 6:4
51. Wilhelm-Benartzi CS, Koestler DC, Karagas MR et al (2013) Review of processing and analysis methods for DNA methylation array data. Br J Cancer 109:1394–1402
52. Thirlwell C, Eymard M, Feber A et al (2010) Genome-wide DNA methylation analysis of archival formalin-fixed paraffin-embedded tissue using the Illumina Infinium HumanMethylation27 BeadChip. Methods 52:248–254
53. Langevin SM, Kelsey KT (2013) The fate is not always written in the genes: epigenomics in epidemiologic studies. Environ Mol Mutagen 54:533–541
54. Costello JF, Fruhwald MC, Smiraglia DJ et al (2000) Aberrant CpG-island methylation has non-random and tumour-type-specific patterns. Nat Genet 24:132–138
55. Esteller M, Corn PG, Baylin SB, Herman JG (2001) A gene hypermethylation profile of human cancer. Cancer Res 61:3225–3229
56. Lee WH, Isaacs WB, Bova GS, Nelson WG (1997) CG island methylation changes near the GSTP1 gene in prostatic carcinoma cells detected using the polymerase chain reaction: a new prostate cancer biomarker. Cancer Epidemiol Biomarkers Prev 6:443–450
57. Goessl C, Muller M, Straub B, Miller K (2002) DNA alterations in body fluids as molecular tumor markers for urological malignancies. Eur Urol 41:668–676
58. Bettegowda C, Sausen M, Leary RJ et al (2014) Detection of circulating tumor DNA in early- and late-stage human malignancies. Sci Transl Med 6:224ra24
59. Jahr S, Hentze S, Englisch S et al (2001) DNA fragments in the blood plasma of cancer patients: quantitations and evidence for their origin from apoptotic and necrotic cells. Cancer Res 61:1659–1665
60. Sun K, Jiang P, Chan KC et al (2015) Plasma DNA tissue mapping by genome-wide methylation sequencing for noninvasive prenatal, cancer, and transplantation assessments. Proc Natl Acad Sci U S A 112:E5503–E5512

61. Pang JM, Deb S, Takano EA et al (2014) Methylation profiling of ductal carcinoma in situ and its relationship to histopathological features. Breast Cancer Res 16:423
62. Belinsky SA, Nikul KJ, Palmisano WA et al (1998) Aberrant methylation of p16(INK4a) is an early event in lung cancer and a potential biomarker for early diagnosis. Proc Natl Acad Sci U S A 95:11,891–11,896
63. Evron E, Dooley WC, Umbricht CB et al (2001) Detection of breast cancer cells in ductal lavage fluid by methylation-specific PCR. Lancet 357:1335–1336
64. Esteller M, Garcia-Foncillas J, Andion E et al (2000) Inactivation of the DNA-repair gene MGMT and the clinical response of gliomas to alkylating agents. N Engl J Med 343:1350–1354
65. Hegi ME, Diserens AC, Gorlia T et al (2005) *MGMT* gene silencing and benefit from temozolomide in glioblastoma. N Engl J Med 352:997–1003
66. Middleton MR, Grob JJ, Aaronson N et al (2000) Randomized phase III study of temozolomide versus dacarbazine in the treatment of patients with advanced metastatic malignant melanoma. J Clin Oncol 18:158–166
67. Schraml P, von Teichman A, Mihic-Probst D et al (2012) Predictive value of the MGMT promoter methylation status in metastatic melanoma patients receiving first-line temozolomide plus bevacizumab in the trial SAKK 50/07. Oncol Rep 28:654–658
68. TCGA (2011) Integrated genomic analyses of ovarian carcinoma. Nature 474:609–615
69. Stefansson OA, Villanueva A, Vidal A et al (2012) BRCA1 epigenetic inactivation predicts sensitivity to platinum-based chemotherapy in breast and ovarian cancer. Epigenetics 11:1225–1229
70. Veeck J, Ropero S, Setien F et al (2010) BRCA1 CpG island hypermethylation predicts sensitivity to poly(adenosine diphosphate)-ribose polymerase inhibitors. J Clin Oncol 28:e563–e564
71. Kane MF, Loda M, Gaida GM et al (1997) Methylation of the hMLH1 promoter correlates with lack of expression of hMLH1 in sporadic colon tumors and mismatch repair-defective human tumor cell lines. Cancer Res 57:808–811
72. Toyota M, Ahuja N, Ohe-Toyota M et al (2001) CpG island methylator phenotype in colorectal cancer. Proc Natl Acad Sci U S A 96:8681–8686
73. Newton K, Jorgensen NM, Wallace AJ, Buchanan DD, Lalloo F, McMahon RF, Hill J, Evans DG (2014) Tumour MLH1 promoter region methylation testing is an effective prescreen for Lynch Syndrome (HNPCC). J Med Genet 51:789–796
74. Loughrey MB, Waring PM, Tan A et al (2007) Incorporation of somatic BRAF mutation testing into an algorithm for the investigation of hereditary non-polyposis colorectal cancer. Fam Cancer 6:301–310
75. Metcalf AM, Spurdle AB (2014) Endometrial tumour BRAF mutations and MLH1 promoter methylation as predictors of germline mismatch repair gene mutation status: a literature review. Fam Cancer 13:1–12
76. Yu M, Carter KT, Makar KW, Vickers K et al (2015) MethyLight droplet digital PCR for detection and absolute quantification of infrequently methylated alleles. Epigenetics 10:803–809

Clinical Flow Cytometry for Hematopoietic Neoplasms

David Wu, Brent L. Wood, and Jonathan R. Fromm

Introduction

Immunophenotypic analysis of hematopoietic neoplasms is commonly required for disease classification and is currently accomplished by immunohistochemistry and flow cytometry [1]. Multi-parameter flow cytometry is an advanced diagnostic technology that is widely used for the clinical evaluation of complex cellular mixtures such as peripheral blood, bone marrow aspirates, body fluids, lymph nodes, and other tissue specimens [2]. The ability of the technique to rapidly and simultaneously characterize the expression of multiple cell surface and cytoplasmic antigens in numerous cellular populations has resulted in the clinical adoption of flow cytometry as "standard-of-care" analysis for the evaluation of hematopoietic processes in the clinical laboratory.

In flow cytometry, a suspension of cells such as that derived from a patient sample is injected into a fluid stream under laminar flow conditions [3] resulting in individual cells being directed through a quartz capillary tube (flow cell). Following illumination by one or more light sources, typically lasers, multiple cellular properties may be simultaneously assessed, including light scatter and the expression of surface and/or cytoplasmic antigens [3, 4] when cells are pre-labeled with fluorochrome-conjugated antibodies directed against specific antigens [5]. Peripheral blood, bone marrow, cerebrospinal fluid, and tissue specimens are routinely analyzed to provide

D. Wu (✉) • B.L. Wood • J.R. Fromm
Department of Laboratory Medicine, University of Washington,
Room G7811, Seattle Cancer Care Alliance, 825 Eastlake Ave E, Seattle, WA 98109, USA
e-mail: dwu2@u.washington.edu; woodbl@u.washington.edu; jfromm@u.washington.edu; jfromm@uw.edu

S.R. Lakhani, S.B. Fox (eds.), *Molecular Pathology in Cancer Research*,
DOI 10.1007/978-1-4939-6643-1_13

unique diagnostic information about hematopoietic populations [2]. In general, flow cytometry permits the evaluation of multiple antigens on cells in a given experiment, typically at least 4–6, and increasingly in up to 10 or more antigens simultaneously [4]. Here, we discuss practical aspects of lymphoma, leukemia, and myeloma diagnosis and classification by flow cytometry. Critical to the approach described herein is the assumption that an abnormal cell population will have an aberrant immunophenotype as compared to the background normal or reactive cells. As such, emphasis is placed on identifying an aberrant immunophenotype of cells by multi-parametric analysis. As there have been numerous articles describing various strategies for flow cytometry [2, 3], the reader is also advised to access other articles in the field to address deficiencies or intentional omissions herein.

Materials

Reagents are generally used as provided from the manufacturer. However, it is important to titer antibodies for optimal signal-to-noise response under the conditions to be used, and this may result in the use of antibodies at concentrations below (or above) that recommended by the manufacturer. The antigens that we target for routine analysis of hematopoietic neoplasms are included in Table 1, with specific antibodies and fluorochromes employed using a modified LSRII flow cytometer (Becton-Dickinson).

Buffers and Cell Staining Reagents

1. PBS-BSA buffer: Dulbecco's Phosphate-Buffered Saline (GIBCO®) containing 3 % bovine serum albumin. PBS contains 2.67 mM KCl, 1.47 mM KH_2PO_4, 137.9 mM NaCl, 8.1 mM Na_2HPO_4.
2. RPMI 1640.
3. Lysing/fixation solution (solution fixes cells and lyses red blood cells): 0.15 mol/L NH_4Cl, pH 7.2 containing 0.25 % ultrapure formaldehyde (Polysciences).
4. Medium A and B for cytoplasmic/nuclear antibody staining are from Invitrogen (Fix & Perm®).

Methods

Sample Preparation

Disaggregation of Tissue Specimens

Tissue specimens are initially disaggregated to create a single cell suspension:

Table 1 Commonly used fluorochrome combinations for immunophenotyping hematopoietic neoplasms at the University of Washington

Antigen	B cells	T cell	TCR V-beta	cHL (9-color)	Plasma cell	B-ALL MRD	T-ALL MRD-1	T-ALL MRD-2	Lineage defining	AML M1	AML M2	AML M4
CD2		FITC										
CD3		PE-Cy7	PE-Cy7				PE-Cy7	APC				
cCD3								FITC	ECD			
CD4		A594	A594				A594				ECD	
CD5	PE-Cy5.5	PE	ECD	APC-Cy7 (or ECD)			PE	PE-Cy7				PE-Cy5
CD7		APC	APC				APC	PE				PE
CD8		BV421	APC-Cy7				BV421					
CD10	APC					PE						
CD13										PE-Cy7	PE-Cy7	
CD14											PE-Cy5.5	
CD15				APC						FITC		
CD16							APC-A700	BV421			APC-A700	
CD19	PE-Cy7				PE-Cy5	PE-Cy7				ECD		
CD20	V450			PE-Cy7		FITC						
CD30		APC-A700		PE								
CD33										PE		PE-Cy7
CD34		ECD				PerCP-Cy5.5	ECD		APC	APC	APC	APC
CD38	A594					A594		A594		A594	A594	A594
CD40				PE-Cy5.5								

(continued)

Table 1 (continued)

Antigen	B cells	T cell	TCR V-beta	cHL (9-color)	Plasma cell	B-ALL MRD	T-ALL MRD-1	T-ALL MRD-2	Lineage defining	AML M1	AML M2	AML M4
CD45	APC-H7	APC-H7		ECD (or APC-H7)	APC-H7	APC-H7	APC-H7	APC-H7	V450	APC-H7	APC-H7	APC-H7
CD48							FITC					
CD56		PE-Cy5	PE-Cy5		PE-Cy7		PE-Cy5	PE-Cy5				A488
CD58						APC						
CD64				FITC							FITC	
CD71				APC-A700						APC-A700		
cCD79a									PE			
CD95				PB								
CD117										PE-Cy5		
CD123											PE	
CD138					APC							
HLA-DR			V450							PB	PB	
Kappa	FITC				FITC[a]							
Lambda	PE				PE[a]							
V-beta isoform reagents			FITC, PE, and FITC and PE									
DAPI					PB							
cMPO									FITC			

Abbreviations: *PB* Pacific Blue, *BV421* brilliant violet421, *FITC* fluorescein isothiocyanate, *A488* AlexaFluor488, *PE* Phycoerythrin, *ECD/PE-TR* PE-Texas Red, *PerCPCy5.5* Peridinin-chlorophyll Cyanine-5.5, *PECy5* PE-Cyanine-5, *PE-Cy5.5* PE-Cyanine-5.5, *PE-Cy7* PE-Cyanine-7, *A594* AlexaFluor 594, *APC* allophycocyanin, *APC-A700* APC-AlexaFluor 700

[a]Light chain antibodies for plasma cell neoplasms are cytoplasmic. Less commonly used tubes are described elsewhere [90]

1. Mince the tissue with a scalpel in RPMI (5 ml).
2. Filter the disaggregated cell suspension through a 40 μm filter.
3. Pellet the cells by centrifugation (550 × *g* for 5 min) and decant the remaining supernatant.
4. Re-suspend the cells (in RPMI), pellet the cells again by centrifugation (550 × *g* for 5 min), and re-suspend in RPMI to a cell count of 10,000 cells/μl or less.
5. Add sufficient cell suspension to deliver up to one million cells in a volume of less than 200 μl.

Cell Surface Labeling of Cell Suspensions

1. Add appropriate fluorescently labeled, titered antibodies to cell suspension derived from disaggregated tissue, bone marrow, blood, etc., typically 5–20 μl of each antibody and mix gently. The antibodies may be combined in a cocktail prior to use.
2. In the dark, incubate the labeled cells for 15 min at room temperature (RT).
3. Add 1.5 mL of lysing/fixation solution.
4. Incubate for 15 min at RT in the dark.
5. Centrifuge the cells (550 × *g* for 5 min) and discard the supernatant.
6. Add 3 mL of PBS-BSA, centrifuge (550 × *g* for 5 min), and decant the supernatant.
7. Re-suspend the cells in 100 μL of PBS-BSA.
8. Collect 150,000 events (if possible) for routine evaluations or up to one million events for minimal residual disease detection.

Cytoplasmic and Surface Labeling of Cell Suspensions (for Cytoplasmic Immunoglobulin, MPO, ZAP-70, and Bcl-2)

1. Add appropriate fluorescently labeled, titered cell surface antibodies to cell suspension and gently mix.
2. Incubate the labeled cells for 15 min at room temperature (RT) in the dark.
3. After washing the cells twice with PBS-BSA, add 100 μl of Medium A and mix well.
4. Incubate for 15 min at RT in the dark.
5. Wash the cells twice with PBS-BSA and add 100 μl of Medium B and mix.
6. Add appropriate amount of cytoplasmic/nuclear antibody, mix, and incubate the labeled cells for 30 min (1 h for MPO) in the dark at room temperature.
7. Wash the cells twice with PBS-BSA and re-suspend the cells in 100 μl of PBS-BSA.

Immunophenotyping of Cells Using DRAQ5

1. Process sample as described in section "Cell Surface Labeling of Cell Suspensions".
2. Prior to analysis on the flow cytometer, add 5 μl of DRAQ5 solution (1:25 dilution of commercially available 5 mM solution (from Enzo Life Sciences)) to the sample followed by incubation in the dark for 10 min.
3. Run on flow cytometer.
4. Note that DRAQ5 emits over a broad range of wavelengths in the red range of the spectrum under both blue and red laser excitation and is not recommended for use with PE-Cy5.5, PE-Cy5, APC, APC-A700, or APC-Cy7 fluorochromes. Additionally, if surface immunophenotyping with CD41 and CD61 is to be performed, washing the cell suspension 3 times with PBS-BSA or RPMI prior to adding the surface antibodies is recommended to reduce platelet adhesion to white cell populations.

For presentation specimens, we typically analyze approximately 150,000 cells per tube in order to allow detection and enumeration of populations at a frequency of 10^3. For Hodgkin lymphoma, the sample processing and immunostaining protocols are the same as those used for non-Hodgkin lymphoma (NHL) with the exception that more events should be analyzed, preferably ~500,000 or more. This classical Hodgkin assay has only been validated on lymph nodes and thus this assay is only recommended for tissue specimens. Likewise, collection of more events (typically 5×10^5 to 1×10^6) is critical for the evaluation of specimens for minimal residual disease (MRD).

Gating Strategies, Data Analysis, and Interpretation

With the exception of Hodgkin lymphoma, the analysis for hematopoietic neoplasms begins with the exclusion coincident or aggregated events, so-called doublets, using a plot of forward scatter area versus forward scatter height. The use of other combinations of area, height, and width may also be used for this purpose. The doublet events represent coincident cells in the flow cell and need to be excluded as the antigenic profile derived from these events may result in an apparent composite immunophenotype due to the combined antigenic expression of two or more cells. Subsequently, non-viable events are excluded using forward and side light scatter gating. As cells degenerate, initially forward scatter decreases while side scatter increases and later both decrease in intensity. These non-viable events can be readily excluded by selective gating using these findings. However, it is important that in some cases the population of interest may show preferential degradation, e.g., Burkitt lymphoma, and evaluation of the low forward scatter events is recommended prior to their exclusion.

B-Cell Analysis

The evaluation for B-cell NHL involves identification of B cells with subsequent evaluation for aberrant immunophenotypes concurrently showing the presence of surface immunoglobulin light chain restriction. We gate lymphocytes with forward and side light scatter, computationally subtract the CD5-positive, CD19-negative T cells, and then isolate the B cells on a plot of CD19 vs side scatter. Data is subsequently evaluated by examining various antigens (Table 1) plotted against each other for both the B cells and lymphocytes. Evaluation of lymphocytes based on forward versus side scatter (in addition to CD19-positive B cells) prevents inadvertent exclusion of B-cell populations that have aberrant loss or decreased expression of CD19. In general, the normal kappa to lambda ratio of the B cells is approximately 1.4; however, sole emphasis on a skewed light chain ratio for identifying an abnormal B-cell population is discouraged due to its poor sensitivity and specificity—such an approach is especially problematic when small clonal B-cell populations are encountered that do not alter the kappa to lambda ratio. Rather, an emphasis is placed on identifying abnormalities (under- or over-expressed antigens relative to a normal population) of other antigens assessed in these studies (CD5, CD10, CD19, CD20, CD38, and CD45), that allow separation of the abnormal and normal populations. Restricted light chain expression should then be present on the gated population. Examples of typical examples of CLL/SLL (Fig. 1), Burkitt lymphoma (Fig. 2), hairy cell leukemia-variant (Fig. 3), minimal residual disease in a case of B-cell lymphoma (Fig. 4), and CD10+ B-cell lymphoma (Fig. 5) are provided. While

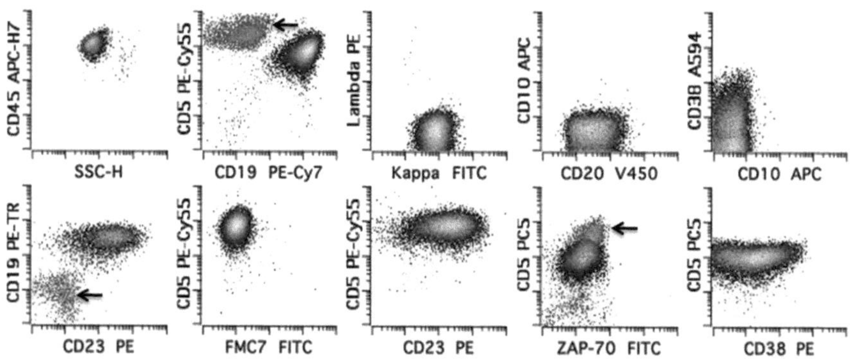

Fig. 1 Flow cytometric characterization of chronic lymphocytic leukemia/small lymphocytic lymphoma (CLL/SLL). The neoplastic population is in *black*; all other events are in *gray*. Where relevant, T cells are indicated with an *arrow*. The CLL/SLL population demonstrates expression of CD45, CD19, CD5, kappa surface light chains, decreased CD20, CD23, and no expression of FMC7 or CD10. The neoplastic population expresses ZAP-70 (positive cells determined by a discriminator which is negative on the normal B cells and positive on the normal T cells) and CD38, features associated with a less favorable prognosis in CLL/SLL [91, 92]. The first dot plot of the *top panel* shows all cells, the second shows lymphocytes, and the last three shows only B cells. In the *bottom panel* of dot plots, all lymphocytes are shown in the first and fourth plots, while the second, third and fifth plots show only B cells

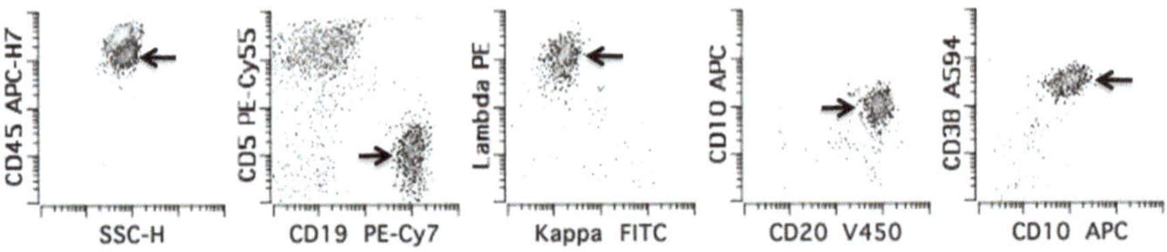

Fig. 2 Flow cytometric characterization of Burkitt lymphoma in a liver biopsy. The neoplastic Burkitt cells (*darker gray*, *arrows*) demonstrate lambda light chain restriction with expression of CD10, CD19, CD20, CD38 (expressed at a level somewhat above normal germinal center B cells), and CD45 without CD5. Fluorescence in situ hybridization (FISH) studies demonstrate the presence of an MYC translocation

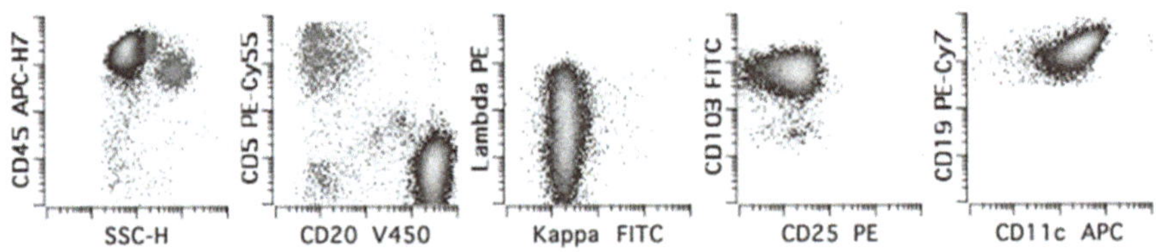

Fig. 3 Immunophenotypic characterization of a case of hairy cell leukemia-variant in the bone marrow. The first panel shows all leukocytes, the second all lymphocytes, and the remaining three panels show only B cells. Neoplastic cells are in *darker gray* or *black*; all other events are *gray*. The neoplastic population demonstrates expression of CD45, bright CD19 and CD20, lambda light chains, CD11c, CD38 (data not shown), and CD103, without CD5, CD10 (data not shown), or CD25. The lack of expression of CD25 is one of the features that distinguish hairy cell leukemia-variant from hairy cell leukemia

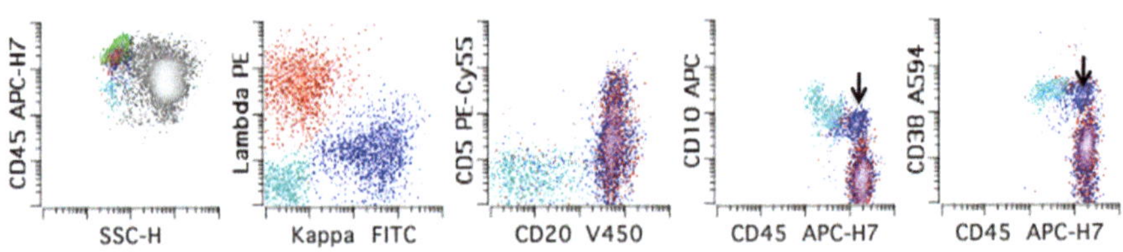

Fig. 4 Minimal residual disease detection of Burkitt lymphoma in the bone marrow. Overall, this bone marrow specimen demonstrates a normal kappa lambda ratio of 1.2 and multiple projections of the data do not demonstrate an abnormal B-cell population. However, a small (1.3 % of leukocytes) kappa restricted B-cell population is noted (*arrows*) with relatively bright expression of CD38 and CD10 can be identified on plots of CD10 and CD38 vs CD45 (last two panels). This immunophenotype is identical to that identified previously in a tissue specimen. The first dot plot shows all white cells, while the remaining four panels show B cells. T cells are colored *green*; kappa and lambda restricted mature B cells are colored *blue* and *red*, respectively; immature, normal B lymphoblasts are colored *cyan*

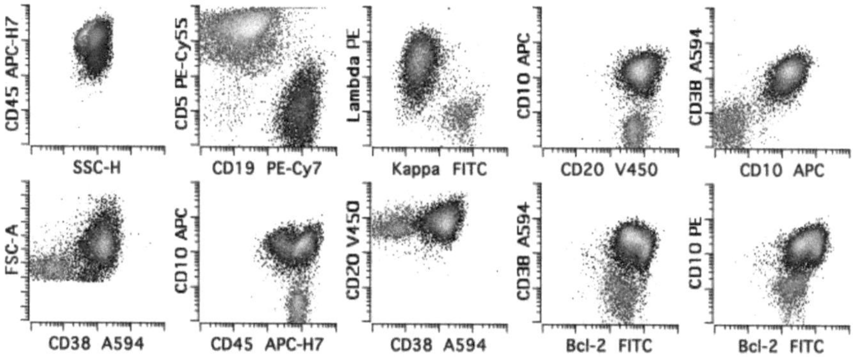

Fig. 5 Immunophenotypic characterization of a case of CD10+ B-cell lymphoma. Components of both diffuse large B-cell lymphoma and follicular lymphoma (grade 3) are present on morphologic evaluation. The first dot plot of the *top panel* shows all leukocytes, the second shows all lymphocytes, and the last three in the *top panel* show B cells. All plots in the *bottom panel* show B cells. Neoplastic B cells are in *dark gray* or *black*; all other events are in *gray*. The neoplastic cells show variable expression of CD45, increased forward light scatter, normal expression of CD19 relatively to the reactive B cells, lambda light chain restriction, and expression CD10, CD20, bcl-2 (relative to the reactive B cells), and CD38 (normal to slight decreased relative to the expression of a normal germinal center population [7]) and no expression of CD5

individual cases vary, a summary of common immunophenotypes seen in typical B-cell lymphomas is provided (Table 2). A more comprehensive discussion of immunophenotypes in B-cell NHL can be found in the review by Craig and Foon [2].

The approach for evaluating specimens for B-cell NHL minimal residual disease (MRD) is identical to that used for evaluating samples at disease presentation, with the obvious difference being that the abnormal population post-therapy is usually very small (often less than 0.1 % of the viable cells). In practice, a series of sequential gates to selectively include and/or exclude various populations may be required to convincingly isolate the abnormal population and evaluate for definite light chain restriction (Fig. 4). Identifying these small population is facilitated by knowledge of the original immunophenotype of the abnormal B-cell population at presentation (when the population is usually present at significantly increased proportion) preferably using the identical reagent combination used to immunophenotype the presentation specimen.

A number of caveats related to the analysis of B cells deserve mention. Some B-cell populations may demonstrate aberrant loss of surface light chains, although occasionally evaluation of cytoplasmic light chains may show monotypic restriction. While hematogones (normal immature B cells in the marrow and rarely in lymphoid tissues) and plasma cells largely lack surface light chains, their absence on mature B-cell populations is usually considered aberrant antigen expression [6]. Note, however, that some normal B-cell populations may exhibit decreased light chain expression (for example, germinal center B cells [1]), and consequently this finding should be interpreted cautiously.

Table 2 Immunophenotypes of common B-cell NHL[a]

B-cell lymphoma	Immunophenotype
Chronic lymphocytic leukemia/ small lymphocytic lymphoma	CD5, CD20 (dim), CD23, CD38 (variable), CD45, and ZAP-70 (variable) positive; mono- or bi-typic dim to absent surface light chain expression; FMC-7 negative
Mantle cell lymphoma	CD5, CD20 (normal), CD45, monotypic surface light chain restriction, FMC-7 positive; CD23 negative
Marginal zone lymphoma	CD19, CD20, and CD45 positive; negative for CD5 and CD10. Occasional cases have expression of CD43
Hairy cell leukemia	CD11c, CD19, CD20 (bright), CD25, monotypic surface light chain expression, and CD103 positive; CD5 and CD10 negative
Burkitt lymphoma	CD10, CD20, CD38 (increased), CD45, and monotypic surface light chain expression positive; BCL2 negative
Follicular lymphoma	CD10, CD19 (decreased), CD20, CD38 (decreased), CD45, BCL-2, and monotypic surface light chain restriction positive
Diffuse large B-cell lymphoma	CD19, CD20, and monotypic surface light chain positive; increased forward and side light scatter; CD10 may or may not be expressed

[a]Unless otherwise stated, these are the most common immunophenotypes for these neoplasms

Some large cell B-cell lymphomas may be missed if analyzed using the gating strategy described above if the lymphocyte gate is too restrictive due to increased light scatter of the neoplastic cells. Consequently, evaluating events that fall outside the normal expected size range for typical small lymphoid cells based on light scatter properties is critical for proper diagnosis and for ensuring that complete sampling has been achieved. Similarly, some B-cell lymphomas may show absent or decreased expression of CD19 [7–9] and/or CD20 [7] and may therefore be inadvertently excluded from analysis. Evaluating multiple B-cell antigens and examining B-cell antigen expression of all lymphocytes gated by light scatter help to avoid this pitfall.

In the last 15 years, it has become increasingly apparent that not every clonal B-cell populations is neoplastic. Small clonal B-cell populations that do not represent B-cell lymphoma are relatively common in normal (most commonly older) individuals and have been described in the literature [10, 11]. Larger clonal but nonmalignant germinal center B-cell populations can occur in reactive states [12], particularly Hashimoto's thyroiditis [13]. For this reason, antigenic abnormality rather than clonality is what provides specificity for these assays and one should not rely on clonality alone in making a diagnosis of lymphoma. Ultimately, flow cytometric evaluation in the context of clinical findings, all laboratory findings (such as peripheral blood counts), and morphology is required for the diagnosis of B-cell lymphoma.

It is important to remember that not all neoplastic populations that express CD19 represent B-cell lymphoma. B lymphoblastic leukemia/lymphoma (B-ALL) and rarely, myeloid neoplasms (namely myeloid neoplasms with t(8;21) [14]) may show expression of CD19. It is the composite immunophenotype that is required for assignment of a population to a particular cell lineage and maturational stage, not the expression of any single antigen.

Finally, dying cells and debris may appear to express B-cell antigens and appear clonal. While the differentiation of debris and authentic abnormal B-cell populations can be challenging, as a general rule, debris binds antibodies non-specifically, often resulting in a correlated (diagonal) relationship between antigens in multiple 2-dimensional projections of the immunophenotypic data. The exclusion of low forward scatter events, e.g., a viability gate, may significantly reduce apparent non-specific binding and simplify analysis. However, neoplasms with a high proliferative rate may show preferential degeneration with loss of forward scatter, e.g., Burkitt lymphoma, so examination of events prior to exclusion by viability gating is important to detect this occurrence.

T-Cell Analysis

In general, the strategy for gating of T cells is similar to that for B cells. Compensated data files are first gated to exclude doublet events and then non-viable events. As with B cells, debris can mimic an authentic abnormal T-cell population. In general, we use two gating strategies to identify T cells. The first approach is by analyzing forward versus side scatter to identify lymphocytes. The second approach is by analyzing CD3 versus side scatter to identify T cells. To identify T cells that have aberrant decreased or absence of CD3 expression, the former approach is helpful. To identify larger cells that may have increased side scatter, the latter approach is helpful. It is important to remember that some large T-cell lymphomas may be missed if analyzed using the routine gating strategy due to increased light scatter. In every case, it is important to consider events that fall outside the normal expected size range for typical small lymphoid cells based on light scatter properties. Evaluating events with increased scatter properties is helpful to ensure complete sampling has been achieved.

Whereas the analysis of the B cells rests on identifying an immunophenotypically abnormal B-cell population that is clonal with respect to surface light chain expression, the evaluation of T cells involves the identification of an immunophenotypically abnormal T-cell population. In this approach, it is critical to recognize that numerous reactive T-cell populations may be otherwise present in any given sample. These T cells may include memory T cells (decreased CD7 expression), gamma-delta T cells (increased CD3, absent CD4, and CD8 (partial)), and large granular lymphocytes (typically CD8+/CD5 dim/absent) and may be identified in any given patient sample, sometimes in increased proportion [2]. Evaluation of T-cell receptor V-beta repertoire may be pursued as a measure of clonality of T-cell population by flow cytometry. Critical to prevent misinterpretation of a reactive population of T cells as a possible malignant clone is the recognition and familiarity with this normal spectrum of antigenic variation of reactive T cells, which is developed during the routine review of many reactive cases. On the other hand, it is also important to recognize that a neoplastic, clonal population of T cells may also have immunophenotypic change similar or identical to these reactive populations. As such, careful clinical and pathologic correlation is required in every case. Lastly, it is important to remember that not all clonal and immunophenotypically aberrant T-cell populations represent T-cell lymphomas [2]. For example, in some situations, clonal populations may arise as part of the normal process of the adaptive immune system, such as in response to cytomegalovirus [15].

Multi-parametric identification of these normal variations in antigenic changes can permit one to reliably identify an abnormal T-cell population, which in conjunction with histologic findings or clinical data can permit subsequent definite classification. In general, it is fortuitous that the majority of T-cell lymphoproliferative disorders will demonstrate some aberrant antigenic expression, such that the level of expression of T-cell associated antigens (CD2, CD3, CD4, CD5, CD7, CD8, and CD45) can be either increased, decreased, or completely absent, as compared to the background, normal reactive T cells [16, 17]. Some examples of the flow cytometry data for mature T-cell lymphomas (adult T-cell leukemia/lymphoma and angioimmunoblastic T-cell lymphoma) are shown in Fig. 6. As another example, flow cytometry of a liver core needle sample from a 69-year-old man with multiple liver lesions (Fig. 7). While individual cases vary, a summary of common immunophenotypes of typical T-cell lymphomas is provided (Table 3). A more comprehensive discussion of immunophenotypes in T-cell NHL can be found elsewhere [2]. Lastly, some T-cell lymphomas may show aberrant expression of CD19 or CD20 [18, 19].

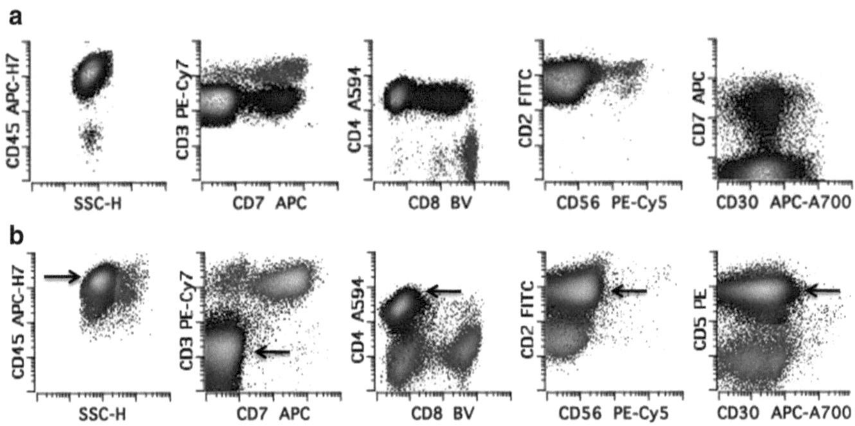

Fig. 6 Examples of mature T-cell lymphomas. (**a**) Adult T-cell leukemia/lymphoma. An abnormal population of T cells (*dark gray*) is identified, comprising 72.3 % of total leukocytes, with aberrant expression of CD2 (slightly decreased), CD3 (decreased), CD4 (slightly decreased), CD5 (absent), CD7 (absent on major subset), CD8 (variable, dim to absent), and CD45 (slightly decreased), without CD34, CD56, TdT, or CD1a. Additional studies show that both CD25 and CD52 are uniformly expressed. Additional serologic studies were positive for HTLV-1, supporting the diagnosis. (**b**) Probable, angioimmunoblastic T-cell lymphoma. An abnormal T-cell population (*dark gray* and identified with *arrows*) comprising 67.6 % of total white cells from this lymph node biopsy has aberrant expression of CD3 (low to absent) and CD7 (absent) with normal expression of CD2, CD4, CD5, and CD45 without CD8, CD34, or CD56

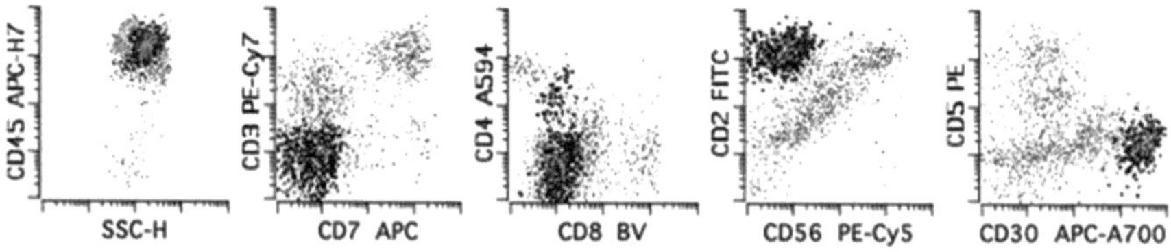

Fig. 7 Example of a CD30+ T-cell lymphoma. An abnormal T-cell population (*dark gray*) is identified comprising 25.5 % of this liver core biopsy with aberrant expression of uniform CD30, dim CD4, and aberrant absence of CD3, CD5, CD7 without CD8 or CD56. Of note, although the expression of CD2 and possible dim expression of CD4 is not considered T-cell lineage specific, subsequent molecular analysis of T-cell receptor gamma gene rearrangement confirms the population is clonal and of T-cell origin

Table 3 Immunophenotypes of common T-cell NHL[a]

T-cell lymphoma	Immunophenotype
ATLL	CD3, CD4, CD7 (decreased), and CD25 positive
T-PLL	CD2 (variable), CD3 (variable), CD4, CD5 (variable), and CD7 (variable) positive, CD8 negative. Occasional cases are CD4+/CD8+ or CD4-/CD8+
T-LGL	CD3, CD5 (decreased), CD7 (decreased), and CD8 positive; CD4 negative
Mycosis fungoides/Sezary syndrome	CD3, CD4, CD5, CD7 (decreased), and CD45 positive
ALCL	CD30, variable loss of T-cells antigens. ALK, and TIA1/granzyme B positive (by immunohistochemistry). A minority of cases are ALK negative
AITL	CD4, CD5, and CD10 positive; surface CD3 negative (but cytoplasmic CD3 positive)
PTCL-NOS	CD4 positive; often negative for CD5 and/or CD7

Abbreviations: *T-PLL* T-cell prolymphocytic leukemia, *AITL* angioimmunoblastic T-cell lymphoma, *PTCL-NOS* peripheral T-cell lymphoma, not otherwise specified, *ATLL* adult T-cell leukemia/lymphoma; *ALCL* anaplastic large cell lymphoma, *T-LGL* T-cell large granular lymphocyte leukemia, *TIA1* T-cell intracellular antigen 1

[a]Unless otherwise stated, these are the most common immunophenotypes for these neoplasms

This should not be inadvertently interpreted as a B-cell neoplasm. It is the composite immunophenotype that provides specificity and analysis of other T-cell antigens (cytoplasmic CD3) and additional B-cell antigens (cytoplasmic CD79a or CD22) may be required.

At presentation, a common question is whether a T-cell population of interest represents a clonal process or alternatively is a reactive oligoclonal or polyclonal expansion of an immunophenotypically distinct subset. To determine if the identified T-cell populations represent a reactive or clonal process, flow cytometric analysis of TCR V-beta repertoire has been used. This is a methodology that uses fluorescently labeled anti-TCR V-beta antibodies to assess clonality [20–23]. As provided by the manufacturer, the IOTest® Beta Test Mark assay is a set of 24 antibodies that covers approximately 70 % of the normal human TCR V-beta repertoire. These V-beta specific antibodies are each conjugated to one of three fluorochromes: FITC, PE, or PE and FITC conjugate. According to the manufacturer's protocol, the assay is run in eight separate tubes, such that three V-beta family specific antibodies (one labeled with FITC, one labeled with PE, and one labeled with FITC and PE) are present in each tube. The T-cell population of interest can be identified through the use of gating reagents (for example, CD3, CD4, or CD8), and then this population can be isolated and evaluated for over-representation of a particular TCR V-beta isoform relative to normal. This finding of TCR V-beta over-representation would strongly be suggestive of clonality. Numerous studies have demonstrated the utility of this approach for assessing putative T-cell clonality in both the clinical and research environment [20–23].

At our institution, we have developed and used a simple permutation of the original protocol for TCR V-beta repertoire analysis (manuscript in preparation). By combining all of the fluorescently labeled TCR V-beta antibodies into a single-tube, we minimize the standard 8-tube analysis intended by the manufacturer. With this approach, we can rapidly determine putative TCR V-beta restricted T-cell population with emphasis on concurrent evaluation for aberrant antigenic expression through the use of an increased number of T-cell specific gating reagents (Table 1). As compared to the standard method, this modified approach can be adopted in any laboratory currently performing routine TCR V-beta analysis, is relatively quick and not labor-intensive, and substantially minimizes the amount of reagent and sample requisite for analysis, thus permitting analysis of samples, such as skin biopsies and cerebrospinal fluids that are typically hypocellular in nature. An example of this modified approach is shown in which an expanded large granular lymphocyte population is identified with expression of CD8 and decreased expression of CD5 and CD7 (Fig. 8). Subsequent assessment of TCR V-beta repertoire analysis using a modified approach in which all V-beta antibodies are combined together permits identification of probable FITC-fluorochrome labeled TCR V-beta isoform. Molecular analysis of TCR gene rearrangement could be subsequently performed to confirm the presumed clonal nature of such a population, if clinically indicated. Further, if determination of the specific TCR V-beta isoform is important, such as for future minimal residual disease monitoring, one could perform the standard TCR V-beta repertoire assay to subsequently identify which TCR V-beta isoform is labeled by the combined anti-V beta reagents.

Hodgkin Cell Analysis

Classical Hodgkin lymphoma (CHL) is an unusual type of B-cell lymphoma [24–26] in which the rare neoplastic Hodgkin and Reed–Sternberg (HRS) cells, less than 1 % of the cells in lymph node [25, 27], are embedded in a reactive infiltrate including reactive lymphocytes, eosinophils, plasma cells, and histiocytes [27]. HRS binds to non-neoplastic T cells, resulting in HRS cell-T-cell rosettes [28–34].

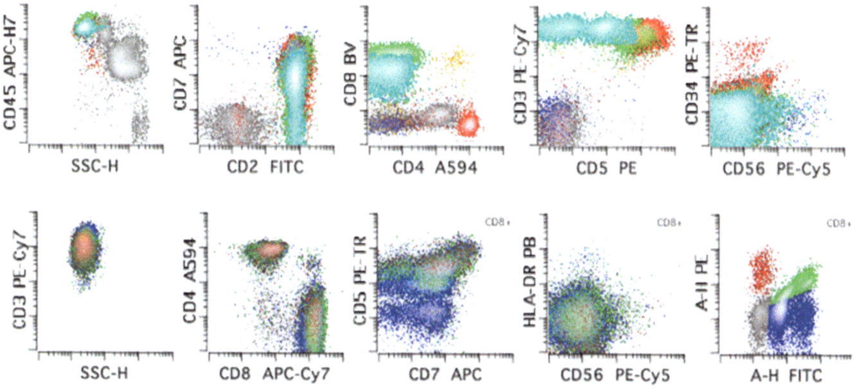

Fig. 8 Example of combined using modified TCR V-beta assay showing clonality of T-LGL population. An abnormal CD8+ T-cell population is identified in this 76-year-old female with decreased CD5 and CD7 expression as compared to background CD4+ and CD8+ T cells. Subsequent analysis of TCR V-beta expression using the modified assay in which all TCR V-beta antibodies are combined together shows the T-LGL population with apparent TCR V-beta restriction limited to an FITC-labeled isoform. By comparison, background, reactive T cells show variable expression of PE (*red*), FITC (*blue*), PE-FITC (*green*), and unlabeled TCR V-beta isoforms (*gray*). In the *top panel*, the neoplastic population is in teal, while reactive CD4 and CD8 positive T cells are colored in *red* and *green*, respectively

Traditionally, CHL has been diagnosed by a combination of morphology and immunohistochemistry (HRS cells demonstrate expression of CD15 and CD30 but lack expression of CD20, CD3, and CD45 [24, 27, 35, 36]). Our work, however, has demonstrated that CHL can be diagnosed by flow cytometry with high clinical sensitivity and specificity [34, 37]. The interaction of T cells and HRS cells (rosetting) can be directly detected by flow cytometry (Fig. 9), as demonstrated by the observation of a composite immunophenotype of the HRS cells and T cells; that is, events with expression of both Hodgkin and T-cell antigens [34]. T-cell-HRS cell rosetting can be abrogated by the addition of unlabeled antibodies that can compete for the binding of the adhesion molecule binding partner ("blocking" antibodies) [32, 34], a practice useful for purifying HRS cells [34]. Blocking of T-cell-HRS cells interactions is not necessary for diagnostic flow cytometry [37], and in fact, observation of these interactions can be a useful diagnostically finding. Reagent combinations are proposed for either 9-color [37] (Table 1) or 6-color [38] flow cytometry platforms. In addition, a reactive T-cell population (CD4+ T-cell population with CD45^{bright}, CD7^{bright}) has been identified in lymph nodes involved by CHL, a finding that can be used to suggest a diagnosis of CHL (see section "Analysis of T Cells in Classical Hodgkin Lymphoma") [39].

The gating strategy to identify HRS cells differs from that for non-Hodgkin lymphomas. The first difference is that because of the relatively large cell size of HRS and rosetted cells, increased side scatter is used to identify these populations (Figs. 9 and 10); HRS cells are then identified by their relatively bright expression of CD30, CD40, and CD95 with absence of strong CD20 expression [37, 38].

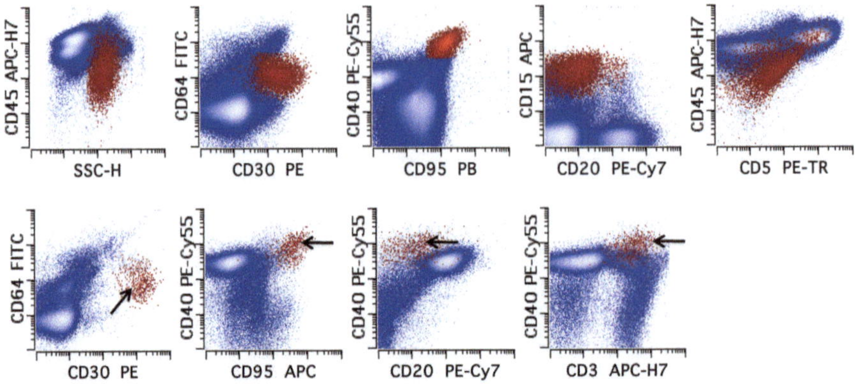

Fig. 9 Representative examples of 9- (*top panel*) and 6-color (*bottom panel*) flow cytometry studies of CHL cases. HRS cells (*arrows*, where needed; also shown in *red* and *emphasized*) are identified by their absence of expression of CD64 (position of negative determined by control experiments, not shown), expression of CD30, CD40, CD95, and increased side light scatter (SSC-H) compared to normal lymphocytes; all remaining viable events are in *blue*. For the 9-color study, the population of neoplastic HRS cells has expression of intermediate CD15, intermediate to bright CD30, intermediate to bright CD40, variable CD71 (data not shown), and intermediate to bright CD95, without expression of CD64 or CD20. The diagonal relationship between CD45 and CD5 is due to the presence of T cells bound to the HRS cells. For the 6-color study, neoplastic HRS cells have expression of intermediate to bright CD30, bright CD40, and intermediate CD95, without expression of CD64 or CD20. CD3 is expressed suggesting some degree of HRS-T-cell rosetting

A HRS population must be present in all four gates shown in Fig. 10 having: 1) increased side scatter as compared to background lymphocytes, 2) expression of CD30 and increased autofluorescence, as assessed using the FITC channel, 3) absence of or decreased expression of CD20 (relative to reactive B cells), and 4) strong expression of CD40 (relative to reactive B cells), and strong expression of CD40 (relative to reactive B cells). CD30+ reactive immunoblasts show minimally increased autofluorescence (relative to small lymphocytes), a finding that is useful in distinguishing these cells from HRS cells. If properly gated, HRS cells may appear as two subset populations, one with apparent T-cell antigen expression (such as CD3 or CD5) and the second with decreased (but not absent) CD45 expression but no expression of T-cell antigens, due to presence and absence, respectively, of T-cell rosetting. This phenomenon can provide support for the diagnosis of CHL and one should attempt to identify the presence of a diagonal relationship on a plot of CD45 versus CD3 or CD5. This diagonal relationship occurs due to the presence of varying numbers of T-cells rosetting of the HRS cells.

While the flow cytometry-derived immunophenotype is characteristic of CHL, there may be slight variance from case to case. For instance, while most HRS cells do not express CD20, some cells may have low-level expression. Further, some HRS populations may lack expression of CD15, a finding that correlates with reported immunohistochemical studies (~20 % of CHL case show lack expression of CD15 [40–42]).

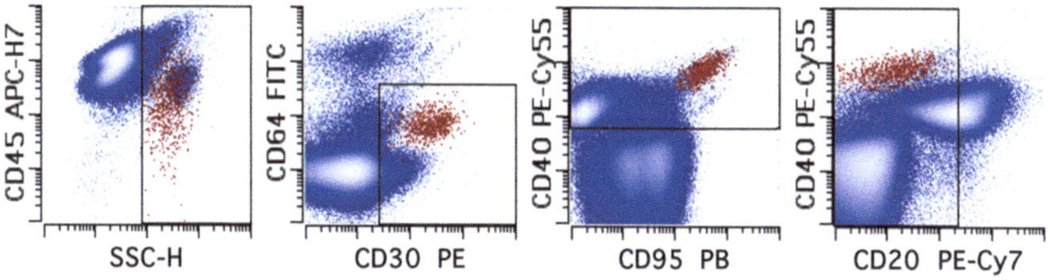

Fig. 10 Gating strategy to identify HRS cells in CHL. HRS cells are in *red* and are *emphasized*, while all other events are in *blue*. The B cells are the large population that is CD20 and CD40+. Putative HRS cells events must fall in all four gates and form a distinct population. Specifically, the cells must have increased side scatter (first plot), be CD30+ and have increased auto-fluorescence in the FITC channel compared to small lymphocytes (second plot), express CD40 at or greater intensity than a reactive B cell (third plot), and express no or low CD20 (fourth plot)

In clinical samples, there may be a population of cells with an apparent immunophenotype that has some resemblance to HRS cells. However, since the immunophenotype does not meet the four basic criteria outlined in Table 4, these events are almost always shown to be a different type of hematologic neoplasm or cellular debris. For example, anaplastic large T-cell lymphoma cells will commonly express CD30 with increased scatter properties. However, its expression of CD40 (if expressed) is usually at a lower level than reactive, background B cells, as also observed in immunohistochemical studies [43, 44]. Diffuse large B-cell lymphoma may occasionally show expression of CD30; however, the level of expression of CD30 in DLBCL (if present) is usually lower than that seen for a typical HRS population. Further, DLBCL cells will have intermediate to bright expression of CD20 (not dim as compared to HRS) with a slight increase in side scatter properties (but not as increased in CHL). Lastly, nodular lymphocyte predominant Hodgkin lymphoma is a type of Hodgkin lymphoma in which the neoplastic cells, referred to now as lymphocyte predominant (LP) cells, express CD20, CD40, CD45 without CD30 or CD15 [45]. Although these cells cannot yet be reliably identified by flow cytometry, the expression of CD20 without CD30 would argue against this population from being a putative HRS population.

One important caveat deserves mention. It is critical to note that HRS cells may occasionally be identified by flow cytometry in patients with non-Hodgkin lymphomas, such as peripheral T-cell lymphoma (PTCL) [46] and chronic lymphocytic leukemia (CLL/SLL) [47–49]. However, all cases for which this phenomenon occurred in our laboratory showed an abnormal B-cell (two cases of CLL/SLL) or T-cell population (two cases of PTCL) in the NHL tubes that were run. Accordingly, the determination of the presence of a HRS population should be considered in the context of all the flow cytometry data. Ultimately, the diagnosis of lymphoma requires the integration of all the clinical, morphologic, and immunophenotypic data.

A summary of the immunophenotypic criteria for identifying a putative HRS population is provided in Table 4 [34, 37, 38].

Table 4 Criteria for identifying a HRS population

1	Have increased forward and side light scatter (compared to background lymphocytes)
2	Express CD30, CD40 and CD95
3	Absence of moderate to bright expression of CD20
4	Absence of expression of CD64
5	Represents a discrete population in multidimensional projections of the immunophenotypic data; increased autofluorescence as compared to CD30+ reactive immunoblasts

The putative HRS population must meet all of the following criteria

Analysis of T Cells in Classical Hodgkin Lymphoma

In the absence of specific analysis for HRS cells, the presence of a characteristic, reactive T-cell population may suggest the diagnosis of CHL in the appropriate clinical and histologic context [50, 51]. We have noted that increased expression of CD2, CD5, CD7, and CD45 on the CD4+ T-cells, increased CD5 and CD45 on the CD8+ T-cells, and decreased expression of CD3 on the CD4 and CD8+ T cells is suggestive of CHL [39]. The increased expression of CD45 and CD7 on CD4+ T cells (Fig. 11) may be the most useful reactive T-cell finding to suggest involvement bya CHL [39]. In addition to suggesting the diagnosis, this observation can also suggest that a specific tube for CHL (above) be run.

Plasma Cell Analysis

Given their derivation from B cells, it is perhaps not surprising that the evaluation of plasma cell neoplasms is similar to that for B-cell NHL. After excluding doublets from compensated flow cytometry data, viable events are identified on a plot of forward scatter versus side scatter. Events that do not produce a signal from the DNA binding dye DAPI (that is, those events that lack DNA) are also excluded. Plasma cells are then identified either by their expression of bright CD38 or CD138 (on plots of CD45 vs. CD38 and CD138 that are computationally summed). Like with the B-cell analysis, one useful feature for plasma cell analysis is identifying cytoplasmic kappa or lambda light chain restricted plasma cells. Plasma cell neoplasms almost invariably have a characteristic immunophenotype (decreased or absent CD19 and/or CD45 with or without expression of CD56 [2]) that allows the ready identification of abnormal populations. Additionally, CD38 is often slightly decreased in plasma cell neoplasms compared to polyclonal plasma cells [52]. When abnormal, DAPI affords the identification of aneuploid plasma cells that increase diagnostic certainty. An example of a typical plasma cell neoplasm is shown in Fig. 12. Analysis of MRD for plasma cell neoplasms is essentially identical to that for B-cell NHL (see above): a given case is evaluated for cytoplasmic kappa or lambda restricted plasma cells with an aberrant immunophenotype identified in multidimensional space.

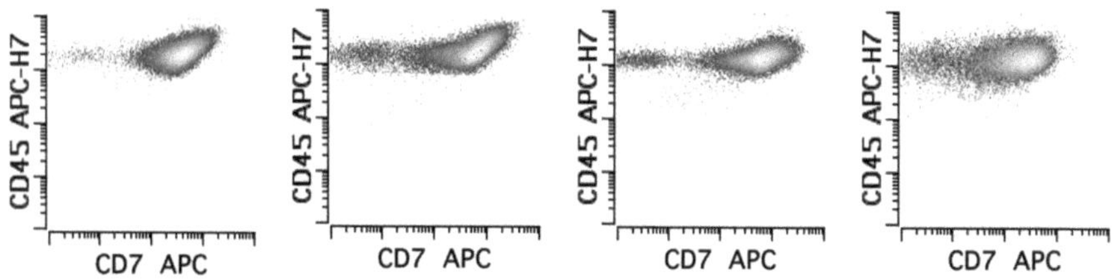

Fig. 11 Reactive T cells in classical Hodgkin lymphoma. CD4+ T cells from classical Hodgkin lymphoma (CHL) demonstrate increased expression of CD45 and CD7 when compared to reactive lymph nodes. First to third dot plots, three examples of CHL cases from three different patients. Fourth dot plot, typical reactive lymph node

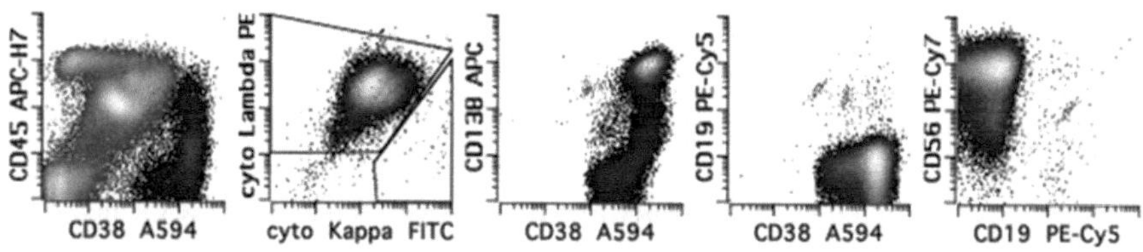

Fig. 12 Example of plasma cell neoplasm. An expanded, abnormal plasma cell population is identified in this 60-year-old male patient, pre-transplant. The abnormal plasma cells (*black*) have aberrant expression of CD19 (absent), CD45 (variable, low to absent), subset CD56, and monoclonal lambda light chain restriction with normal expression of CD38 and CD138. The first plot shows all cells, while the last four plots show only plasma cells. Plasma cells are identified by expression of CD38 and CD138 (see text). Gates in the second panel show very few cytoplasmic kappa and many cytoplasmic lambda restricted plasma cells

A few caveats regarding the evaluation of plasma cells deserve mention. CD138 (syndecan-1) is shed and consequently decreased expression of this antigen can be frequently identified [53]. It is recommended at presentation that evaluation for a plasma cell neoplasm includes evaluation of the B cells, as a subset of B-cell non-Hodgkin's lymphomas will demonstrate evidence of plasmacytic differentiation and consequently may show an abnormal plasma cell population. Importantly, the plasma cell component of a B-cell neoplasm, in contrast to plasma cell neoplasms, often does not show decreased expression of CD45 and CD19 and does not show expression of CD56 [54, 55].

Acute Myeloid Leukemia Analysis

The diagnosis and classification of AML require a combination of diagnostic modalities including morphology, immunophenotyping (most commonly by flow cytometry), cytogenetics, and (increasingly) molecular studies [56]. This section is focused on immunophenotyping AML by flow cytometry; the reader is advised to consult reviews on the other diagnostic modalities for a more complete description of AML diagnosis [57, 58].

Our evaluation of a putative case of AML begins with evaluation of the blast population with two myeloid, B cell, and T-cell tubes (Table 1), the foci being blast identification and lineage assignment. Analysis of myeloid blasts by flow cytometry begins within an inclusive blast gate drawn from a plot of CD45 vs side scatter. From this gate, myeloid blasts can subsequently be identified either by their expression of CD34 and/or CD117 in conjunction with low side scatter and B-cell progenitors can be subsequently excluded based on their expression of CD19. Erythroid cells are identified by their expression of bright CD71 and lack of expression of CD15 and CD34, and reduced CD45. Monocytes are initially identified in the first myeloid tube by gating all myelomonocytic cells on a plot of CD45 vs side scatter, identifying those cells with expression of variable HLA-DR and bright expression of CD33, and then refining on a plot of CD45 vs side scatter. In the second myeloid tube, monocytes are identified first by their expression of bright CD64, exclusion of mature neutrophils expressing CD16 without HLA-DR, usual maturational expression of CD14 and HLA-DR, and finally gated on their characteristic side scatter properties; myeloid cells are typically identified as the majority of the events in the myelomonocytic gate that are not monocytes. Eosinophils can easily be identified in the second myeloid tube by their increased expression of CD45 compared to other myeloid cells, generally higher side scatter, and lack of expression of CD16. Eosinophils also have increased autofluorescence, easily demonstrated in the Pacific Blue channel and may be useful for their identification. As basophils and dendritic cells fall in the blast gate and demonstrate high-level expression of CD123, these cells can be readily identified in the second myeloid tube; basophils have bright expression of CD123 but lack expression of HLA-DR, while dendritic cells have bright expression of CD123 and also express HLA-DR.

Expression of abnormal T-cell antigens on the myeloid blasts is facilitated by the inclusion of CD34 in the usual T-cell tube (Table 1; blasts gated on a plot of CD45 vs side scatter), allowing CD34+ blasts with aberrant T-cell antigen expression to be recognized (Fig. 13). Likewise, blasts (identified on a plot of CD45 vs side scatter) can be evaluated in the B-cell tube (Table 1) to identify any expression of B-cell associated antigens (CD19, CD20, or CD10).

At presentation, myeloid blasts in AML are markedly expanded and demonstrated an abnormal immunophenotype. The abnormal blast population by flow cytometry usually represents greater than 35–40 % of the leukocytes in a given blood or bone marrow specimen. As the formal definition of AML requires the presence of 20 % blasts on a morphologic count that includes nucleated red blood cells (cells that often are excluded during acquisition or analysis of flow cytometry data), we recommended against formally diagnosing "AML" by flow cytometry, if the blast population is less than 35 % in a bone marrow specimen. In peripheral blood specimens, this issue is less problematic and a diagnosis of AML can be established if the blast percentage is more than 20 % of the white blood cells in the blood.

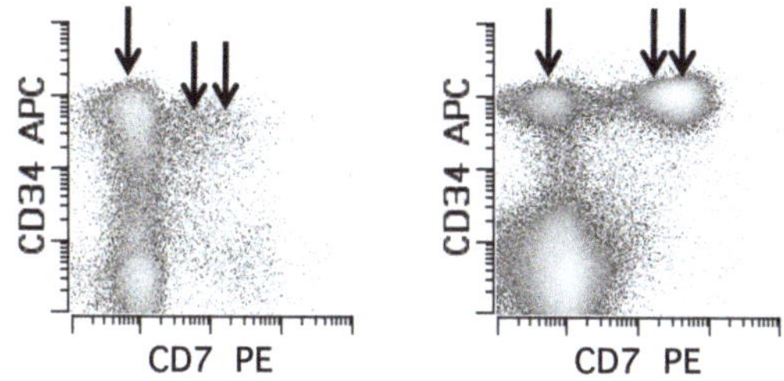

Fig. 13 T-cell antigen expression on myeloid blasts. The *first panel* demonstrates a normal fraction of CD34-positive cells expressing CD7. The *second panel* demonstrates a population with uniform expression of CD7 (an immunophenotypic abnormality). CD34-positive blasts represent approximately 7 % of the leukocytes in both cases. Blasts were initially identified on a plot of CD45 vs side scatter; CD34-positive, CD7-negative cells are identified with a *single arrow*; CD34-positive, CD7-positive cells are identified with a *double arrow*

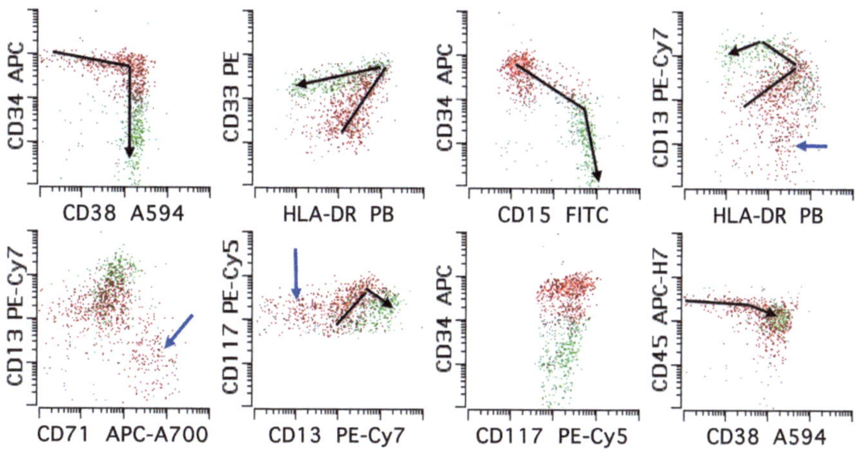

Fig. 14 Normal patterns of myeloid blast maturation. Blasts are first identified in a blast gate drawn on a plot of CD45 vs side scatter (not shown). CD34-positive blasts (*red*) are shown maturing toward immature myeloid cells (*green*), and the path is described by *black arrows*. At the tail of the marrow, the most immature blasts (stem cell population) are present. Where indicated, a *blue arrow* notes the presence of CD34-positive cells maturing to erythroid cells. CD34-positive blasts represent approximately 1.2 % of the leukocytes

Analysis of blast populations requires knowledge of normal myeloid progenitor maturation (Fig. 14) and comparison of a given patient's blast population to the normal patterns of maturation for that cell lineage to determine if immunophenotypic abnormalities are present. Such an approach is particularly important as expanded, but reactive myeloid blast population can occasionally be encountered in the setting of bone marrow regeneration and growth factor therapy when AML minimal residual disease detection (MRD; see below) is performed or during evaluation of myeloid progenitors in other myeloid stem cell disorders (myelodysplastic syndromes and myeloproliferative neoplasms; see below). When viewing progenitor maturation, all events in the blast gate are displayed, allowing maturation of the progenitors to monocytes, myeloid cells, basophils, and dendritic cells to be identified. An initial orientation to early progenitor maturation is provided by examination of the expression of CD34 and CD38, the putative hematopoietic stem cells being those with bright CD34 and low CD38 expression. Once lineage commitment has occurred the level of CD38 is uniform and moderate in intensity and a progressive decline in CD34 expression occurs. Concurrently, the subsequent emergence of specific cell lineages can be seen by the acquisition of lineage-associated antigens or patterns of antigenic expression in conjunction with CD34 or CD117, e.g., CD19 for B cells, bright CD71 for erythroid cells, and CD15 for myelomonocytic cells. Early myelomonocytic cells may be further delineated in early myeloid and monocytic lineages based on differential expression of CD13, being high on early promyelocytes and low on promonocytes. Additionally, identification of immunophenotypic abnormalities on progenitor "stem cell" populations can be facilitated by gating CD34-positive progenitors and displaying each antigen of interest vs CD38.

Immunophenotypic abnormalities can be broadly defined as one of four types: (1) aberrant expression of antigens not normally seen on cells of that type (for example, uniform expression of CD7 on a myeloid blast population; Fig. 13); (2) increased or decreased expression of antigens normally present on that cell type; (3) homogeneous antigen expression (suggesting an arrest in maturation often associated with clonal expansion); and (4) asynchronous co-expression of antigens (two or more antigens that do not show normal co-expression in maturation). In general, as the number and severity of immunophenotypic abnormalities increase, the probability that the blast population in question is abnormal also increases. Characterization and grading of every immunophenotypic abnormality is beyond the scope of this manuscript; however, Fig. 15 demonstrates selected abnormalities that are, in isolation, likely represent an abnormal blast population.

Lineage assignment for most cases of acute leukemia does not require evaluation for cytoplasmic antigen expression (see below). In general, the surface expression of CD117 and other myeloid-associated antigens (CD13, CD33), in the absence of expression of intermediate to bright T-cell antigens (most importantly CD3 and to a lesser extent CD5 and CD7) and the absence of significant expression of B-cell antigens (CD10, CD19, and CD20), is sufficient to imply myeloid lineage. In difficult cases, more definitive assessment of lineage can be performed by evaluating for cytoplasmic antigen expression [59] (Table 1). While criteria are not formally defined in the literature, the presence of at least 10 % of the cells expressing a lineage-defining antigen suggests lineage. Cytoplasmic MPO or monocytic antigen expression (CD64, CD14, and/or CD11c) define myeloid lineage. Surface or cytoplasmic CD3 defines T-cell lineage. Expression of strong CD19 defines B-cell lineage as does expression of more than one of cytoplasmic CD79a, cytoplasmic CD22, CD10, and/or CD20.

Some specific AML subtypes (defined by the French–American–British (FAB) classification (AML-M1 to M7) and the 2008 WHO [60]) are amenable to flow

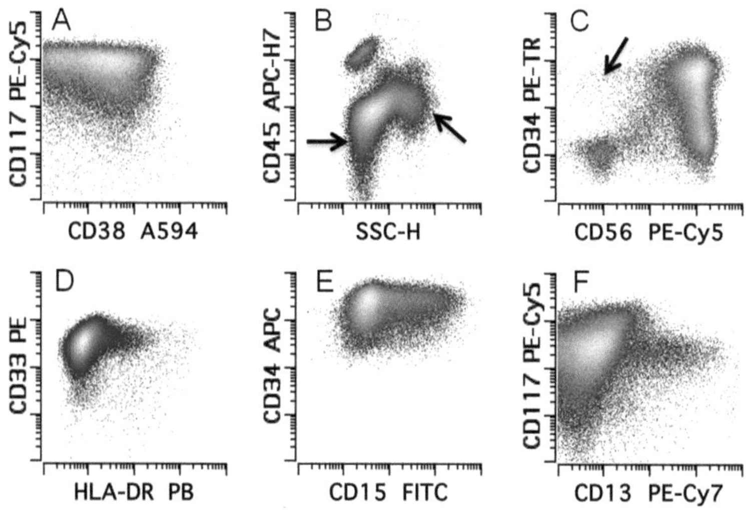

Fig. 15 Immunophenotypic abnormalities that in isolation are suggestive of an abnormal myeloid blast population. Except for panel **b** (which shows all events), blasts were initially identified on a plot of CD45 vs side scatter. (**a**) Bright CD117 (50.2 % of the leukocytes) expression on a CD38-negative (stem cell) population. This population should be at least 1 log lower in antigen intensity for CD117. (**b**) Blasts (69.3 % of the leukocytes and identified with a *horizontal arrow*) have aberrantly low expression of CD45, best seen relative to the granulocytes (*oblique arrow*). (**c**) CD34-positive blasts (61.5 % of the leukocytes) show uniform increased expression of CD56. A very small population of normal CD34-positive, CD56-negative blasts is identified with an *arrow*. An example of uniform expression of CD7 as an immunophenotypic abnormality is shown in Fig. 13. (**d**) CD117-positive blasts (92.7 % of the leukocytes) show uniform and strong expression of CD33 without expression of HLA-DR. This should compared to Fig. 14. (**e**) CD34-positive blasts (56.4 % of the leukocytes) show a subset with aberrant expression of CD15. CD34 expression is also aberrantly increased. (**f**) CD34/CD117-positive blasts (93 % of the leukocytes) show aberrant loss of CD13. Other antigens (not shown) suggest these cells do not have differentiation toward erythroid cells and consequently are abnormal

cytometric diagnosis. Acute myeloid leukemia with minimal differentiation or without maturation (AML M0 and M1) and AML with maturation (AML M2) do not have distinguishing features by flow cytometry (other than an expanded myeloid blast population), and are not described further. Likewise, acute myelomonocytic leukemia (AML M4) demonstrates evidence of monocytic differentiation (below) often with a non-descript myeloid blast population. Acute promyelocytic leukemia (AML M3) can be strongly suggested by flow cytometry with the progenitor population typically expressing CD117, CD13 (variable), and CD33 (bright) but without HLA-DR, CD34, or CD15 (in contrast to normal promyelocytes that show expression of CD15). These abnormal promyelocytes often show increased side scatter compared to a normal CD34-positive progenitor population and can express CD2 [61], CD64, and CD56 [62] (example of APL in Fig. 16). The hallmark of acute monoblastic/monocytic leukemia (AML M5) is the presence of greater than 80 % monocyte forms and an expanded population of

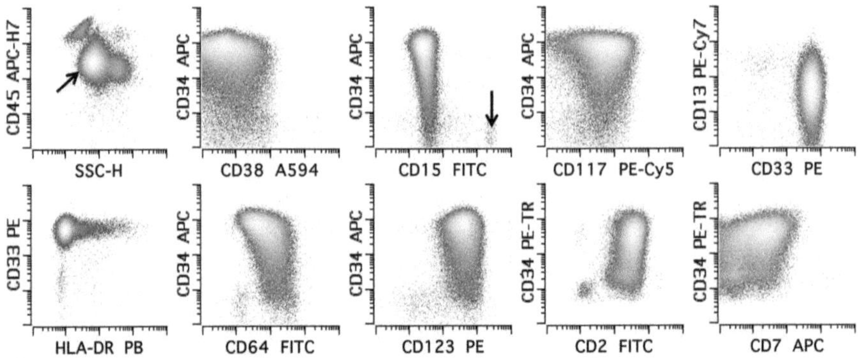

Fig. 16 Example of acute promyelocytic leukemia (APL). The first panel of the *top row* shows all events. All other panels show events in the blast gate (drawn by CD45 vs SS). The blasts have expression of CD2, CD7 (low), CD13 (variable), CD33 (bright), CD34 (more than most cases of APL), CD64, CD117, and CD123 (bright) without CD15, CD38, or significant HLA-DR. The *oblique* and *vertical arrows* identify the blast population and a small population of normal immature myeloid cells, respectively

immature monoblasts/promonocytes that show expression of "monocyte markers" CD4, HLA-DR, CD64 without expression of CD14 (mature monocytes show expression of CD14). Abnormal immature forms (blast equivalent) are usually expanded and, in contrast to normal immature monocytes, often form a distinct subset with minimal maturation and immunophenotypic abnormalities (for example, increased CD64, CD56, and/or CD15) compared to normal monocytes (Fig. 17). Acute erythroid leukemia is defined by two subtypes: erythroleukemia (erythroid/myeloid; AML M6a) (on morphologic evaluation, erythroid cells must comprise greater than 50 % of the nucleated cells and blasts must represent greater than 20 % of the non-erythroid cells) and pure erythroid leukemia (AML M6b). Because of the nature of the diagnosis of erythroleukemia (erythroid/myeloid), flow cytometry will allow for the identification of an abnormal myeloid blast population, but often at a percentage below that typically seen in other types of AML. Pure erythroleukemia is diagnosed morphologically when blasts committed to the erythroid lineage represent greater than 80 % of the nucleated cells. While flow cytometry often underestimates nucleated red blood cells due to exclusion during acquisition and analysis, early immature erythroid cells generally are expanded. The erythroid cells may express CD235a (Glycophorin A) and CD117 without CD34, with expression of bright CD71 being more consistently seen. Recent studies have suggested that E-cadherin may be useful for identifying immature erythroid cells [63]; however, currently our laboratory has not used E-cadherin diagnostically.

Acute megakaryoblastic leukemia (AML M7), as its name would suggest, expresses early megakaryocytic markers CD41 and CD61 (Fig. 18). Evaluation of such antigens requires a special method of specimen preparation to both minimize

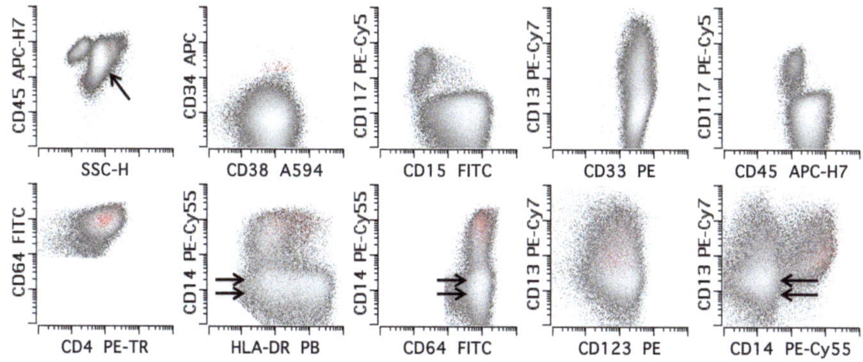

Fig. 17 Example of acute monocytic leukemia (AML M5). The first panel of the *top row* shows all events. All other panels show events in the blast gate (drawn by CD45 vs SS). The events in the blast gate (monocytes) show expression of CD4, CD13 (variable), CD15 (major subset positive), CD38, CD45 (variable), CD64, CD117 (small subset positive), and CD123 without CD34. Importantly, the immature monocyte population (blast equivalents—promonocytes or monoblasts) shows no expression of CD14 (*double arrows*) whereas the mature monocyte population (not a blast equivalent) shows the presence of CD14. The immature monocytes show the small subset with CD117 with lower CD45. The *oblique arrow* in the first panel (*top row*) identifies the blast population

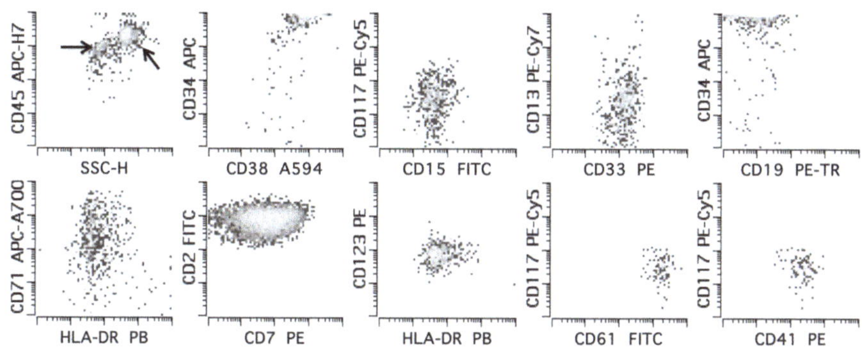

Fig. 18 Example of acute megakaryoblastic leukemia in tissue (AML M7 immunophenotype as myeloid sarcoma). The first panel of the *top row* shows all events. All other panels show events in the blast gate (drawn by CD45 vs SS). The blast population is shown with the *oblique arrow*; reactive lymphocytes are identified with a *horizontal arrow* in the first panel. The neoplastic megakaryoblastic population (67 % of the leukocytes) expresses CD2, CD7, CD13 (dim), CD34 (very bright), CD33, CD38, CD41, CD45 (increased), CD61, CD71 (variable), CD117 (dim), and CD123, without CD15, CD19, or HLA-DR. These cells were also negative for surface CD3, CD5, CD20, and CD64. Bright expression of CD34 and CD45 is unusual for megakaryoblasts

white cell-platelet aggregates and exclude platelet–platelet aggregates. The former is improved by washing the unprocessed sample multiple times prior to further processing, the latter by use of a permeant DNA binding dye, e.g., DRAQ5 (see above), to ensure identification of only nucleated cells. These blasts are often negative for progenitor antigens including CD34 and/or CD117 and are negative for cytoplasmic

MPO and CD235a. Other rare entities (acute basophilic leukemia, acute pan myelosis with myelofibrosis) are not described here. Readers should consult specialized texts on the subject.

A number of types of AML with recurrent genetic abnormalities demonstrate characteristic immunophenotypes (including AML with t(15;17) or acute promyelocytic leukemia, see above). AML with t(8;21)(q22;q22) frequently shows expression of B-cell antigens CD19 and CD79a, bright CD34 and occasionally CD56 [14, 64]. Identification of this immunophenotype is critical to (1) avoid suggesting the diagnosis of mixed phenotype acute leukemia and (2) suggest evaluation for the t(8:21) by conventional cytogenetics or fluorescence in-situ hybridization (FISH).

AML Minimal Residual Disease Detection

Evaluation of MRD attempts to assess therapeutic response for a given patient's neoplasm (example Fig. 19). Flow cytometry can often identify abnormal myeloid blast populations with sensitivity of 10^{-4} to 10^{-5}[57]. When evaluating a bone marrow specimen for AML MRD, a critical question that must be evaluated is whether the myeloid blast population observed is normal. The approach is similar to determining whether the blast population at presentation in AML is normal, with the obvious exception that the MRD populations are typically much smaller. The evaluation for AML MRD is significantly easier if the laboratory has previously immunophenotyped the patient's neoplastic myeloid blast population at presentation, as

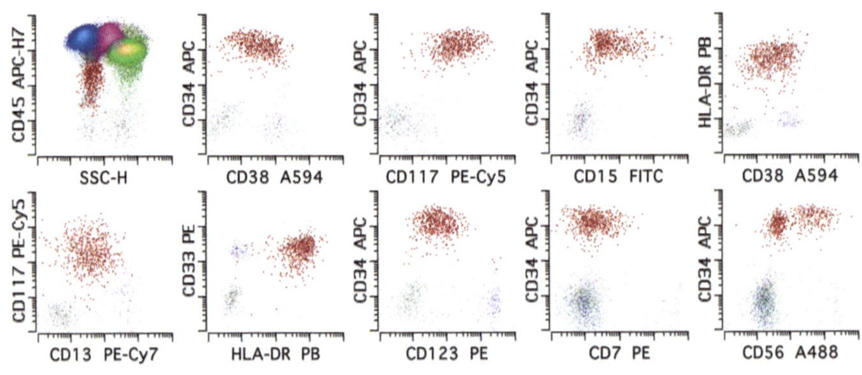

Fig. 19 Example of a positive AML MRD specimen. The first panel of the *top row* shows all events (myeloid blasts, *red*; lymphocytes, *blue*; monocytes, fuchsia; and granulocytes, *green*). All other panels show events in the blast gate (drawn by CD45 vs SS). The abnormal blasts represent 0.16 % of the white cells and express CD7 (low), CD13 (low), CD15, CD33, CD34 (bright), CD38 (low), CD45 (variable), CD56 (subset positive), CD117, CD123, and HLA-DR (increased). Evidence that the blasts are abnormal include increased expression of CD34, expression of CD15, increased expression of HLA-DR, and subset positivity of CD56; this is essentially identical to the abnormal blast population seen before in this patient

many (but frequently not all) of the immunophenotypic abnormalities observed at presentation will be present in the MRD specimen.

A number of important caveats regarding AML MRD deserve mention: (1) active bone marrow regeneration may show a number of immunophenotypic alterations on progenitors, e.g., the expression of low-level CD5 and CD7 or increases in CD123, CD56, and CD33, and these findings should be interpreted with caution; (2) occasionally, immunophenotypic abnormalities are encountered where it is not clear that the changes observed are sufficient for the diagnosis of an abnormal myeloid blast population. In these cases, it is prudent to note the ambiguity and suggest close clinical follow-up with a repeat bone marrow specimen be evaluated further removed from therapy; (3) immunophenotypic drift (changes in immunophenotype after chemotherapy) can result in abnormal blast populations with unexpected immunophenotypes that differ from those seen at diagnosis; (4) some leukemic blast populations have only minimal immunophenotypic abnormalities with usual clinical reagent panels. While at presentation acute myeloid leukemia may be relatively easy to diagnose even with the relatively normal immunophenotype (given the markedly increased number of myeloid blasts), AML MRD in this setting can be exceedingly difficult; and (5) no unified standards have been adopted between laboratories, making comparison of MRD assays between laboratories difficult.

Flow Cytometry for Other Myeloid Stem Cell Neoplasms

The evaluation of other myeloid stem cell neoplasms (myelodysplastic syndrome and myeloproliferative neoplasms) is essentially identical to the approach seen with AML MRD. Maturation of myeloid progenitors, maturing myeloid forms and monocytes is compared to normal maturational patterns for each lineage in multidimensional space. While most cases of myelodysplasia exhibit immunophenotypic abnormalities [65–68], experience suggests that certain myeloproliferative neoplasms are less likely to be detected by flow cytometry. For example, changes in the myeloblasts in essential thrombocythemia or polycythemia vera are rarely identified, while immunophenotypic abnormalities on the myeloblasts are more common in primary myelofibrosis. Ultimately, correlation of morphologic, cytogenetic, molecular, and flow cytometric data is required for diagnosis.

Lymphoblastic Leukemia/Lymphoma Analysis

Lymphoblastic leukemia/lymphoma or acute lymphoblastic leukemia (ALL) is an aggressive immature lymphoid neoplasm that affects both adult and pediatric patients. Over the last couple of decades, there has been substantial improvement in disease outcomes for these patients as a whole [69]. In large part, this gain has been achieved through the systematic application of therapeutic clinical trials and to a lesser extent to an enhanced understanding of the genomic and cytogenetics underlying different subtypes of acute lymphoid leukemias. However, contributing to these advances has also been the systematic identification of patients requiring additional therapy based on the assessment of minimal residual disease by flow cytometry [70, 71] and molecular methods [72, 73]. Here, we discuss the utility of flow cytometry for the initial diagnosis of B- and T-lymphoid acute lymphoblastic leukemia/lymphoma and also provide an overview of the utility of multi-parametric flow cytometry for the assessment of minimal residual disease to guide continuing patient therapy [74].

Lineage Assignment in Acute Lymphoblastic Leukemia

At presentation, leukemic lymphoblasts will typically comprise more than 20 % of the total leukocyte cellularity. In these scenarios, gating and identification of the lymphoblasts will generally be easily performed on a plot of CD45 versus side scatter with lymphoblasts typically having decreased expression levels of both CD45 as compared to mature lymphocytes and slightly decreased side scatter as compared to myeloid progenitors. Sole reliance, however, on the CD45 versus side scatter gate to identify blasts may be problematic in cases with low blast counts, as previously considered [75]. Generally, an abnormal lymphoblast population is defined primarily by the identification of aberrant antigen expression and less commonly by the relative size of the population, as some individuals may have exuberant, non-neoplastic lymphoblastic proliferations that may occur within the marrow [76], peripheral blood [77], and even in tissues [78].

B lymphoblastic leukemia/lymphoma is usually positive for CD19 with usually strong CD10 [79]. At presentation, these cases are immunophenotyped with standard lineage screening tubes (Table 1) and B cells identified in a manner similar to that used for B-cell NHL as above. For most cases of B-ALL, the composite immunophenotype with expression of CD19, CD10, and/or CD20 is sufficient to suggest B-cell lineage, although problematic cases may require evaluation of cytoplasmic CD79a or CD22 to define lineage (Table 1). According to the WHO classification scheme, B lymphoblasts should also lack evidence of common acute myeloid leukemia-defining translocations, such as t(8;21)(q22;q22); *RUNX1/RUNXT1* translocation as this myeloid neoplasm can frequently have B-lineage antigen expression, see above.

T-ALL is also immunophenotyped at presentation and T cells evaluated in a manner similar to that described above for T-cell NHL. T lineage acute lymphoblastic leukemia/lymphoma is usually positive for strong expression of CD7 with variably dim to absent surface CD3 without myeloperoxidase or CD19 and dim to absent CD79a [80]. There is often variable expression of T lymphoid antigens, such as CD2,

CD4, CD5, and CD8, as well as antigens characteristic of immaturity, including TdT, CD1a, CD99, and/or CD34, with CD56 being present on a subset of cases.

In the absence of definite features indicating B- or T-cell lineage, the current WHO classification scheme permits the diagnosis of mixed phenotype acute leukemia if there is substantial co-expression of antigens indicating two or more lineages. These, however, represent the minority of cases [81]. The reader is advised to consult the section of the WHO 2008 classification on acute leukemia with ambiguous lineage [59].

Unique Immunophenotypic Correlates with Genetic Subtypes

There are several unique genetic subtypes of B- and T-lineage acute lymphoblastic that may be important clinically to identify, in part to ensure that adequate cytogenetic testing is performed. Most of these subtypes have more aggressive behavior than standard-risk ALL, so identifying cases could be of substantial clinical importance for patient and their care-provider.

B Lymphoblastic Leukemia with t(9;22) BCR-ABL1 Translocation

This B-lineage acute leukemia is typically positive for expression of CD10, CD19, and TdT with concurrent, aberrant expression of myeloid-associated genes, CD13 and CD33. Patients with t(9;22) translocation typically have worse clinical prognosis and benefit from tyrosine kinase inhibitor therapy. Clinicians may follow treatment response by either flow cytometry and/or molecular analysis of the *BCR-ABL1* fusion gene.

B Lymphoblastic Leukemia/Lymphoma with t(v;11q23); MLL Rearranged

MLL-rearranged B-ALL is an aggressive subtype of B-lineage lymphoblastic leukemia that is usually CD10-negative and CD24-negative with a pro-B immunophenotype when the translocation occurs with chromosome 4. Patients with *MLL-rearranged* tumors tend to have aggressive disease and poor prognosis. Interestingly, in our experience, some MLL-rearranged tumors that have B lineage differentiation may show dramatic immunophenotypic switch to more myeloid differentiation during the course of therapy.

Early Thymic Precursor T-ALL

The early thymic precursor subtype of T-ALL is an aggressive variant, estimated to involve ~10–12 % of all T-ALL cases [82]. These neoplasms are identified by an immunophenotype that is characterized by the absence of CD1a and CD8 with dim to absent CD5 and expression of one or more markers of myeloid and immature cells, such as HLA-DR, CD34, CD33, CD117, CD13, CD33 or CD11b,and CD65. Recent genetic sequencing studies have shown that this subtype has genetic lesions that overlap with acute myeloid leukemia, raising the hypothesis that treatment with myeloid-active regimens may be useful.

ALL Minimal Residual Disease Detection

Evaluation of minimal residual disease is critical for identifying patients who may require more therapy such as dose-intensification and/or alternative treatments such as stem cell treatments. In order to confidently identify neoplastic lymphoblastic populations, it is necessary to have an appreciation for the normal maturational sequence of B- and T-cell progenitors [83] (Figs. 20 and 21, respectively). Normal B- and T-cell progenitors derive from a pluripotent precursor within the marrow that is CD34+/CD38− and lacks lineage-specific markers. These progenitors show a tightly regulated gain and loss of many antigens as early lineage committed precursors differentiate toward mature B and T lymphocytes.

For B-cell progenitors, the immature stem cell will mature to become pro-B cells with expression of CD34, TdT, and CD22. These cells then acquire bright CD10 expression and CD19 and at the pre-B stage show cytoplasmic expression of IgM. CD20 expression is acquired concurrent with surface expression of IgM and decreased expression of CD10. Naïve, resting mature B cells subsequently express no CD10. During this process, levels of expression of various antigens, including CD19 and CD45, increase while expression of CD34 is lost (Fig. 20).

T lymphoblast maturation begins within the bone marrow as a pluripotent progenitor cell that migrates to the thymus where subsequent T-cell maturation occurs. The prothymocyte shows expression of HLA-DR, CD34, TdT, CD2, CD7, and cytoplasmic CD3. At the immature thymocyte stage, T lymphoblasts will gain expression of CD25 with concurrent rearrangement of T-cell receptor genes. By the

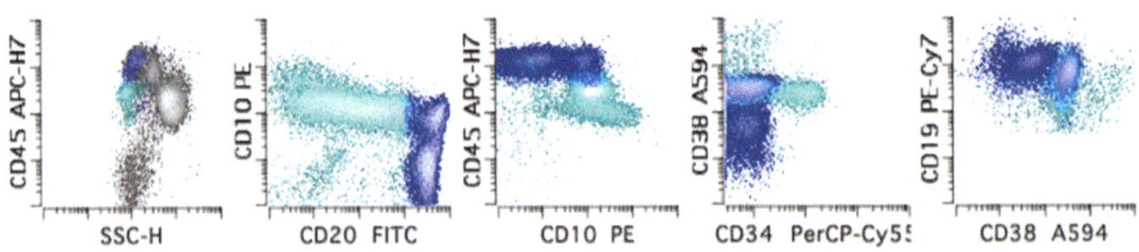

Fig. 20 Normal maturational sequence for B lymphoblasts (hematogones). Precursor B cells (*light blue*) are derived from an immature stem cell within the marrow. The most immature B-cell precusors have expression of CD10, CD34, CD38, without CD20. On maturing, the cells quickly lose CD34, subsequently acquire expression of CD20 while simultaneously gaining expression of CD45 and losing expression of CD38. Mature B cells (*blue*) have bright CD20 and CD45 without CD10

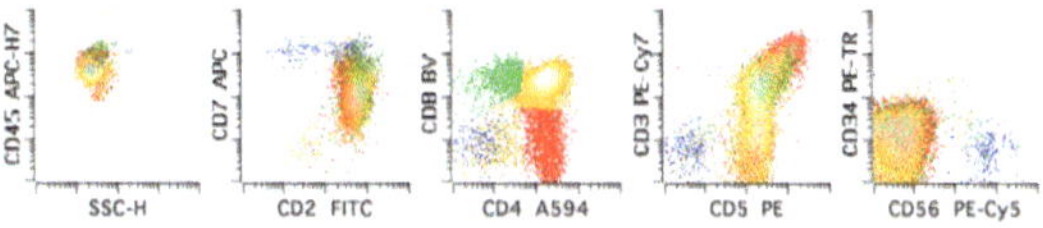

Fig. 21 Normal maturational sequence for T lymphoblasts. Immature T cells are derived initially from an immature stem cell within the marrow, but then subsequently mature in the thymus. T lymphoblasts start as double-negative T cells, become double-positive at the common precursor T lymphoblast and then subsequently express either CD4 or CD8

common thymocyte stage, there is surface expression of the pre-T-cell receptor, including dim expression of CD3. At this point, the cell is double-positive for CD4 and CD8 with expression of CD2, CD5, and CD7 and the common thymocyte antigen, CD1a. As the thymocyte continues to mature, it loses expression of either CD4 or CD8 to give rise to cytotoxic (CD8+) and helper (CD4+) mature T-cell subsets (Fig. 21).

The assessment of MRD begins with the identification of bulk B- or T-cell populations likely to contain any residual leukemia, CD19 or CD7-positive populations gated vs. side scatter, respectively, being useful for this purpose. As the maturational expression sequence of both B and T lymphoblasts is well-defined (at least for the antigens commonly utilized in clinical diagnostic flow cytometry), it is possible to readily identify neoplastic lymphoblast populations when there is deviation from the normal immunophenotypic expression for detection of minimal residual disease [84]. Additionally, evaluation for T-ALL MRD is facilitated by evaluation for immature T-cell immunophenotypes that normally do not occur outside the thymus, e.g., expression of CD1a, bright CD99 or TdT, although following therapy these markers of immaturity may be lost on the leukemia population [85]. A related strategy is the identification of T cells having reduced or absent surface CD3 expression in the presence of cytoplasmic CD3. Since a subset of NK cells may also have this immunophenotype, the use of NK cell-associated antigens such as CD16 and CD56 are helpful to exclude this population. Characterization of the surface immunophenotype alone is similar to the evaluation of mature T cells (see above), save for CD48, an antigen that is decreased to absent on immature T cells but highly expressed on mature T cells (BLW, unpublished results).

As mentioned, an important consideration is that the immunophenotype of B and T lymphoblasts may change substantially during the course of therapy. In B-ALL, Chen and colleagues showed that in the majority of cases of B-ALL, at least one aberrant antigen was lost in the post-treatment period [86]. However, fortuitously for MRD detection by flow cytometry, at least 80 % of abnormalities in the original lymphoblast population were retained [86]. Such changes in immunophenotype are in part related to the use of steroids in the early phases of therapy, which appear to induce maturational progression [87]. Accordingly, enhanced detection of MRD by flow cytometry is best achieved through the use of multidimensional gating strategies that isolate the abnormal population based on its devi-

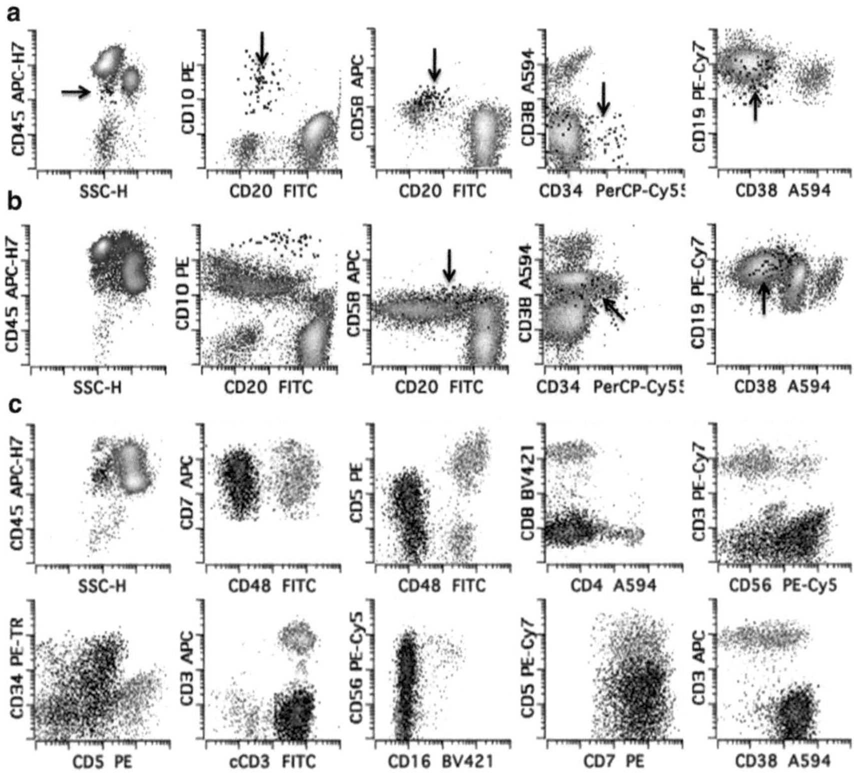

Fig. 22 Utility of flow cytometry for assessment of minimal residual disease in ALL. Assessment of minimal residual disease is critical for guiding patient-specific chemotherapeutic decisions. (**a**) Peripheral blood from a 6-year-old boy with B-ALL at day 8 post-therapy. The blasts (*black* and identified with *arrows*) represented 0.5 % of the white cells (0.04 % of mononuclear cells) and abnormally express CD34, CD38 (absent), CD58 (increased) with normal expression of CD10, CD19, and CD45 without CD20. (**b**) 14-year-old girl with history of B-ALL. Abnormal B lymphoblasts (*black* and identified selectively with *arrows*) in this case comprise only 0.009 % of the white cells (0.015 % of mononuclear cells). Due to characteristically bright expression of CD10, the blasts are readily identified in a background of normal hematogones. The abnormal blasts and abnormally expressed CD10 (bright), CD20 (increased), CD34 (variable), CD38 (decreased), and CD58 (slightly increased) with normal expression of CD19 and CD45. While the original diagnostic flow cytometry analysis was not performed in our laboratory, we did analyze prior MRD samples that had higher proportions of blasts thus permitting establishment of the immunophenotype to consider for high-sensitivity MRD detection. (**c**) 23-year-old male with T-ALL. Abnormal T lymphoblasts (*black*), comprising 0.62 % of total white cells (6.7 % of total mononuclear cells), have aberrant expression of CD3 (absent on surface, present in cytoplasm), CD5 (decreased), CD38 (slightly increased), CD48 (absent), CD34 (variable), CD45 (decreased), and CD56 with normal expression of CD7 without CD4, CD8, or CD16. These abnormalities easily separate the population from background reactive, mature T cells

ance from background normal cells and not solely based on the assessment of fixed-gates based on pre-treatment immunophenotypes [88]. Given the importance of identifying MRD in the post-treatment setting and the challenges with standardizing MRD measurement, investigators continue to seek out novel markers that may be useful in MRD monitoring [89]. Examples of MRD detection in ALL are shown in Fig. 22.

Acknowledgements The authors thank Anju Thomas and the other medical technologists in the Hematopathology Laboratory at the University of Washington for their expert technical assistance.

References

1. Jaffe ES, Harris NL, Stein H, Campo E, Pileri SA, Swerdlow SH (2008) Introduction and overview of the classification of lymphoid neoplasms. In: Swerdlow SH, Campos E, Harris NL, Jaffe ES, Pileri SA, Stein H, Thiele J, Vardiman JW (eds) WHO classification of tumours of haematopoietic and lymphoid tissues, World Health Organization classification of tumors. IARC Press, Lyon, pp 158–166
2. Craig FE, Foon KA (2008) Flow cytometric immunophenotyping for hematologic neoplasms. Blood 111:3941–3967
3. Wood BL Borowitz, MJ (2007) The flow cytometric evaluation of hematopoietic neoplasia. In: McPherson RA, Pincus, MR (eds) Henry's clinical diagnosis and management by laboratory methods. Saunders Elsevier, Philadelphia, pp 599–616
4. Givan AL (2011) Flow cytometry: an introduction. Methods Mol Biol 699:1–29
5. Redelman D (2000) Flow cytometric analyses of cell phenotypes. In: Stewart CC, Nicholson JKA (eds) Immunophenotyping. Wiley-Liss, New York
6. Li S, Eshleman JR, Borowitz MJ (2002) Lack of surface immunoglobulin light chain expression by flow cytometric immunophenotyping can help diagnose peripheral B-cell lymphoma. Am J Clin Pathol 118:229–234
7. Mantei K, Wood BL (2009) Flow cytometric evaluation of CD38 expression assists in distinguishing follicular hyperplasia from follicular lymphoma. Cytometry B Clin Cytom 76:315–320
8. Yang W, Agrawal N, Patel J, Edinger A, Osei E, Thut D, Powers J, Meyerson H (2005) Diminished expression of CD19 in B-cell lymphomas. Cytometry B Clin Cytom 63:28–35
9. Ray S, Craig FE, Swerdlow SH (2005) Abnormal patterns of antigenic expression in follicular lymphoma: a flow cytometric study. Am J Clin Pathol 124:576–583
10. Rawstron AC, Green MJ, Kuzmicki A, Kennedy B, Fenton JA, Evans PA, O'Connor SJ, Richards SJ, Morgan GJ, Jack AS, Hillmen P (2002) Monoclonal B lymphocytes with the characteristics of "indolent" chronic lymphocytic leukemia are present in 3.5% of adults with normal blood counts. Blood 100:635–639
11. Rawstron AC, Shanafelt T, Lanasa MC, Landgren O, Hanson C, Orfao A, Hillmen P, Ghia P (2010) Different biology and clinical outcome according to the absolute numbers of clonal B-cells in monoclonal B-cell lymphocytosis (MBL). Cytometry B Clin Cytom 78(Suppl 1):S19–S23
12. Kussick SJ, Kalnoski M, Braziel RM, Wood BL (2004) Prominent clonal B-cell populations identified by flow cytometry in histologically reactive lymphoid proliferations. Am J Clin Pathol 121:464–472
13. Chen HI, Akpolat I, Mody DR, Lopez-Terrada D, De Leon AP, Luo Y, Jorgensen J, Schwartz MR, Chang CC (2006) Restricted kappa/lambda light chain ratio by flow cytometry in germinal center B cells in Hashimoto thyroiditis. Am J Clin Pathol 125:42–48

14. Hurwitz CA, Raimondi SC, Head D, Krance R, Mirro J Jr, Kalwinsky DK, Ayers GD, Behm FG (1992) Distinctive immunophenotypic features of t(8;21)(q22;q22) acute myeloblastic leukemia in children. Blood 80:3182–3188
15. Rodriguez-Caballero A, Garcia-Montero AC, Barcena P, Almeida J, Ruiz-Cabello F, Tabernero MD, Garrido P, Munoz-Criado S, Sandberg Y, Langerak AW, Gonzalez M, Balanzategui A, Orfao A (2008) Expanded cells in monoclonal TCR-alphabeta+/CD4+/NKa+/CD8-/+dim T-LGL lymphocytosis recognize hCMV antigens. Blood 112:4609–4616
16. Gorczyca W, Weisberger J, Liu Z, Tsang P, Hossein M, Wu CD, Dong H, Wong JY, Tugulea S, Dee S, Melamed MR, Darzynkiewicz Z (2002) An approach to diagnosis of T-cell lymphoproliferative disorders by flow cytometry. Cytometry 50:177–190
17. Jamal S, Picker LJ, Aquino DB, McKenna RW, Dawson DB, Kroft SH (2001) Immunophenotypic analysis of peripheral T-cell neoplasms. A multiparameter flow cytometric approach. Am J Clin Pathol 116:512–526
18. Rahemtullah A, Longtine JA, Harris NL, Dorn M, Zembowicz A, Quintanilla-Fend L, Preffer FI, Ferry JA (2008) CD20+ T-cell lymphoma: clinicopathologic analysis of 9 cases and a review of the literature. Am J Surg Pathol 32:1593–1607
19. Rizzo K, Stetler-Stevenson M, Wilson W, Yuan CM (2009) Novel CD19 expression in a peripheral T cell lymphoma: a flow cytometry case report with morphologic correlation. Cytometry B Clin Cytom 76:142–149
20. Beck RC, Stahl S, O'Keefe CL, Maciejewski JP, Theil KS, Hsi ED (2003) Detection of mature T-cell leukemias by flow cytometry using anti-T-cell receptor V beta antibodies. Am J Clin Pathol 120:785–794
21. Langerak AW, van Den Beemd R, Wolvers-Tettero IL, Boor PP, van Lochem EG, Hooijkaas H, van Dongen JJ (2001) Molecular and flow cytometric analysis of the Vbeta repertoire for clonality assessment in mature TCRalphabeta T-cell proliferations. Blood 98:165–173
22. Morice WG, Kimlinger T, Katzmann JA, Lust JA, Heimgartner PJ, Halling KC, Hanson CA (2004) Flow cytometric assessment of TCR-Vbeta expression in the evaluation of peripheral blood involvement by T-cell lymphoproliferative disorders: a comparison with conventional T-cell immunophenotyping and molecular genetic techniques. Am J Clin Pathol 121:373–383
23. Tembhare P, Yuan CM, Xi L, Morris JC, Liewehr D, Venzon D, Janik JE, Raffeld M, Stetler-Stevenson M (2011) Flow cytometric immunophenotypic assessment of T-cell clonality by Vbeta repertoire analysis: detection of T-cell clonality at diagnosis and monitoring of minimal residual disease following therapy. Am J Clin Pathol 135:890–900
24. Chan WC (2001) The Reed-Sternberg cell in classical Hodgkin's disease. Hematol Oncol 19:1–17
25. Stein H, Hummel M (1999) Cellular origin and clonality of classic Hodgkin's lymphoma: immunophenotypic and molecular studies. Semin Hematol 36:233–241
26. Marafioti T, Hummel M, Foss HD, Laumen H, Korbjuhn P, Anagnostopoulos I, Lammert H, Demel G, Theil J, Wirth T, Stein H (2000) Hodgkin and Reed-Sternberg cells represent an expansion of a single clone originating from a germinal center B-cell with functional immunoglobulin gene rearrangements but defective immunoglobulin transcription. Blood 95:1443–1450
27. Stein H, Delsol G, Pileri SA, Weiss LM, Poppema S, Jaffe ES (2008) Classical Hodgkin lymphoma, introduction. In: Swerdlow SH, Campos E, Harris NL, Jaffe ES, Pileri SA, Stein H, Thiele J, Vardiman JW (eds) WHO classification of tumours of haematopoietic and lymphoid tissues, World Health Organization classification of tumors. IARC Press, Lyon, pp 326–329
28. Dorreen MS, Habeshaw JA, Stansfeld AG, Wrigley PF, Lister TA (1984) Characteristics of Sternberg-Reed, and related cells in Hodgkin's disease: an immunohistological study. Br J Cancer 49:465–476
29. Kadin ME, Newcom SR, Gold SB, Stites DP (1974) Letter: origin of Hodgkin's cell. Lancet 12:167–168
30. Payne SV, Jones DB, Wright DH (1977) Reed-Sternberg-cell/lymphocyte interaction. Lancet 2:768–769

31. Payne SV, Newell DG, Jones DB, Wright DH (1980) The Reed-Sternberg cell/lymphocyte interaction: ultrastructure and characteristics of binding. Am J Pathol 100:7–24
32. Sanders ME, Makgoba MW, Sussman EH, Luce GE, Cossman J, Shaw S (1988) Molecular pathways of adhesion in spontaneous rosetting of T-lymphocytes to the Hodgkin's cell line L428. Cancer Res 48:37–40
33. Stuart AE, Williams AR, Habeshaw JA (1977) Rosetting and other reactions of the Reed-Sternberg cell. J Pathol 122:81–90
34. Fromm JR, Kussick SJ, Wood BL (2006) Identification and purification of classical Hodgkin cells from lymph nodes by flow cytometry and flow cytometric cell sorting. Am J Clin Pathol 126:764–780
35. Harris NL (1999) Hodgkin's disease: classification and differential diagnosis. Mod Pathol 12:159–175
36. Schmitz R, Stanelle J, Hansmann ML, Kuppers R (2009) Pathogenesis of classical and lymphocyte-predominant Hodgkin lymphoma. Annu Rev Pathol 4:151–174
37. Fromm JR, Thomas A, Wood BL (2009) Flow cytometry can diagnose classical Hodgkin lymphoma in lymph nodes with high sensitivity and specificity. Am J Clin Pathol 131:322–332
38. Fromm JR, Wood BL (2010) A six-color flow cytometry tube for immunophenotyping classical Hodgkin lymphoma in lymph nodes. Cytometry B Clin Cytom 78B:395
39. Fromm JR, Thomas A, Wood BL (2010) Increased expression of T cell antigens on T cells in classical Hodgkin lymphoma. Cytometry B Clin Cytom 78:387–388
40. Hsu SM, Jaffe ES (1984) Leu M1 and peanut agglutinin stain the neoplastic cells of Hodgkin's disease. Am J Clin Pathol 82:29–32
41. Stein H, Mason DY, Gerdes J, O'Connor N, Wainscoat J, Pallesen G, Gatter K, Falini B, Delsol G, Lemke H et al (1985) The expression of the Hodgkin's disease associated antigen Ki-1 in reactive and neoplastic lymphoid tissue: evidence that Reed-Sternberg cells and histiocytic malignancies are derived from activated lymphoid cells. Blood 66:848–858
42. Stein H, Uchanska-Ziegler B, Gerdes J, Ziegler A, Wernet P (1982) Hodgkin and Sternberg-Reed cells contain antigens specific to late cells of granulopoiesis. Int J Cancer 29:283–290
43. Carbone A, Gloghini A, Gattei V, Aldinucci D, Degan M, De Paoli P, Zagonel V, Pinto A (1995) Expression of functional CD40 antigen on Reed-Sternberg cells and Hodgkin's disease cell lines. Blood 85:780–789
44. Carbone A, Gloghini A, Pinto A (1996) CD40: a sensitive marker of Reed-Sternberg cells. Blood 87:4918–4919
45. Poppema S, Delsol G, Pileri SA, Stein H, Swerdlow SH, Warnke RA, Jaffe ES (2008) Nodular lymphocyte predominant Hodgkin lymphoma. In: Swerdlow SH, Campos E, Harris NL, Jaffe ES, Pileri SA, Stein H, Thiele J, Vardiman JW (eds) WHO classification of tumours of haematopoietic and lymphoid tissues. IARC Press, Lyon, pp 323–325
46. Quintanilla-Martinez L, Fend F, Moguel LR, Spilove L, Beaty MW, Kingma DW, Raffeld M, Jaffe ES (1999) Peripheral T-cell lymphoma with Reed-Sternberg-like cells of B-cell phenotype and genotype associated with Epstein-Barr virus infection. Am J Surg Pathol 23:1233–1240
47. Mao Z, Quintanilla-Martinez L, Raffeld M, Richter M, Krugmann J, Burek C, Hartmann E, Rudiger T, Jaffe ES, Muller-Hermelink HK, Ott G, Fend F, Rosenwald A (2007) IgVH mutational status and clonality analysis of Richter's transformation: diffuse large B-cell lymphoma and Hodgkin lymphoma in association with B-cell chronic lymphocytic leukemia (B-CLL) represent 2 different pathways of disease evolution. Am J Surg Pathol 31:1605–1614
48. Momose H, Jaffe ES, Shin SS, Chen YY, Weiss LM (1992) Chronic lymphocytic leukemia/small lymphocytic lymphoma with Reed-Sternberg-like cells and possible transformation to Hodgkin's disease. Mediation by Epstein-Barr virus. Am J Surg Pathol 16:859–867
49. Ohno T, Smir BN, Weisenburger DD, Gascoyne RD, Hinrichs SD, Chan WC (1998) Origin of the Hodgkin/Reed-Sternberg cells in chronic lymphocytic leukemia with "Hodgkin's transformation". Blood 91:1757–1761

50. Seegmiller AC, Karandikar NJ, Kroft SH, McKenna RW, Xu Y (2009) Overexpression of CD7 in classical Hodgkin lymphoma-infiltrating T lymphocytes. Cytometry B Clin Cytom 76:169–174
51. Bosler DS, Douglas-Nikitin VK, Harris VN, Smith MD (2008) Detection of T-regulatory cells has a potential role in the diagnosis of classical Hodgkin lymphoma. Cytometry B Clin Cytom 74:227–235
52. Paiva B, Almeida J, Perez-Andres M, Mateo G, Lopez A, Rasillo A, Vidriales MB, Lopez-Berges MC, Miguel JF, Orfao A (2010) Utility of flow cytometry immunophenotyping in multiple myeloma and other clonal plasma cell-related disorders. Cytometry B Clin Cytom 78:239–252
53. Jourdan M, Ferlin M, Legouffe E, Horvathova M, Liautard J, Rossi JF, Wijdenes J, Brochier J, Klein B (1998) The myeloma cell antigen syndecan-1 is lost by apoptotic myeloma cells. Br J Haematol 100:637–646
54. San Miguel JF, Vidriales MB, Ocio E, Mateo G, Sanchez-Guijo F, Sanchez ML, Escribano L, Barez A, Moro MJ, Hernandez J et al (2003) Immunophenotypic analysis of Waldenstrom's macroglobulinemia. Semin Oncol 30:187–195
55. Morice WG, Chen D, Kurtin PJ, Hanson CA, McPhail ED (2009) Novel immunophenotypic features of marrow lymphoplasmacytic lymphoma and correlation with Waldenstrom's macroglobulinemia. Mod Pathol 22:807–816
56. Vardiman JW, Brunning RD, Arber DA, Le Beau MM, Porwit A, Tefferi A, Bloomfield CD, Thiele J (2008) Introduction and overview of the classification of the myeloid neoplasms. In: Swerdlow SH, Campos E, Harris NL, Jaffe ES, Pileri SA, Stein H, Thiele J, Vardiman JW (eds) WHO classification of tumours of haematopoietic and lymphoid tissues. World Health Organization classification of tumors. IARC Press, Lyon, pp 18–30
57. Buccisano F, Maurillo L, Del Principe MI, Del Poeta G, Sconocchia G, Lo-Coco F, Arcese W, Amadori S, Venditti A (2012) Prognostic and therapeutic implications of minimal residual disease detection in acute myeloid leukemia. Blood 119:332–341
58. Morrissette JJ, Bagg A (2011) Acute myeloid leukemia: conventional cytogenetics, FISH, and moleculocentric methodologies. Clin Lab Med 31:659–686
59. Borowitz MJ, Bene M-C, Harris NL, Porwit A, Matutes E (2008) Acute leukemias of ambiguous lineage. In: Swerdlow SH, Campos E, Harris NL, Jaffe ES, Pileri SA, Stein H, Thiele J, Vardiman JW (eds) WHO classification of tumours of haematopoietic and lymphoid tissues. World Health Organization classification of tumors. IARC Press, Lyon, pp 150–155
60. Arber DA, Brunning RA, Orazi A, Prowit A, Peterson L, Thiele J, Le Beau MM (2008) Acute myeloid leukemia, not otherwise specified. In: Swerdlow SH, Campos E, Harris NL, Jaffe ES, Pileri SA, Stein H, Thiele J, Vardiman JW (eds) WHO classification of tumours of haematopoietic and lymphoid tissues. IARC Press, Lyon, pp 130–139
61. Exner M, Thalhammer R, Kapiotis S, Mitterbauer G, Knobl P, Haas OA, Jager U, Schwarzinger I (2000) The "typical" immunophenotype of acute promyelocytic leukemia (APL-M3): does it prove true for the M3-variant? Cytometry 42:106–109
62. Ferrara F, Morabito F, Martino B, Specchia G, Liso V, Nobile F, Boccuni P, Di Noto R, Pane F, Annunziata M et al (2000) CD56 expression is an indicator of poor clinical outcome in patients with acute promyelocytic leukemia treated with simultaneous all-trans-retinoic acid and chemotherapy. J Clin Oncol 18:1295–1300
63. Liu W, Hasserjian RP, Hu Y, Zhang L, Miranda RN, Medeiros LJ, Wang SA (2011) Pure erythroid leukemia: a reassessment of the entity using the 2008 World Health Organization classification. Mod Pathol 24:375–383
64. Kita K, Nakase K, Miwa H, Masuya M, Nishii K, Morita N, Takakura N, Otsuji A, Shirakawa S, Ueda T et al (1992) Phenotypical characteristics of acute myelocytic leukemia associated with the t(8;21)(q22;q22) chromosomal abnormality: frequent expression of immature B-cell antigen CD19 together with stem cell antigen CD34. Blood 80:470–477
65. Kussick SJ, Fromm JR, Rossini A, Li Y, Chang A, Norwood TH, Wood BL (2005) Four-color flow cytometry shows strong concordance with bone marrow morphology and cytogenetics in the evaluation for myelodysplasia. Am J Clin Pathol 124:170–181

66. Ogata K, Kishikawa Y, Satoh C, Tamura H, Dan K, Hayashi A (2006) Diagnostic application of flow cytometric characteristics of CD34+ cells in low-grade myelodysplastic syndromes. Blood 108:1037–1044
67. Kern W, Bacher U, Schnittger S, Alpermann T, Haferlach C, Haferlach T (2013) Multiparameter flow cytometry reveals myelodysplasia-related aberrant antigen expression in myelodysplastic/myeloproliferative neoplasms. Cytometry B Clin Cytom 84:194–197
68. Wood BL (2007) Myeloid malignancies: myelodysplastic syndromes, myeloproliferative disorders, and acute myeloid leukemia. Clin Lab Med 27(551–575):vii
69. Pui CH, Evans WE (2006) Treatment of acute lymphoblastic leukemia. N Engl J Med 354:166–178
70. Bruggemann M, Raff T, Kneba M (2012) Has MRD monitoring superseded other prognostic factors in adult ALL? Blood 120:4470–4481
71. Campana D (2010) Minimal residual disease in acute lymphoblastic leukemia. Hematology Am Soc Hematol Educ Program 2010:7–12
72. Faham M, Zheng J, Moorhead M, Carlton VE, Stow P, Coustan-Smith E, Pui CH, Campana D (2012) Deep-sequencing approach for minimal residual disease detection in acute lymphoblastic leukemia. Blood 120:5173–5180
73. Wu D, Sherwood A, Fromm JR, Winter SS, Dunsmore KP, Loh ML, Greisman HA, Sabath DE, Wood BL, Robins H (2012) High-throughput sequencing detects minimal residual disease in acute T lymphoblastic leukemia. Sci Transl Med 4:134ra63
74. McGregor S, McNeer J, Gurbuxani S (2012) Beyond the 2008 World Health Organization classification: the role of the hematopathology laboratory in the diagnosis and management of acute lymphoblastic leukemia. Semin Diagn Pathol 29:2–11
75. Harrington AM, Olteanu H, Kroft SH (2012) A dissection of the CD45/side scatter "blast gate". Am J Clin Pathol 137:800–804
76. McKenna RW, Asplund SL, Kroft SH (2004) Immunophenotypic analysis of hematogones (B-lymphocyte precursors) and neoplastic lymphoblasts by 4-color flow cytometry. Leuk Lymphoma 45:277–285
77. Kroft SH, Asplund SL, McKenna RW, Karandikar NJ (2004) Haematogones in the peripheral blood of adults: a four-colour flow cytometry study of 102 patients. Br J Haematol 126:209–212
78. Ohgami RS, Zhao S, Ohgami JK, Leavitt MO, Zehnder JL, West RB, Arber DA, Natkunam Y, Warnke RA (2012) TdT+ T-lymphoblastic populations are increased in Castleman disease, in Castleman disease in association with follicular dendritic cell tumors, and in angioimmunoblastic T-cell lymphoma. Am J Surg Pathol 36:1619–1628
79. Seegmiller AC, Kroft SH, Karandikar NJ, McKenna RW (2009) Characterization of immunophenotypic aberrancies in 200 cases of B acute lymphoblastic leukemia. Am J Clin Pathol 132:940–949
80. Patel JL, Smith LM, Anderson J, Abromowitch M, Campana D, Jacobsen J, Lones MA, Gross TG, Cairo MS, Perkins SL (2012) The immunophenotype of T-lymphoblastic lymphoma in children and adolescents: a Children's Oncology Group report. Br J Haematol 159:454–461
81. Han X, Bueso-Ramos CE (2007) Precursor T-cell acute lymphoblastic leukemia/lymphoblastic lymphoma and acute biphenotypic leukemias. Am J Clin Pathol 127:528–544
82. Coustan-Smith E, Mullighan CG, Onciu M, Behm FG, Raimondi SC, Pei D, Cheng C, Su X, Rubnitz JE, Basso G et al (2009) Early T-cell precursor leukaemia: a subtype of very high-risk acute lymphoblastic leukaemia. Lancet Oncol 10:147–156
83. McKenna RW, Washington LT, Aquino DB, Picker LJ, Kroft SH (2001) Immunophenotypic analysis of hematogones (B-lymphocyte precursors) in 662 consecutive bone marrow specimens by 4-color flow cytometry. Blood 98:2498–2507
84. Gorczyca W, Tugulea S, Liu Z, Li X, Wong JY, Weisberger J (2004) Flow cytometry in the diagnosis of mediastinal tumors with emphasis on differentiating thymocytes from precursor T-lymphoblastic lymphoma/leukemia. Leuk Lymphoma 45:529–538

85. Roshal M, Fromm JR, Winter S, Dunsmore K, Wood BL (2010) Immaturity associated antigens are lost during induction for T cell lymphoblastic leukemia: implications for minimal residual disease detection. Cytometry B Clin Cytom 78:139–146
86. Chen W, Karandikar NJ, McKenna RW, Kroft SH (2007) Stability of leukemia-associated immunophenotypes in precursor B-lymphoblastic leukemia/lymphoma: a single institution experience. Am J Clin Pathol 127:39–46
87. Gaipa G, Basso G, Aliprandi S, Migliavacca M, Vallinoto C, Maglia O, Faini A, Veltroni M, Husak D, Schumich A et al (2008) Prednisone induces immunophenotypic modulation of CD10 and CD34 in nonapoptotic B-cell precursor acute lymphoblastic leukemia cells. Cytometry B Clin Cytom 74:150–155
88. Griesinger F, Piro-Noack M, Kaib N, Falk M, Renziehausen A, Troff C, Grove D, Schnittger S, Buchner T, Ritter J et al (1999) Leukaemia-associated immunophenotypes (LAIP) are observed in 90% of adult and childhood acute lymphoblastic leukaemia: detection in remission marrow predicts outcome. Br J Haematol 105:241–255
89. Coustan-Smith E, Song G, Clark C, Key L, Liu P, Mehrpooya M, Stow P, Su X, Shurtleff S, Pui CH et al (2011) New markers for minimal residual disease detection in acute lymphoblastic leukemia. Blood 117:6267–6276
90. Wu D, Wood BL, Fromm JR (2013) Flow cytometry for non-Hodgkin and classical Hodgkin lymphoma. Methods Mol Biol 971:27–47
91. Crespo M, Bosch F, Villamor N, Bellosillo B, Colomer D, Rozman M, Marce S, Lopez-Guillermo A, Campo E, Montserrat E (2003) ZAP-70 expression as a surrogate for immunoglobulin-variable-region mutations in chronic lymphocytic leukemia. N Engl J Med 348:1764–1775
92. Damle RN, Wasil T, Fais F, Ghiotto F, Valetto A, Allen SL, Buchbinder A, Budman D, Dittmar K, Kolitz J et al (1999) Ig V gene mutation status and CD38 expression as novel prognostic indicators in chronic lymphocytic leukemia. Blood 94:1840–1847

Bioinformatics Analysis of Sequence Data

Anthony T. Papenfuss, Daniel Cameron, Jan Schroeder, and Ismael Vergara

Introduction

Bioinformatics is a relatively young, rapidly evolving discipline, which can be broadly defined as the application of mathematics, statistics and computer science to the analysis of biological data. Information technology and software engineering skills are also important, particularly in molecular pathology. Bioinformatics is about deriving insight from biological data. For the outsider, understanding what bioinformatics is and engaging with practitioners is complicated by the different types, multiple specialties and rapid development of bioinformatics. In this chapter, we aim to provide an overview of

A.T. Papenfuss (✉)
Bioinformatics Division, Walter and Eliza Hall Institute of Medical Research,
Parkville, Victoria 3052, Australia

Computational Cancer Biology Program, Peter MacCallum Cancer Centre, Victorian Comprehensive Cancer Centre Building, 305 Grattan Street, Melbourne,
Victoria 3000, Australia

Department of Medical Biology, University of Melbourne,
Parkville, Victoria 3010, Australia

Department of Mathematics and Statistics, University of Melbourne,
Victoria 3010, Australia

Sir Peter MacCallum Department of Oncology, University of Melbourne,
Victoria 3010, Australia
e-mail: anthony.papenfuss@petermac.org; papenfuss@wehi.edu.au

D. Cameron
Bioinformatics Division, Walter and Eliza Hall Institute of Medical Research,
Parkville, Victoria 3052, Australia

Department of Medical Biology, University of Melbourne, Parkville, Victoria 3010, Australia

S.R. Lakhani, S.B. Fox (eds.), *Molecular Pathology in Cancer Research*,
DOI 10.1007/978-1-4939-6643-1_14

bioinformatics to help readers engage with bioinformaticians and, since the data and methods used in bioinformatics change rapidly, present key concepts of most relevance to molecular pathology and point readers in the direction of the major tools.

Types of Bioinformatician

There are several different types of bioinformatician:

Production bioinformatician is associated with sequencing facilities and involves operation of Laboratory Information Management System (LIMS) and ensuring data gets from sequencer or other platform to database/disk storage.

Infrastructure bioinformatician involves database development and maintenance, and tool development.

Service bioinformatician typically performs on a fee-for-service basis, often, in a core facility.

Research bioinformatician may be collaborative, but will typically also have a focus on methods development and data analysis. Research bioinformaticians will need to write papers and bring in grants. The term *computational biologist* is sometimes used synonymously with research bioinformatician. If one needs to draw a distinction, then a research bioinformatician may be more focused on methods development, while a computational biologist is focused on drawing biological insight from data and modelling.

In molecular pathology, arguably a new type of bioinformatician is needed. The *clinical bioinformatician* needs to work in close partnership with clinicians and pathologists, to develop and maintain pipelines in an environment of continuously evolving tools, to test tools using carefully designed validation datasets, and to control change and to meet the needs of diagnostic certification. The clinical bioinformatician needs a good understanding of software engineering principles, practices and tools, including version control software, but also be capable of solving analysis

J. Schroeder
Bioinformatics Division, Walter and Eliza Hall Institute of Medical Research, Parkville, Victoria 3052, Australia

Department of Computing and Information Systems, University of Melbourne, Victoria 3010, Australia

I. Vergara
Bioinformatics Division, Walter and Eliza Hall Institute of Medical Research, Parkville, Victoria 3052, Australia

Computational Cancer Biology Program, Peter MacCallum Cancer Centre, Victorian Comprehensive Cancer Centre Building, 305 Grattan Street, Melbourne, Victoria 3000, Australia

Sir Peter MacCallum Department of Oncology, University of Melbourne, Victoria 3010, Australia problems, as there are still many unsolved problems of relevance to molecular pathology, such as analysis of copy number from targeted sequencing data.

Bioinformatics has evolved rapidly and its development is closely linked to the emergence of new technologies for generating -'omics data. Many specialty areas have already emerged, for example, sequence analysis, analysis of gene expression and gene regulation, analysis of proteomics data, and some would allow statistical genetics including genome-wide association studies. Typically, analyses of genomics data from new technologies start out rough, methods and tools improve over time and eventually become standardised and even commoditized. In recent years, the importance and impact of bioinformatics has grown substantially due to the rapid increase in the rate at which we can generate data, particularly using massively parallel sequencing (MPS) or next-generation sequencing (NGS).

In the remainder of this chapter, key bioinformatics analyses relevant to molecular pathology are introduced. We cover copy number analysis using single nucleotide polymorphism (SNP) arrays, MPS, the primary analysis of this sequencing data, prediction of single nucleotide variants (SNVs) and small indels, sequencing-based copy number analysis, and prediction of genomic rearrangements.

Copy Number Analysis Using Single Nucleotide Polymorphism Arrays

A number of DNA hybridization technologies have been applied to the detection of copy number variants. Here, we describe the use of SNP arrays. These arrays consist of a set of probes that cover SNPs. For example, the Illumina HumanOmni2.5-8 array covers about 2.5 million polymorphic markers. For each marker, there are probes for the A and B alleles with A and B defined by the manufacturer and independent of allele population frequency.

The DNA is fragmented, fluorescently labelled and hybridised to the array. The fluorescent intensity of each probe is then measured and the intensity is approximately proportional to the copy number of the allele. Typically, the intensities, A and B, are corrected for background and cross-talk between alleles and normalised, then expressed as the $\log_2$-relative ratio (LRR):

$$LRR = \log_2\left(\frac{A+B}{m}\right)$$

and the B-allele frequency (BAF):

$$BAF = \frac{B}{A+B}$$

where m is the median total intensity across all markers. Values of LRR and BAF are returned for each SNP. The same relationships hold for the allele-specific copy numbers (n_A and n_B) and median copy number (or average ploidy n_m).

$$LRR = \log_2\left(\frac{n_A + n_B}{n_m}\right)$$

$$BAF = \frac{n_B}{n_A + n_B}$$

LRR is zero when the total copy number is equal to the median ploidy, while BAF is 0 for homozygous A alleles (AA), ½ for heterozygous (AB), 1 for homozygous B alleles (BB), and may take other fractional values where there is allelic imbalance (e.g. 1/3 for AAB). In reality, these relationships can be further complicated by the normalisation process used by the manufacturer and other factors. If the average ploidy is known, then these could be solved for the allelic copy numbers. However, more typically the ploidy is unknown and somatic samples may be impure or heterogeneous.

Unavoidably, there is a loss of information when performing array-based copy number estimation on a sample of tissue. The data collected is relative to the average signal, which results from the overall ploidy of the sample and this is usually unknown. There may be contamination by normal tissue and heterogeneity. These factors make the mathematical problem underdetermined (i.e. there are more unknowns to estimate than data points). Several methods attempt to infer the cellularity (purity) and average ploidy (see, for example, [1]). Another approach, which is attractive, but not widely used, is to independently estimate copy number in selected regions.

To profile somatic copy number changes in a tumour, DNA from the tumour and the patient germline are usually hybridised to arrays. The germline array allows us to subtract out the germline copy number profile, so only somatic mutations are considered. It can also help to reduce noise in the data through normalisation methods like CalMaTe [2] and identify somatic Loss of Heterozygosity (LOH).

Figure 1 shows the LRR and BAF for a tumour, where each dot in the scatter plot represents one SNP. This is a useful way to represent copy number profiles.

The LRR represents the total relative copy number. The centre of the LRR distribution is centred on zero and corresponds to the average ploidy of the sample. Regions that show a decrease in LRR either focally or at the chromosomal arm or whole chromosome level correspond to deletions; increases correspond to copy number gain events. As the LRR is relative and frequently we do not know if the tumour is diploid, it can be difficult to assign an absolute copy number to each region.

The BAF shows two, three or four bands of dots (Fig. 1). Three bands of dots occur when a region is in allelic balance (i.e. an equal number of copies of each allele are present). The bottom band corresponds to homozygous A alleles, the top band corresponds to homozygous B alleles and the central band corresponds to heterozygosity. This is the normal situation when a genome is diploid, in which case the bands correspond to the allelic states AA (bottom), AB (middle) and BB (top). However, 3 bands arise if both chromatids are duplicated, with corresponding allelic states AAAA, AABB and BBBB, or more generally occur in allelic balance. Four bands

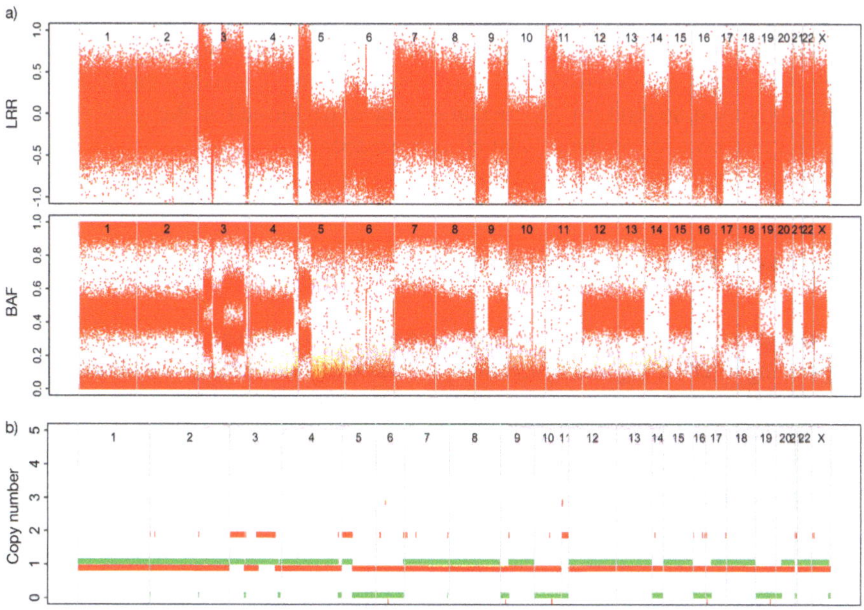

Fig. 1 Log relative ratio (LRR) and B-allele frequency (BAF) plots from a single nucleotide polymorphism (SNP) array of a tumour sample

occur if there is allelic imbalance (i.e. one allele is amplified), leading to the splitting of the heterozygous band. If there is a gain of 1 copy in a region of a clonal tumour, the corresponding allelic states of the middle bands are AAB and ABB. More generally the separation of these bands tells us about the relative allelic copy number. The larger the allelic imbalance, the further the bands are apart. For example, in a pure tumour sample, in a region with copy number 3, the central bands will sit at 1/3 (AAB) and 2/3 (ABB). If the copy number is 4 with allelic imbalance, the bands will sit at 1/4 (AAAB) and 3/4 (ABBB). In other words, the closer the bands are to 1/2, the closer the region is to allelic balance, while the further they are apart, the greater the allelic imbalance. Finally, 2 bands at a BAF of 0 and 1 in a pure tumour correspond to LOH. To determine if this coincides with a deletion or is copy number neutral, one must look at the LRR. A deletion will be apparent from a drop in LRR.

Since tumours frequently have normal cells contaminating them, the tumour DNA is rarely pure. This can be expressed as the cellularity, purity or contamination of the tumour. The impact of cellularity is to squeeze the paired heterozygous bands due to allelic imbalance (or LOH) back towards a BAF of 1/2. This effect impacts every chromosome.

A variety of tools exist to estimate the average ploidy and cellularity of samples (e.g. qpure estimates cellularity only [3]). Once these are known, the copy number can be estimated. The tool ASCAT predicts the cellularity, ploidy and the allele-specific copy number.

A typical analysis performed by most CNV callers is to segment regions of constant copy number and identify the positions in the genome where the copy number state jumps. Several approaches exist. Circular binary segmentation (CBS) is a commonly used method [4]. Segmentation can be performed on the LRR and BAF, or jointly on both, or on the estimated copy number.

Massively Parallel Sequencing

To sequence a genome to 30 times (30×) coverage using 100 nt long read paired-end sequencing of DNA fragments, 450 million reads ((30-fold coverage × 3 billion nt haploid genome)/(2 × 100 nt reads)) are required. The figure of 30× is the estimated coverage at which one theoretically detects >99.5 % of heterozygous variants in a diploid genome [5]. Empirical analyses have shown that at an average coverage of 30×, 99.15 % of SNPs (both heterozygous and homozygous) are correctly identified [6].

Read Alignment

In re-sequencing projects (as opposed to de novo assembly or other analyses), analysis of NGS data typically begins with alignment of reads to the reference genome. To align the massive quantities of data generated by NGS platforms, new alignment algorithms were necessary. NGS aligners (or mappers) are much faster than generic aligners (such as BLAST, BLAT and exonerate). This is achieved by: reducing the alignment problem to global alignment (the whole read must be aligned) or simpler types of local alignment (a single contiguous section of the read must be aligned, allowing the start and end only to be clipped); limiting how different the read can be from the reference genome; and most importantly by introducing acceleration techniques (such as the Burrows–Wheeler Transform in BWA [7]).

Reads are generally aligned to the most recent version of the human reference genome. However older versions are sometimes used to ensure compatibility either with previous analysis, or an analysis package or database not yet updated to the latest version. In some cases, the reference can be augmented with additional sequences such as Merkel cell polyomavirus or other integrated viruses.

Prediction of Single Nucleotide Variants and Small Indels

SNVs are the most common form of genetic diversity in the human population, with a germline mutation rate of approximately one per 100 million base pairs per generation [8]. Somatic mutation rates can be much higher [9]. SNP arrays can detect a large number of known polymorphisms included in the array, but genome-wide detection of novel variants requires a different technology, with MPS being well suited to this task. A typical SNV detection pipeline using GATK software toolkit performs the following steps:

1. *Sorting*: Following read alignment, reads are sorted according to their aligned genomic position, allowing for computationally efficient processing in subsequent steps.
2. *Duplicate removal*: If a library preparation protocol involving PCR amplification is used, multiple reads originating from a single source fragment may appear. This is especially a problem in conjunction with small starting amounts of DNA. This can create biases in downstream analyses such as copy number analysis and propagate nucleotide sequencing errors into incorrectly called variants. To overcome this, duplicate reads which have the same alignment start and end coordinates for each read pair are flagged or removed. The Picard tools (http://broadinstitute.github.io/picard/) subprogram MarkDuplicates is the most widely used duplicate removal tool for SNV calling pipelines and identifies duplicates based on matching alignment position and strand orientation.
3. *Indel realignment*: As each read is aligned independently, small insertions or deletions occurring in the middle of a read will result in correct alignment. Indels occurring near either end of a read however result in spurious SNVs adjacent to the indel. Indel realignment removes these artefacts by performing targeted local realignment against an alternative indel consensus sequence and, if sufficiently improved over the reference sequence alignments, adjusting the alignment of all reads supporting the alternate consensus.
4. *Base quality score recalibration:* For most sequencing runs, the base quality scores generated from the sequencer do not match the empirical distribution of base mismatches implied by the alignment of the reads to the reference genome. Tools such as GATK correct for this by adjusting the quality score for known sequencing chemistry effects such as the reported quality score, position within the read and the dinucleotide context.
5. *Variant calling*: For each genomic position, the base calls for each read mapping to that location are compared. Low-quality bases and reads with low mapping quality are filtered, and each of the four nucleotides in the remaining bases are counted. This is sometimes performed explicitly via samtools mpileup (http://www.htslib.org/), while some methods generate these counts on the fly directly from the BAM file (e.g. MuTect [10]). SNVs are then called by either applying a series of heuristic thresholds (e.g. VarScan [11–13]) or a statistical model (e.g. GATK, MuTect) to the nucleotide counts.

   Most variant callers are run from a Unix command-line and take as input one or more SAM or BAM read alignment files. Output is a human-readable file format VCF (Variant Call Format), or the compressed binary equivalent BCF. Each line in a VCF corresponds to a single variant call and includes a variant identifier, the genomic location of the variant, the reference and alternate alleles, the estimated probability of all samples being homozygous reference allele (variant quality score), the genotypes for all samples and any additional informational fields written by the calling software. Genome visualisation software such as IGV can be used to view the location, and supporting evidence for the called variants (Fig. 2).

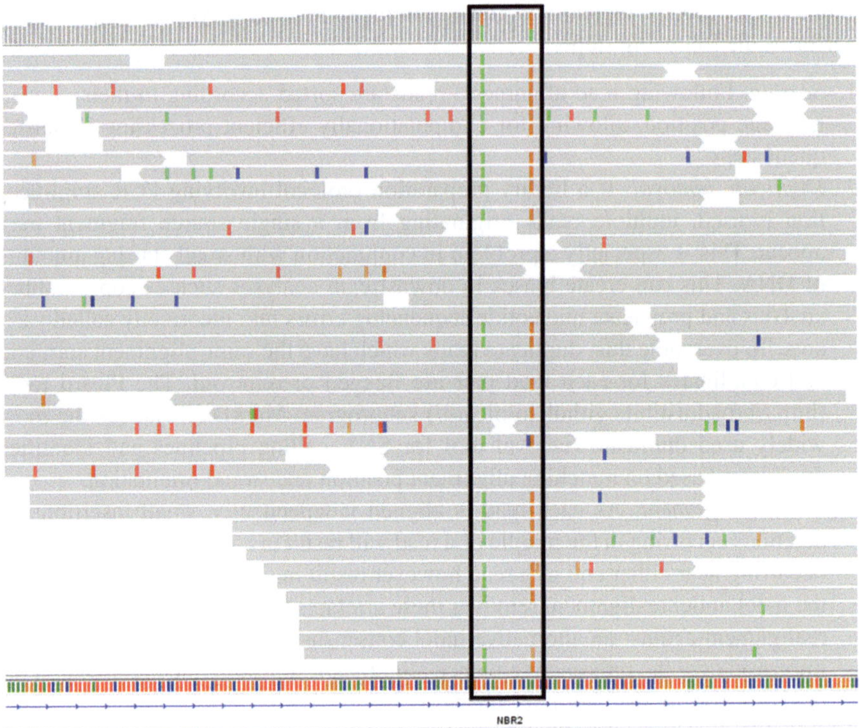

Fig. 2 Single nucleotide variants (SNVs) visualised using IGV. Read bases matching the reference are *grey*; mismatched base are *coloured*. The SNVs (highlighted in the *box*) are part of a single haplotype as reads contain either both or neither of the SNVs

For applications requiring a high-confidence list of putative variants, the results of multiple variant callers can be combined to form a consensus call set. Applications of this approach have shown improved sensitivity and specificity over any of the single variant callers used to form the consensus [14].

Although usage of mature variant callers such as samtools/bcftools and GATK is widespread, specialised callers have become more prevalent. Of particular interest to cancer researchers are somatic variant callers such as VarScan2, SomaticSniper and MuTect. These callers take a germline and somatic sample and classify variants according to their source: germline, somatic, or somatic LOH (loss of heterozygosity).

6. *Variant filtering*: Once putative variants have been called, low-quality variants and variants likely due to artefacts are filtered. Common filters include:

- Quality score: variants below a threshold quality score are filtered
- Coverage: variants with insufficient coverage are filtered
- Strand bias: variants showing a strong strand bias are indicative of sequencing or alignment artefacts

- BAF: variants with a BAF below a threshold value are filtered
- Blacklisted regions: variants occurring in known problematic regions are filtered

7. *Small Indels:* The read alignment techniques used to identify SNVs can, with only minor modifications, be used to detect small insertions and deletions. Unfortunately, there is no consensus in the literature as to what constitutes a small indel, though 'small' is typically taken as a synonym for 'detectable by the software used' when applied to SNV and small indel variant callers. The most common range used is 1–50 bp insertions or deletions as this is the range of indel submissions accepted by dbSNP. However a smaller or larger size range is sometimes used depending on the read length of the dataset (longer read lengths increase the maximum indel size that most variant callers can detect), or type of analysis performed.
8. *Annotation and phenotypic effect*: Once somatic variants have been identified in a cancer, their functional impact needs to be determined. Previously reported variants can be found in a number of human cancer databases, the largest being COSMIC (Catalogue Of Somatic Mutations In Cancer). When known, attributes describing the functional impact of listed variants are included in variant annotation databases. However due to the very high heterogeneity of cancers, many novel somatic variants can be found. The type of variant strongly determines the phenotypic impact of the mutation. As well as classifying the SNV/indel according to the mutation type (synonymous, nonsense, missense, insertion, deletion and frameshift) and location (exonic, splicing, ncRNA, UTR5, UTR3, intronic, upstream, downstream and intergenic), software such as Polyphen2, LRT, PhyloP and MutationTaster assign scores based on different in silico approaches that indicate how damaging a particular variant is likely to be. Emerging technologies, such as deep mutational scanning, may eventually provide unbiased empirical evidence for whether a particular variant has a significant functional impact on a protein [15, 16].

Prediction of Copy Number Variants from Sequence Data

The primary approach to using sequencing data to identify copy number changes is through analysis of read depth. Sequencing-based copy number methods follow the same principles of those based on techniques such as array Comparative Genomic Hybridization (aCGH) and SNP arrays (described previously). There are typically four stages to read depth analyses. The first stage corresponds to the definition of a window size over which the number of reads or median read depth is calculated on the samples across all covered regions. The main assumption is that this value is proportional to the copy number of the sequenced sample, and hence the LRR of these values for a case sample and a control/reference set should be indicative of the relative presence/absence of CNVs. Importantly, pooling of cells and sequencing results in loss of absolute copy number information and the results are relative.

Single cell sequencing overcomes this shortcoming. As with arrays, statistical models can be used to infer ploidy and cellularity. The second stage corrects the read depth profiles for technical biases such as GC content and mappability. The third stage corresponds to the segmentation of the LRR by merging adjacent windows of similar LRR while detecting change points between segments; common methods used are CBS [4] and hidden Markov model (HMMs)-based approaches [e.g. 17]. The last stage implemented in some methods involves the classification of the segments into gains and losses, copy number states, LOH and allele-specific CNVs. Several reviews have revisited and discussed in detail the main approaches used for most read depth-based methods for the steps listed above [18–22].

While there are a large (and growing) number of tools developed for read depth-based CNV detection, they can be broadly separated according to their: (1) type of sequencing and (2) type of control/reference set.

Type of Sequencing

While it still suffers from biases due to GC and mappability, detection of CNVs using whole genome sequencing (WGS) benefits from a more uniform distribution of the reads across the genome compared to capture-based enrichment methods, for example, whole exome sequencing (WES) and targeted re-sequencing (TRS), making the proportionality between read depth and copy number clearer. Additionally, WGS data allows for accurate detection of breakpoints, if the depth of sequencing is sufficient. This allows for refinement of the edges of CNVs and single nucleotide resolution. In WES and TRS, the reduced proportion of the genome being sequenced (1–3 % in the case of WES, and much lower for TRS), the non-uniform distribution of exons along the chromosomes and additional technical biases such as differences in hybridization capture efficiency across regions and probe concentration challenges this assumption and makes the accurate detection of copy number changes more difficult. A number of tools have been developed specifically for WES and TRS data (e.g. ADTEx [23] and CONTRA [24]), as well as amplicon-based TRS (oncoCNV [25]) and long-range TRS (cnvCapSeq [26]).

Type of Control/Reference

Methods for read depth-based CNV detection can be grouped into those that use paired case–control data and those that use pooled data to build a reference set. The former type requires each case sample to have a matched control against which the LRR is built. This is preferable for detection of somatic CNVs. The latter builds a reference from pooled samples that is used to measure LRR against each case sample. The pooled data approach is useful when matched controls are not available for all samples.

Two popular tools that can be used for the detection of CNVs include BIC-seq [27] and CONTRA [24]. BIC-seq is a tool designed for WGS paired data. In the BIC-seq algorithm, mixed bins of uniquely aligned tumour and normal reads (sorted by their genomic coordinates and with removed amplification bias—see Section "Prediction of Copy Number Variants from Sequence Data" above) are merged iteratively according to similarity. The similarity is calculated with the Bayesian Information Criterion (BIC), which includes a term to represent how well the model fits the data and another to avoid overfitting of complex models. The model is the joint likelihood of the reads as a function of their sample of origin and their coordinates. BIC for neighbouring bins is computed and those with a BIC difference lower than zero are merged. This is repeated until no BIC difference is less than zero with a final merging step of three or more bins in order to improve BIC. The copy ratios of the segments are calculated and breakpoints have an assigned confidence interval. The bin size is explicitly assigned by the user and can be as low as 1 bp, enhancing accurate breakpoint localization [18].

CONTRA is designed for WES and TRS and can be used with paired data and pooled data. The first step for detecting small region-level CNVs is the computation of a base-level log ratio of the adjusted coverage (excluding regions with coverage lower than a threshold and scaled by the geometric mean of the library size between case and control). The region-level log ratio (RLR) is then computed as the mean of the base-level log ratios across the region and corrected for bias due to unequal library size between case and control. The RLR is modelled with a normal distribution and an adjusted two-tailed p-value is computed for each region. Larger CNV regions from the RLR can be obtained with a heuristic approach based on CBS.

The interest in sensitive and specific CNV callers (especially in the WES setting due to its low cost) has sparked a number of comparative studies [18, 19, 21, 28]. Unfortunately, results from comparative assessments on these tools utilise different metrics, types of dataset (e.g. simulated, primary tumour and cell lines), gold standards (e.g. aCGH, SNP arrays and WGS) and tools, making them difficult to compare and highly variable among them. Additionally, tools are usually benchmarked with default/recommended settings, without further exploration of the space of parameters that may yield better performance on the validation set. These make the decision on the best methodology for a given dataset of interest hard to make and CNV calling from WES and targeted sequencing remains an open research question.

Cell Admixture, Baseline Ploidy and Subclonal Heterogeneity

Separate technical issues which complicate CNV prediction include: the presence of cell admixture in the sequenced case samples, which can dilute the signal of a CNV; genomes of somatic cases may not be diploid, thus affecting the interpretation of the baseline LRR; and the sample may include an unknown number of subclones with different CNV profiles that may be relevant for the disease

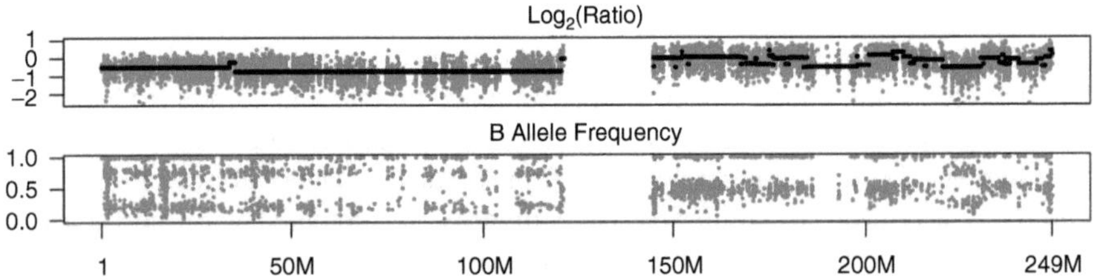

Fig. 3 Copy number from whole exome sequencing (WES) data. The LRR and SNP BAF are shown for one chromosome from a tumour sample. Allelic imbalance is apparent from the split in BAF on the p-arm and a focal change on part of the q-arm

under study (e.g. subclones that develop resistance to disease). Thus, it is important that the pathologist is aware of any prior knowledge that could give insights into these. For example, inspection of BAF and Mutant Allele Frequency (MAF) plots can be informative regarding the presence of cell admixture and potential heterogeneity. Tools like Control-FREEC [25] estimate cell admixture, but it requires the user to specify an input ploidy value (and further recommends trying different values). ADTEx [23] and Sequenza [29] on the other hand allow the user to estimate both ploidy and normal cell contamination, whereas SomatiCA [30] estimates cell admixture and subclonal heterogeneity, while bypassing the estimation of ploidy (Fig. 3).

Formalin Fixed Paraffin Embedded Samples

Routine storage of samples by formalin fixation results in degradation of the genetic material and deamination. While previous studies using fresh frozen versus matched Formalin Fixed Paraffin Embedded (FFPE) samples for MPS [31, 32] have shown that the detection of FFPE-based CNVs is feasible and informative, the damaged genetic material results in additional sequencing noise. A recent study [33] has shown that informative depth of coverage (DOC) profiles can be obtained from shallow WGS of fresh frozen and FFPE samples and provides a package—QDNASeq—that performs simultaneous GC content and mappability correction via LOESS on the profiles.

Genomic Rearrangements

Genomic rearrangements or structural variations are large-scale changes to chromosomes defined by one or more double-strand break points. Typically, these refer to translocations, inversions, and tandem or inverted duplications, large insertions and deletions (>50 nt), which may also be detected as copy number changes. Genomic rearrangements may involve large-scale gains or losses or be copy number neutral. Unlike copy number variation detection, the search for structural variants is aimed at identifying breakpoints of genomic fusions rather than amplified or deleted regions. Breakpoints may also be associated with finescale deletions or insertion of untemplated sequence.

FISH and spectral karyotyping are low-resolution methods of detecting genomic rearrangements. Here we focus on the use of high-throughput sequencing data.

The methods to search sequencing data for variants in aligned read data generally utilise one or more of the following techniques:

Discordant Paired-End Methods

If paired-end (PE) sequencing of DNA fragments is performed, a cluster of read pairs that align discordantly or anomalously to the reference genome may provide support for a genomic rearrangement. For example, a number of independent read pairs that map to the same neighbourhoods of two different chromosomes provide evidence for an interchromosomal genomic fusion. Figure 4b illustrates the concept for a single fragment mapping discordantly. The average distance between PE reads on a single chromosome can also be utilised to detect rearrangements. Since the PE library preparation creates fragments with some size distribution, read pairs that map significantly closer or farther than this expected distance to each other can reveal the presence of an insertion or deletion event. Figure 4c–d sketches both these cases. Again, clusters of fragments with such anomalous mappings increase the likelihood of a real event having taken place.

In general, discordant read pair methods do not achieve single nucleotide resolution, nor can they identify untemplated insertions at the breakpoint. However, they can achieve high sensitivity if the physical coverage of the genome by DNA fragments is high, that is if the average fragment size is substantially larger than the read length.

BreakDancer is a method for genomic rearrangement prediction that uses a pure paired-end analysis tool [30]. After identifying clusters of discordantly aligned read pairs, BreakDancer applies a statistical test to filter out false positives. Despite this, it is generally reported to have a high false positive rate [34].

Split-Read Methods

Since genomic rearrangements cause fusions of non-contiguous segments of the genome (on the same or different chromosomes), it is possible for reads to be sampled in such a manner that one part of a read is aligned on the 'left side' of the fusion and the other part on the 'right'. Figure 4f displays such a scenario. Such reads are referred to as split-reads (SR).

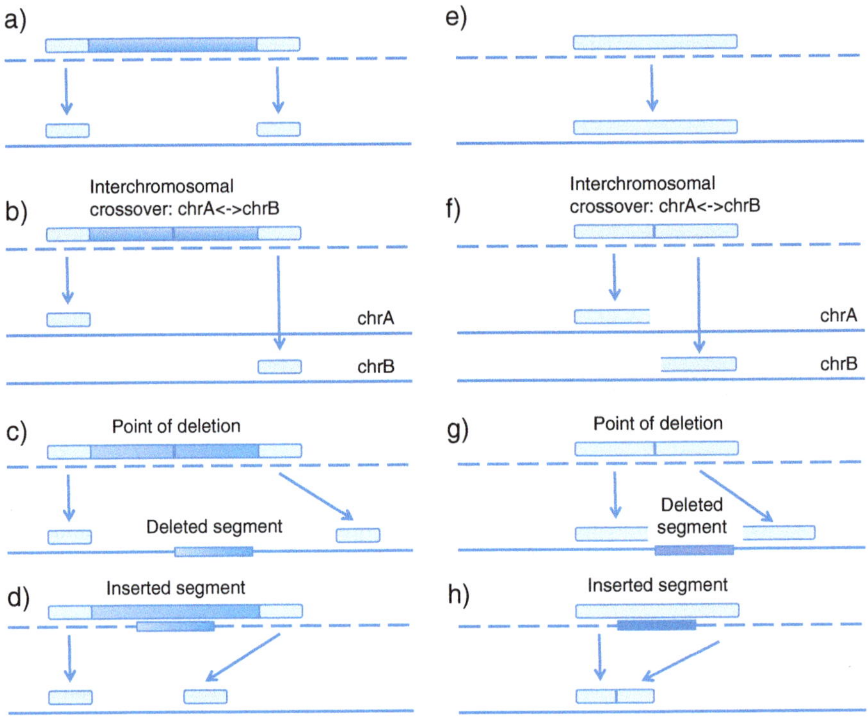

Fig. 4 Different types of genomic rearrangements and supporting evidence. Details of evidence for genomic rearrangements from paired-end and split-reads. The donor genome (from which read fragments have been sequenced) is drawn as a *dashed line* in each panel and the reference genome as a *solid line*. Read fragments are indicated on the donor genome as *solid blocks with light* ends indicating the actually sequenced reads (**a–d**). Single reads (or parts thereof) are displayed as *larger solid blocks* in panels (**e–h**). (**a**) A DNA fragment from a homologous stretch of donor genome; both reads are aligned to the same position in concordant distance. (**b**) A chromosome crossover (for example, a translocation) in the donor genome causes the two reads to be mapped to two different chromosomes in the reference genome. (**c**) The donor genome is lacking a stretch of DNA from the reference. Read pairs that span the point of deletion cause reads to be mapped further apart in the reference genome than the average fragment length. (**d**) An insertion in the donor genome (shorter than the fragment length) causes the read pairs spanning this segment to map closer to each other than the average fragment length. (**e**) A read sampled from a homologous region of DNA from the donor gets placed at the same position in the reference. (**f**) Interchromosomal fusions in the donor can create split-reads that partially align to two different chromosomes. (**g**) A deletion in the donor genome results in split-reads that partially align to two distant points in the reference genome. (**h**) A short insertion in the donor genome (shorter than the read length) can result in a read whose beginning and ending, but not the middle part, map to the reference genome

Split-reads cause a particular type of signal in aligned sequencing data. Correctly configured aligners allow for partially mapped reads. Such reads align from the read start for some length, but then no longer match the genome sequence and enter a special clipped state. Only certain aligners support such soft clipped alignment (e.g.

bowtie2). The two halves of a split-read can then be associated with each other by realigning the clipping portions of reads, or by assembling them into longer segments of DNA together with other clipped reads in the vicinity.

One method that uses split-read detection is Socrates [34], which uses standard aligners (such as bowtie2) to realign unmapped portions of reads, and then clusters sets of split-reads into genomic fusions. Socrates is not as dependent on the library design as BreakDancer above, but benefits from longer reads. Socrates provides single nucleotide resolution, as well as highlighting micro-homologies and untemplated sequence at the fusion site. Other evidence including discordantly aligned reads can be integrated post hoc.

An integrative approach to structural variation detection that uses both discordant read pairs and split-reads is Delly [35]. This algorithm searches for anomalously aligned PE reads (much like BreakDancer does), but then refines the fusion by also investigating the surrounding reads for evidence of split-reads. A consequence of this approach is that the sensitivity of Delly is similar to paired-end methods, but obtains single nucleotide resolution.

Finally, several methods now also include de novo assembly as part of the algorithm to improve the specificity of calls [e.g. 22, 36].

Summary

In this chapter, we have introduced the area of bioinformatics, covering the basic analyses of germline and somatic DNA sequence data using NGS and SNP arrays for accurate copy number calling. The field of bioinformatics is complex and fast moving; ideally specialist bioinformaticians should be engaged to undertake analyses. The experience in research is that most new datasets require some level of bioinformatics methods development, or at least exploration. Once established, pipelines make analyses more efficient, and in pathology, where routine assays are more likely, the majority of analyses can be undertaken this way. However, a trained eye is needed to monitor the emergence of better approaches or problematic datasets for which default methods fail. These latter edge cases will always come up from time to time.

References

1. Van Loo P, Nordgard SH, Lingjaerde OC, Russnes HG, Rye IH, Sun W, Weigman VJ, Marynen P, Zetterberg A, Naume B et al (2010) Allele-specific copy number analysis of tumors. Proc Natl Acad Sci U S A 107(39):16910–16915
2. Ortiz-Estevez M, Aramburu A, Bengtsson H, Neuvial P, Rubio A (2012) CalMaTe: a method and software to improve allele-specific copy number of SNP arrays for downstream segmentation. Bioinformatics 28(13):1793–1794
3. Song S, Nones K, Miller D, Harliwong I, Kassahn KS, Pinese M, Pajic M, Gill AJ, Johns AL, Anderson M et al (2012) qpure: A tool to estimate tumor cellularity from genome-wide single-nucleotide polymorphism profiles. PLoS One 7(9), e45835

4. Olshen AB, Venkatraman ES, Lucito R, Wigler M (2004) Circular binary segmentation for the analysis of array-based DNA copy number data. Biostatistics 5(4):557–572
5. Bentley DR, Balasubramanian S, Swerdlow HP, Smith GP, Milton J, Brown CG, Hall KP, Evers DJ, Barnes CL, Bignell HR et al (2008) Accurate whole human genome sequencing using reversible terminator chemistry. Nature 456(7218):53–59
6. Rieber N, Zapatka M, Lasitschka B, Jones D, Northcott P, Hutter B, Jager N, Kool M, Taylor M, Lichter P et al (2013) Coverage bias and sensitivity of variant calling for four whole-genome sequencing technologies. PLoS One 8(6), e66621
7. Li H, Durbin R (2009) Fast and accurate short read alignment with Burrows-Wheeler transform. Bioinformatics 25(14):1754–1760
8. Campbell CD, Chong JX, Malig M, Ko A, Dumont BL, Han L, Vives L, O'Roak BJ, Sudmant PH, Shendure J et al (2012) Estimating the human mutation rate using autozygosity in a founder population. Nat Genet 44(11):1277–1281
9. Alexandrov LB, Nik-Zainal S, Wedge DC, Aparicio SA, Behjati S, Biankin AV, Bignell GR, Bolli N, Borg A, Borresen-Dale AL et al (2013) Signatures of mutational processes in human cancer. Nature 500(7463):415–421
10. Cibulskis K, Lawrence MS, Carter SL, Sivachenko A, Jaffe D, Sougnez C, Gabriel S, Meyerson M, Lander ES, Getz G (2013) Sensitive detection of somatic point mutations in impure and heterogeneous cancer samples. Nat Biotechnol 31(3):213–219
11. Koboldt DC, Chen K, Wylie T, Larson DE, McLellan MD, Mardis ER, Weinstock GM, Wilson RK, Ding L (2009) VarScan: variant detection in massively parallel sequencing of individual and pooled samples. Bioinformatics 25(17):2283–2285
12. Koboldt DC, Larson DE, Wilson RK (2013) Using VarScan 2 for germline variant calling and somatic mutation detection. Curr Protoc Bioinformatics 44:15.14.11–15.14.17
13. Koboldt DC, Zhang Q, Larson DE, Shen D, McLellan MD, Lin L, Miller CA, Mardis ER, Ding L, Wilson RK (2012) VarScan 2: somatic mutation and copy number alteration discovery in cancer by exome sequencing. Genome Res 22(3):568–576
14. Goode DL, Hunter SM, Doyle MA, Ma T, Rowley SM, Choong D, Ryland GL, Campbell IG (2013) A simple consensus approach improves somatic mutation prediction accuracy. Genome Med 5(9):90
15. Fowler DM, Fields S (2014) Deep mutational scanning: a new style of protein science. Nat Methods 11(8):801–807
16. Araya CL, Fowler DM (2011) Deep mutational scanning: assessing protein function on a massive scale. Trends Biotechnol 29(9):435–442
17. Ivakhno S, Royce T, Cox AJ, Evers DJ, Cheetham RK, Tavare S (2010) CNAseg—a novel framework for identification of copy number changes in cancer from second-generation sequencing data. Bioinformatics 26(24):3051–3058
18. Alkodsi A, Louhimo R, Hautaniemi S (2015) Comparative analysis of methods for identifying somatic copy number alterations from deep sequencing data. Brief Bioinform 16:242–254
19. Kadalayil L, Rafiq S, Rose-Zerilli MJ, Pengelly RJ, Parker H, Oscier D, Strefford JC, Tapper WJ, Gibson J, Ennis S et al (2015) Exome sequence read depth methods for identifying copy number changes. Brief Bioinform 16:380–392
20. Liu B, Morrison CD, Johnson CS, Trump DL, Qin M, Conroy JC, Wang J, Liu S (2013) Computational methods for detecting copy number variations in cancer genome using next generation sequencing: principles and challenges. Oncotarget 4(11):1868–1881
21. Tan R, Wang Y, Kleinstein SE, Liu Y, Zhu X, Guo H, Jiang Q, Allen AS, Zhu M (2014) An evaluation of copy number variation detection tools from whole-exome sequencing data. Hum Mutat 35(7):899–907
22. Zhao M, Wang Q, Wang Q, Jia P, Zhao Z (2013) Computational tools for copy number variation (CNV) detection using next-generation sequencing data: features and perspectives. BMC Bioinformatics 14 Suppl 11:S1
23. Amarasinghe KC, Li J, Hunter SM, Ryland GL, Cowin PA, Campbell IG, Halgamuge SK (2014) Inferring copy number and genotype in tumour exome data. BMC Genomics 15:732

24. Li J, Lupat R, Amarasinghe KC, Thompson ER, Doyle MA, Ryland GL, Tothill RW, Halgamuge SK, Campbell IG, Gorringe KL (2012) CONTRA: copy number analysis for targeted resequencing. Bioinformatics 28(10):1307–1313
25. Boeva V, Popova T, Lienard M, Toffoli S, Kamal M, Le Tourneau C, Gentien D, Servant N, Gestraud P, Rio Frio T et al (2014) Multi-factor data normalization enables the detection of copy number aberrations in amplicon sequencing data. Bioinformatics 30(24):3443–3450
26. Bellos E, Kumar V, Lin C, Maggi J, Phua ZY, Cheng CY, Cheung CM, Hibberd ML, Wong TY, Coin LJ et al (2014) cnvCapSeq: detecting copy number variation in long-range targeted resequencing data. Nucleic Acids Res 42(20), e158
27. Xi R, Hadjipanayis AG, Luquette LJ, Kim TM, Lee E, Zhang J, Johnson MD, Muzny DM, Wheeler DA, Gibbs RA et al (2011) Copy number variation detection in whole-genome sequencing data using the Bayesian information criterion. Proc Natl Acad Sci U S A 108(46):E1128–E1136
28. Guo Y, Sheng Q, Samuels DC, Lehmann B, Bauer JA, Pietenpol J, Shyr Y (2013) Comparative study of exome copy number variation estimation tools using array comparative genomic hybridization as control. Biomed Res Int 2013:915636
29. Favero F, Joshi T, Marquard AM, Birkbak NJ, Krzystanek M, Li Q, Szallasi Z, Eklund AC (2015) Sequenza: allele-specific copy number and mutation profiles from tumor sequencing data. Ann Oncol 26(1):64–70
30. Chen K, Wallis JW, McLellan MD, Larson DE, Kalicki JM, Pohl CS, McGrath SD, Wendl MC, Zhang Q, Locke DP et al (2009) BreakDancer: an algorithm for high-resolution mapping of genomic structural variation. Nat Methods 6(9):677–681
31. Schweiger MR, Kerick M, Timmermann B, Albrecht MW, Borodina T, Parkhomchuk D, Zatloukal K, Lehrach H (2009) Genome-wide massively parallel sequencing of formaldehyde fixed-paraffin embedded (FFPE) tumor tissues for copy-number- and mutation-analysis. PLoS One 4(5), e5548
32. Van Allen EM, Wagle N, Stojanov P, Perrin DL, Cibulskis K, Marlow S, Jane-Valbuena J, Friedrich DC, Kryukov G, Carter SL et al (2014) Whole-exome sequencing and clinical interpretation of formalin-fixed, paraffin-embedded tumor samples to guide precision cancer medicine. Nat Med 20(6):682–688
33. Scheinin I, Sie D, Bengtsson H, van de Wiel MA, Olshen AB, van Thuijl HF, van Essen HF, Eijk PP, Rustenburg F, Meijer GA et al (2014) DNA copy number analysis of fresh and formalin-fixed specimens by shallow whole-genome sequencing with identification and exclusion of problematic regions in the genome assembly. Genome Res 24(12):2022–2032
34. Schroder J, Hsu A, Boyle SE, Macintyre G, Cmero M, Tothill RW, Johnstone RW, Shackleton M, Papenfuss AT (2014) Socrates: identification of genomic rearrangements in tumour genomes by re-aligning soft clipped reads. Bioinformatics
35. Rausch T, Zichner T, Schlattl A, Stutz AM, Benes V, Korbel JO (2012) DELLY: structural variant discovery by integrated paired-end and split-read analysis. Bioinformatics 28(18):i333–i339
36. Wong K, Keane TM, Stalker J, Adams DJ (2010) Enhanced structural variant and breakpoint detection using SVMerge by integration of multiple detection methods and local assembly. Genome Biol 11(12):R128

Forgotten Resources – The Autopsy

Deborah Smith, Amy McCart Reed, and Sunil R. Lakhani

Historical value of the autopsy in research
The autopsy procedure and preservation of specimens
Research benefits of the autopsy
- Investigating the biology of malignancy
- Evaluate the effects of medical and surgical therapy
- Ensuring accurate epidemiological data

What makes a good research autopsy program work?

Historical Value of the Autopsy in Research

An autopsy, also known as a postmortem, is the medical examination of the deceased. It is a careful and detailed examination of the body and internal organs, in order to determine the cause of death and answer any clinical questions. The term "autopsy" means "to see for oneself" and refers to the fact that manifestations of disease are directly observed rather than relying solely on clinical findings and

D. Smith
Anatomical Pathology Mater Health Services, Mater Adult Hospital, Brisbane, Queensland, Australia

A.M. Reed
The University of Queensland, UQ Centre for Clinical Research, The Royal Brisbane and Women's Hospital, Herston, Queensland, Australia

S.R. Lakhani (✉)
The University of Queensland, UQ Centre for Clinical Research, The Royal Brisbane and Women's Hospital, Herston, Queensland, Australia

Pathology Queensland, Royal Brisbane Women's Hospital, Herston, Queensland, Australia

The University of Queensland, School of Medicine, Herston, Queensland, Australia
e-mail: s.lakhani@uq.edu.au

S.R. Lakhani, S.B. Fox (eds.), *Molecular Pathology in Cancer Research*,
DOI 10.1007/978-1-4939-6643-1_15

investigations [1, 2]. Modern medical understanding of disease originated when autopsies were first used to examine the structure and function of normal tissue. This could then be compared to the alterations seen in disease [3, 4]. Autopsies contributed to Virchow's theories of cellular pathology in 1876, and to Osler's great advancements of medical knowledge in the early 1900s [2]. Throughout the twentieth century autopsies played a key role in the explosion of medical knowledge. In recent times, autopsy research has contributed to the understanding of diseases as diverse as oesophageal adenocarcinoma, sudden cardiac death in young people, avian H1N1 influenza, and Creutzfeldt–Jakob disease [5–8].

Sadly, many believe there is no longer a place for autopsies in modern medical research; that an autopsy cannot provide specimens that are adequate for modern research techniques, and that all we need to know about a patient and his disease can be derived from premortem clinical investigation and imaging. So is there any place for autopsies in modern research programs? The answer to this is a resounding "YES"! Autopsies can be used to obtain large quantities of tissue for research, assess response to therapy, map the distribution of metastases, highlight rare complications, provide feedback for quality assurance assessment of protocols and procedures, and provide reliable cause of death information.

The decline in hospital based autopsies and the more rigorous ethical standards and consent requirements for tissue retention can make research using autopsy data and tissue difficult [3, 9, 10]. However despite these hurdles, there is a recent resurgence of interest in the benefits of utilising autopsies for research, including obtaining tissue for molecular studies [10–12].

The Autopsy Procedure and Preservation of Specimens

The body is examined externally, then the internal organs are dissected, examined macroscopically, and tissue taken for histologic examination. An autopsy may be authorised by the state to establish a cause of death (coronial autopsy). Alternatively one may be performed at the request of clinicians or families to answer a clinical question, as part of a quality assurance measure, or to obtain tissue for research (hospital based autopsy). Coronial autopsies are a legal requirement to establish if the death was due to natural causes. Hence, there are additional legal and ethical standards that must be met if these autopsies are to be used for research. In contrast, hospital based autopsies are performed after a death certificate has been issued, and an autopsy is only performed with the consent of the patient (obtained prior to death) of the family.

An autopsy can be targeted, it can exclude designated organs (often the brain), or be a detailed examination of the whole body [1]. Appropriate consent for the autopsy, and in many countries the retention of tissue samples or organs, must be obtained. Autopsy research programs may also require research ethics approval, although not all countries mandate this [13]. Tissue and fluids retained for research purposes can be subject to a wide range of interrogation that includes histology, biochemistry, microbiological studies, immunohistochemistry, histomorphometric analysis, and molecular and biochemical studies [2, 14, 15].

"Rapid" autopsies are performed as soon after death as possible, specifically to obtain high quality tissue for research. The autopsy is ideally performed within 4 h of death to minimise postmortem tissue degradation [16]. Rapid autopsies are performed within an established research program. They require hospital infrastructure through which patients can be recruited and give consent prior to death. A team of pathology and laboratory staff experienced in biobanking need to be available on call and able to respond within a few hours of the tissue donor's death [17, 18]. To expedite the removal and processing of tissue samples, clinical and radiologic information can be used to indicate which sites are to be biopsied or organs examined [17]. The extent of autopsy examination varies, with some programs continuing with a full autopsy protocol [18], while others simply obtain tissue using core needle biopsies of sites of interest. Of course, if a complete autopsy is not performed unexpected findings are much less likely to be identified.

Adequacy of Autopsy Tissue for Molecular and Biochemical Research

It is a common misconception that autopsy tissue is not good enough for molecular and biochemical studies [9, 19]. Although there is no doubt that degradation of nucleic acids can be an issue, this may be much less significant than many believe. For many research questions, the disadvantages are well outweighed by the availability of ample tumour and control tissue.

Several research groups, mostly using tissue from rapid autopsy programs, have successfully performed genomic copy number analysis of primary and metastatic tumours. The methods used include comparative genomic hybridisation [20, 21], fluorescent in situ hybridisation [22], and single nucleotide polymorphisms (SNP) analysis [20]. Using high-resolution genome-wide SNP arrays, Liu et al. [20] investigated 58 metastatic prostate cancer samples obtained at rapid autopsy from 14 subjects. They found subject specific data clustering of the 58 samples, suggesting a common origin of the metastatic cells [20]. Zarghooni et al. [40] also used SNP based DNA microarrays to analyse 11 diffuse intrinsic pontine gliomas (DIPG), which are lethal paediatric brainstem tumours. Nine of their cases were obtained at postmortem, with a postmortem interval range of 9–40 h. They found the genomic alterations were different to paediatric supratentorial tumours and identified two novel potential biological targets. To control for any postmortem alterations the samples were matched with their own normal brain tissue, and compared with the two available surgical biopsies [23]. This particular study was not from a rapid autopsy program, and several of their subjects died during terminal care at home.

Adequate preservation and evaluation of RNA is important for functional genomic studies. RNA is known to deteriorate with increasing time from death to autopsy. To consistently obtain high quality RNA requires an organised program that can efficiently obtain and process tissue. The median RIN (RNA Integrity Number) of a rapid autopsy program with a median postmortem interval of 3 h has been recorded as 8.9 for brain and 7.0 for body tissues [24]. Tissue samples are considered to be of high quality if the RIN $\geq$ 6.5, while samples with RIN $\geq$ 8.0 are considered suitable for all downstream molecular techniques [25]. Studies with a longer postmortem interval do show more significant deterioration in RNA quality [26]. Messenger RNA (mRNA) levels have also been determined from brain tissue obtained at autopsy, using reverse transcription followed by real-time polymerase chain reaction (PCR). While this showed a general decline in measured mRNA levels in the autopsy tissue, when the measured mRNA level was adjusted according to a reference gene mRNA level, most genes evaluated were not affected by the postmortem status. One gene did have significantly decreased adjusted mRNA levels. The results suggest that overall the pattern of gene expression in postmortem tissues is similar to surgical biopsy tissue, but carefully chosen controls are required. Factors that may alter RNA expression in postmortem tissue include both individual variation in gene expression and reduced production at the time of death, and possibly relate to the mode of death rather than the postmortem delay [9, 27].

Tissue obtained after a long postmortem interval may have partially degraded RNA, but this can still be utilised for PCR amplification of smaller fragments, so that tissue need not be wasted [28]. Increasingly, new technologies are being developed that can tolerate lower quality RNA samples, for example, NanoString® technology, meaning that gene expression can still potentially be evaluated [29]. DNA can also be obtained from postmortem tissue, including formalin-fixed and paraffin embedded tissue; however, larger DNA fragments are more prone to degradation than in surgically obtained formalin-fixed and paraffin embedded tissue [30].

Studies using proteins are more difficult and complex, as there is significant variation in degradation that is not predictable. Each study must therefore commence with an evaluation of the preservation of that particular protein [9, 31].

Expansion of Autopsy Derived Material: Resource Generation and Applications

Viable tumour tissue can be harvested to establish both in vitro and in vivo models, allowing in-depth studies of both primary and metastatic tumours [18, 32]. Several studies have confirmed the viability of growing fibroblasts from autopsy tissue, which can be reprogrammed into induced pluripotent stem cells [33, 34]. While mouse xenografts of cultured human tumour cell lines have been in the researcher's toolkit for many decades, patient-derived xenograft (PdX) resources are becoming increasingly sought after to provide greater clinically predictive insights. The PdX benefits pre-clinical research by preserving both tumour heterogeneity and tissue architecture, and by facilitating the modeling of specific stages of disease progression (for example, local metastasis, distant metastasis, and/or broad disease dissemination) in the absence of the clonal selective pressures of culture in monolayer [35]. Furthermore, clinical trials are routinely undertaken in cohorts with advanced disease, PdX models generated from metastatic deposits collected in rapid autopsies are certainly a more relevant pre-clinical model as opposed to the use of surgical resections of primary tumours [36–38].

Research Benefits of the Autopsy

Investigate the Biology of Malignancy

Use of autopsy tissue for research is particularly valuable for rare malignancies, those that are not managed by surgical excision, tumours that are frequently disseminated at the time of diagnosis, and metastases. For many of these special groups obtaining enough tissue for research studies can be problematic. If biopsies are sufficient for clinical diagnosis and management, taking more tissue may be unethical. However, biopsies may not be large enough for both clinical diagnosis and research studies. If subsequent treatment does not include surgical excision, further tissue may never be obtained. Malignancies such as pancreatic carcinoma are often disseminated at the time of diagnosis, consequently those patients will often not undergo surgical resection. Metastatic or recurrent malignancy may not undergo repeat biopsy, especially if the tumour is deep seated or difficult to biopsy, hence comparison with the primary may never occur.

In contrast, tissue samples of both primary and metastatic tumours can be obtained at autopsy. Multiple metastases can be sampled, including those in surgically inaccessible sites. The true extent of disease can be determined, including any metastases not detected antemortem. Tissue for controls can easily be obtained, unlike surgical biopsies that target diseased tissue. Research involving rare malignancies also benefits from collection of tumour tissue at autopsy, because larger amounts of tissue can often be obtained and retained for more extensive investigation [10].

DIPG for example, are diagnosed based on clinical and radiologic findings, and biopsy is often not performed at all. The current treatment is ineffectual and they are uniformly lethal. Minimal surgically obtained tissue is available for research, so autopsies can provide material that is critical to understanding the underlying tumour biology [39]. Parents of children with DIPG have been actively encouraging other parents to consider an autopsy, via DIPG cancer support networks. Recent studies using postmortem tissue have finally begun unraveling the molecular alterations present, hopefully allowing more targeted treatments to be identified [40].

With the use of endocrine, chemotherapy, and targeted therapies, the survival for a number of cancer types such as breast cancer has improved dramatically over the last two decades. While survival is good for localised disease, the outcome remains poor once metastases develop [10]. Currently, most metastatic deposits are not biopsied and the treatment of metastatic disease is based on the phenotype (including molecular phenotype) of the primary tumour. There is now compelling evidence that this may be inappropriate. Changes in biomarkers between primary and metastatic sites such as oestrogen receptor (ER), progesterone receptor (PR), and the oncogene HER2 have been demonstrated in breast cancer [16, 41]. In fact, the American Society of Clinical Oncology Clinical Practice Guidelines recommend the use of the ER, PR, and HER2 status of the metastasis to direct therapy, if supported by the clinical scenario and patient's goals for care [42]. Studies investigating metastatic pancreatic carcinoma have found reduced expression of DPC4 in tumours that are widely disseminated as compared to localised, surgically amenable tumours, suggesting an important role for this tumour suppressor gene [43]. Indeed, whole exome sequencing of metastatic pancreatic ductal adenocarcinoma sampled at autopsy has gone some way towards illuminating the oncogenic drivers of this lethal disease progression [38]. Similarly, studies in prostate carcinoma have begun to document the clonal evolution and molecular changes from primary to metastatic sites. Rapid autopsy derived metastatic deposits from lethal castration-resistant prostate cancer were utilised to describe the discordance in *ERG* gene rearrangements and ERG protein expression between tumour sites in heavily treated patients [44]; a subset of these samples have been exome sequenced, identifying recurrent mutations in androgen receptor transcriptional cofactors [45].

These approaches will hopefully shed light on the mutations required to metastasise and the "genomic archeology" of multiple metastatic sites [20, 21, 31]. Understanding the biology of metastatic disease will become increasingly important in order to develop targeted therapies [46] understand why treatment fails and find biomarkers of aggressive disease [10].

Evaluate the Effects of Medical and Surgical Therapies

Autopsy research can provide valuable insights onto the effectiveness of both surgical and medical treatment of cancer [17, 47–49]. This is particularly important for new and rapidly evolving areas, such as transplant medicine [47] and stereotactic surgery [50]. Autopsy studies can reveal the effects of treatment on malignancy, providing information on both responders and non-responders and exposing "privileged sites" not reached by systemic therapies [51]. Autopsy allows the most aggressive disease to be sampled, and for samples to be obtained when treatment has failed. The genetic makeup of distant metastases following treatment failure in patients with breast cancer has been shown to be different to that of local lymph node metastases sampled during primary surgical treatment [31]. Toxic effects on adjacent normal tissue and the spectrum of side effects can also be documented [52]. Many survivors, particularly of paediatric and early adulthood malignancies, now live long enough to develop complications from their oncological treatment. Complications such as cirrhosis and bronchiolitis obliterans can be severe and lead to further morbidity and mortality [51]. A thorough understanding of the range of possible complications and their relative incidence is therefore required in determining treatment protocols.

Within drug therapeutic trials, autopsy examination can be used to accurately differentiate between deaths due to treatment (the so-called toxic death), disease progression, and deaths from unrelated causes [53, 54]. In 1997 a survey of clinical research papers published in the British Medical Journal, Lancet, Annals of Internal Medicine, and New England Journal of Medicine indicated that less than a quarter used autopsy to evaluate the cause of death [54]. A review in 2012 of studies conducted within the European Organisation for Research and Treatment of Cancer (EORTC) showed autopsies had been performed in just 26 treatment related deaths, from a total of 255. Of the 26 cases that underwent autopsy, 46 % had a final diagnosis that was discrepant with the clinical diagnosis. The reviewers also felt a further 64 cases which did not undergo autopsy had a clinical course which did not fit with the reported cause of death [53]. The vast majority of deaths were considered not treatment related, and no information is available on these.

Discrepancy between clinical diagnoses and autopsy findings are well documented [14, 15, 47, 48, 53, 55–63], and involve all levels of clinical practice from community hospitals to intensive care units [56]. The rate of major errors, where a principle underlying disease or cause of death is missed is approximately 30 %, and ranges from 5.5 % to over 45 % [53, 62]. Only a few studies show demonstrable improvement in the major error rate over the past decades [46, 64]. Other studies suggest the discrepancy rate has not changed significantly, but the type of unexpected pathology found has [58, 61]. This shift in the conditions that are most likely to be missed is attributed to both changing diagnostic criteria and changing treatments resulting in novel complications. When clinicians are more certain of their diagnoses the discrepancy rate is somewhat lower, but still significant, being 25 % in a large study of 1152 cases [15]. The lowest discrepancy rates (from 5.5 to 7 %) are reported from centres where the autopsy rate remains consistently over 50 % [62, 65].

The diagnosis of neoplastic disease may have a lower discrepancy rate when compared to other disease categories [58, 66], although some cultural factors and mental health disorders may result in under-diagnosis [67, 68]. Misdiagnosis of treatment complications is more problematic; opportunistic infections and cardiac complications are the most frequently missed diagnoses in cancer autopsy series [53, 63]. For example, invasive mycotic infection in patients following stem cell transplant was missed clinically in half of the cases in one study, despite being investigated with cultures, antigen testing, and high-resolution CT scans [69]. In a case described by Allan et al [47] a man with upper gastrointestinal bleeding thought to be secondary to graft versus host disease died despite appropriate therapy. At autopsy he was discovered to have instead succumbed to severe fungal infection. The patient had been investigated with liver and rectosigmoid biopsies, and the clinical diagnoses were compatible with the biopsy results. As this case demonstrates, clinical history and investigations may appear consistent with a particular diagnosis but that doesn't mean the diagnosis is correct.

Autopsy data should be an essential part of clinical research protocols, particularly in the early stages of patient safety assessment. If autopsy following death during clinical trials is neglected, an under-reporting bias may be present that preferentially favours the death being due to disease, reducing credibility.

In addition, without autopsies the errors that may occur from misplaced clinical bias or suboptimal test performance cannot be documented and learnt from, and unexpected events may not be detected.

Ensuring Accurate Epidemiological Data

Accurate epidemiologic information is required when determining the significance within a given population of specific cancer types, and whether screening or treatment protocols are effective. For cancers that have the potential to remain occult, epidemiological data is not accurate unless it includes a survey of presumed normal subjects. Autopsies of unselected patients provide very accurate epidemiological data as tissue from organs presumed to be normal can be obtained and extensively examined. Prostate carcinoma is one such disease; without autopsy examination an accurate prevalence is unknown. With accurate prevalence data in hand, researchers can better determine the actual effects of prostate screening, and focus their attention on separating the more aggressive carcinomas from those that are indolent [70].

Autopsies also provide comprehensive information about the distribution of metastatic disease, which may be much more widespread than clinical records suggest [40]. A recent review has noted the difficulty in ascertaining the true incidence of brain metastases given the marked reduction in autopsies [46].

While newer imaging techniques may improve detection of metastases, like all diagnostic tests, false positive, and false negative results occasionally occur [71]. Even new, sensitive modalities may not detect disease that is present. For instance, positron emission tomography is one of the most sensitive imaging modalities clinically available and has a lower limit of 10 mm when imaging lung nodules [72]. Over-diagnosis may also occur, with positive scans resulting from active inflammatory nodules or the so-called metabolic flare reaction after chemotherapy [73, 74]. Histology on autopsy samples can be much more sensitive, and may detect tiny residual foci of malignancy, missed by imaging studies [47, 48].

As a definitive and detailed examination of the deceased, autopsies play a vital role in determining the incidence of cancer and proximate cause of death. This will naturally affect population statistics of disease incidence. Given the discrepancies documented in all studies, the reliability of death certificates has been questioned [15, 74]. Accurate population health records are also essential for assessing screening program effectiveness and developing evidence based public health policy [68, 74].

What Makes a Good Research Autopsy Program Work?

Although many types of research can utilise autopsy data, prospective autopsy research programs are the most valuable as fresh and frozen tissue can be retained for molecular studies and future research [11]. Some oncology protocols such as from the Children's Oncology Group (COG) provide a facility for storage and dissemination of tissue from specific malignancies [12]. Postmortem brain banks have been an integral part of neuropathology research for decades, and increasingly, similar banks for malignancies are being established within academic medical centres [75, 76]. Biobanking of normal tissue is also valuable, and recently the United States National Cancer Institute published recommendations regarding the postmortem recovery of such specimens for research [77].

To be successful, a research autopsy program requires good collaboration between clinicians, pathologists, and researchers in order that sufficient autopsy consents are obtained, and that autopsies are performed to a high standard. Much of the focus in the literature is on how to achieve a higher autopsy rate and the most sensitive method of obtaining consent. Other significant factors that are less frequently examined are the problems of funding and geographic issues.

The consent rate for autopsy research varies greatly between studies from 47 to 98 % [23, 78]. This variance may reflect the difference between obtaining consent to perform research on autopsies that are mandated (Coronial) versus requesting an autopsy specifically for research purposes, with the latter having a much lower consent rate. Some studies suggest that higher autopsy rates can be maintained within specialised programs [47]. Others emphasise the role that a good pathologist–clinician relationship plays [49]. Autopsy request can be part of the end of life discussion when treatment has failed, ideally through the treating oncologist who has already established a relationship with the patient and family [18, 23]. Feedback to the families can also be arranged through the oncological team [23], and this may provide closure and answer any lingering questions [10].

Obtaining consent for the use of tissue for research requires explicit consent in many countries [79]. Although clinicians who have a close relationship with the deceased's family are often considered to be in the best position to request tissue samples, very high consent rates for obtaining tissue for research (96–98 %) have been obtained by nurse practitioners contacting bereaved families by telephone [7, 78, 80].

A common theme that emerges is that one of the major barriers to obtaining autopsies is the reluctance by medical staff to ask families for consent [12]. It is suggested that the response of families to requests for an autopsy is much more positive than medical professionals assume [78], and that when doctors ask, the autopsy rate increases [9]. A survey of parents of children who had died from cancer found that 93 % indicated they would have agreed to donate tissue for research if asked. Of those same parents, only half had been given the opportunity to do so [12]. Families of research participants are often positive about being given the opportunity to contribute to an area of knowledge that caused suffering for a loved one [7, 23, 78, 80]; in exceptional cases tissue donation has been initiated by parents [28]. Discussing possibly autopsy prior to death can allow for decisions to be made away from the grief of death, although sensitivity is clearly required [81]. Involving patient network and advocacy groups may allow researchers to understand and respond to potential concerns, as well as disseminate information [13].

The reasons for refusal are varied, and include emotional distress, religious and cultural issues, the feeling a loved one has suffered enough, and time pressures [14, 23]. Time pressures are one of the most commonly cited reasons, due to the additional delay imposed by an autopsy [82]; this may be alleviated by a rapid autopsy program. While families may refuse consent, few have indicated dissatisfaction with being asked [78]. Patients may have terminal care at home or in a hospice, so arrangements to transport the body to the mortuary after death will be required [18]. Organisation of the transportation issues and associated costs in advance was found to be helpful [23], and removes an otherwise significant barrier to participation. Some research groups have successfully coordinated external non-academic centres to perform the autopsy and obtain tissue to overcome geographic barriers [28, 32].

Education of both medical staff and families on the value of obtaining tissue at autopsy for research is necessary. Medical staff must be made aware of the presence of research protocols that use autopsy tissue, and the value of tissue donations. Families need better information regarding the potential benefit of donating tissue for research and the process of tumour banking. Education regarding the practical aspects of the autopsy procedure is also important [12]. In addition, tailoring request protocols according to the specific needs of racial and cultural minorities may improve representation of those groups within clinical studies [83].

Conclusion

The autopsy is an essential component of clinical audit as well as cancer research, but remains under-appreciated by many medical researchers. Autopsies can provide large quantities of high quality tissue suitable for modern research methods, as well as providing accurate information on extent of disease, treatment response, and cause of death. Major barriers to obtaining tissue from autopsies for research include a lack of awareness of both current research protocols and the potential value of autopsies, and reluctance to approach family members. However, patients and families are often positive about donating tissue for research, provided consent requests are carefully considered. Time pressures and transportation costs are some of the potential barriers that can be ameliorated [84]. Funding may also be an obstacle, as the infrastructure required is expensive and the benefits may take some years to be realised.

References

1. Burton JL, Underwood J (2007) Clinical, educational, and epidemiological value of autopsy. Lancet 369(9571):1471–1480
2. Burton JL, Rutty NG (eds) (2010) The Hospital Autopsy, 3rd edn. Hodder Arnold, Great Britain
3. McManus JF (1968) Types of research and the postmortem. Bull N Y Acad Med 44(7):799–807
4. Angrist A (1969) Experimental research and the autopsy. Bull N Y Acad Med 45(1):3–9
5. Ormsby AH, Kilgore SP, Goldblum JR, Richter JE, Rice TW, Gramlich TL (2000) The location and frequency of intestinal metaplasia at the esophagogastric junction in 223 consecutive autopsies: implications for patient treatment and preventive strategies in Barrett's esophagus. Mod Pathol 13(6):614–620

6. Nakajima N, Sato Y, Katano H, Hasegawa H, Kumasaka T, Hata S et al (2012) Histopathological and immunohistochemical findings of 20 autopsy cases with 2009 H1N1 virus infection. Mod Pathol 25(1):1–13
7. Millar T, Lerpiniere C, Walker R, Smith C, Bell JE (2008) Postmortem tissue donation for research: a positive opportunity? Br J Nurs 17(10):644–649
8. Larsen MK, Nissen PH, Berge KE, Leren TP, Kristensen IB, Jensen HK et al (2012) Molecular autopsy in young sudden cardiac death victims with suspected cardiomyopathy. Forensic Sci Int 219(1-3):33–38
9. Kretzschmar H (2009) Brain banking: opportunities, challenges and meaning for the future. Nat Rev Neurosci 10(1):70–78
10. Spunt SL, Vargas SO, Coffin CM, Skapek SX, Parham DM, Darling J et al (2012) The clinical, research, and social value of autopsy after any cancer death: a perspective from the Children's Oncology Group Soft Tissue Sarcoma Committee. Cancer 118(12):3002–3009
11. Kleiner DE, Emmert-Buck MR, Liotta LA (1995) Necropsy as a research method in the age of molecular pathology. Lancet 346(8980):945–948
12. Alabran JL, Hooper JE, Hill M, Smith SE, Spady KK, Davis LE et al (2013) Overcoming autopsy barriers in pediatric cancer research. Pediatr Blood Cancer 60:204–209
13. Pentz RD, Cohen CB, Wicclair M, DeVita MA, Flamm AL, Youngner SJ et al (2005) Ethics guidelines for research with the recently dead. Nat Med 11(11):1145–1149
14. Sirkia K, Saarinen-Pihkala UM, Hovi L, Sariola H (1998) Autopsy in children with cancer who die while in terminal care. Med Pediatr Oncol 30(5):284–289
15. Roulson J, Benbow EW, Hasleton PS (2005) Discrepancies between clinical and autopsy diagnosis and the value of post mortem histology; a meta-analysis and review. Histopathology 47(6):551–559
16. Subhawong AP, Nassar H, Halushka MK, Illei PB, Vang R, Argani P (2010) Heterogeneity of Bcl-2 expression in metastatic breast carcinoma. Mod Pathol 23(8):1089–1096
17. Morrissey C, Roudier MP, Dowell A, True LD, Ketchanji M, Welty C et al (2013) Effects of androgen deprivation therapy and bisphosphonate treatment on bone in patients with metastatic castration resistant prostate cancer: results from the University of Washington rapid autopsy series. J Bone Miner Res 28:333–340
18. Rubin MA, Putzi M, Mucci N, Smith DC, Wojno K, Korenchuk S et al (2000) Rapid ("warm") autopsy study for procurement of metastatic prostate cancer. Clin Cancer Res 6(3):1038–1045
19. Start RD, Firth JA, Macgillivray F, Cross SS (1995) Have declining clinical necropsy rates reduced the contribution of necropsy to medical research? J Clin Pathol 48(5):402–404
20. Liu W, Laitinen S, Khan S, Vihinen M, Kowalski J, Yu G et al (2009) Copy number analysis indicates monoclonal origin of lethal metastatic prostate cancer. Nat Med 15(5):559–565
21. Robbins CM, Tembe WA, Baker A, Sinari S, Moses TY, Beckstrom-Sternberg S et al (2011) Copy number and targeted mutational analysis reveals novel somatic events in metastatic prostate tumors. Genome Res 21(1):47–55
22. Garcia Parra-Perez FA, Zavala-Pompa A, Pacheco-Calleros J, Cortes-Gutierrez EI, Cerda-Flores RM, Lara-Miranda S et al (2012) Monosomy of chromosome 8 could be considered as a primary preneoplastic event in breast cancer: a preliminary study. Oncol Lett 3(2):445–449
23. Angelini P, Hawkins C, Laperriere N, Bouffet E, Bartels U (2011) Post mortem examinations in diffuse intrinsic pontine glioma: challenges and chances. J Neurooncol 101(1):75–81
24. Beach TG, Adler CH, Sue LI, Serrano G, Shill HA, Walker DG et al (2015) Arizona Study of Aging and Neurodegenerative Disorders and Brain and Body Donation Program. Neuropathology 35:354–389
25. Kap M, Oomen M, Arshad S, de Jong B, Riegman P (2014) Fit for purpose frozen tissue collections by RNA integrity number-based quality control assurance at the Erasmus MC tissue bank. Biopreserv Biobank 12(2):81–90

26. van der Linden A, Blokker BM, Kap M, Weustink AC, Robertus JL, Riegman PH et al (2014) Post-mortem tissue biopsies obtained at minimally invasive autopsy: an RNA-quality analysis. PLoS One 9(12), e115675
27. Cummings TJ, Strum JC, Yoon LW, Szymanskim MH, Hulette CM (2001) Recovery and expression of messenger RNA from postmortem human brain tissue. Mod Pathol 14(11):1157–1161
28. Broniscer A, Baker JN, Baker SJ, Chi SN, Geyer JR, Morris EB et al (2010) Prospective collection of tissue samples at autopsy in children with diffuse intrinsic pontine glioma. Cancer 116(19):4632–4637
29. Reis PP, Waldron L, Goswami RS, Xu W, Xuan Y, Perez-Ordonez B, Gullane P, Irish J, Jurisica I, Kamel-Reid S (2011) mRNA transcript quantification in archival samples using multiplexed, color-coded probes. BMC Biotechnol 11:46
30. Macoska JA, Benson PD, Turkeri LN, Haas GP, Sakr W (1993) PCR-based genetic analysis of DNA from autopsied prostate tissue. Genome Res 2(4):354–355
31. Singhi AD, Cimino-Mathews A, Jenkins RB, Lan F, Fink SR, Nassar H et al (2012) MYC gene amplification is often acquired in lethal distant breast cancer metastases of unamplified primary tumors. Mod Pathol 25(3):378–387
32. Embuscado E, Laheru D, Ricci F, Yun K, Boom Witzel S, Seigel A, Flickinger K, Hidalgo M, Bova S, Lacobuzio-Donahue C (2005) Immortalizing the complexity of cancer metastasis genetic features of lethal metastatic pancreatic cancer obtained from rapid autopsy. Cancer Biol Ther 4(5):548–554
33. Bliss LA, Sams MR, Deep-Soboslay A, Ren-Patterson R, Jaffe AE, Chenoweth JG et al (2012) Use of postmortem human dura mater and scalp for deriving human fibroblast cultures. PLoS One 7(9), e45282
34. Hjelm BE, Rosenberg JB, Szelinger S, Sue LI, Beach TG, Huentelman MJ et al (2011) Induction of pluripotent stem cells from autopsy donor-derived somatic cells. Neurosci Lett 502(3):219–224
35. Siolas D, Hannon GJ (2013) Patient-derived tumor xenografts: transforming clinical samples into mouse models. Cancer Res 73:5315–5319
36. Aparicio S, Hidalgo M, Kung AL (2015) Examining the utility of patient-derived xenograft mouse models. Nat Rev Cancer 15(5):311–316
37. Majumder B, Baraneedharan U, Thiyagarajan S, Radhakrishnan P, Narasimhan H, Dhandapani M et al (2015) Predicting clinical response to anticancer drugs using an ex vivo platform that captures tumour heterogeneity. Nat Commun 6:6169
38. Xie T, Musteanu M, Lopez-Casas PP, Shields DJ, Olson P, Rejto PA, Hidalgo M (2015) Whole exome sequencing of rapid autopsy tumors and xenograft models reveals possible driver mutations underlying tumor progression. PLoS One 10(11):e0142631
39. Jansen MH, van Vuurden DG, Vandertop WP, Kaspers GJ (2012) Diffuse intrinsic pontine gliomas: a systematic update on clinical trials and biology. Cancer Treat Rev 38(1):27–35
40. Zarghooni M, Bartels U, Lee E, Buczkowicz P, Morrison A, Huang A et al (2010) Whole-genome profiling of pediatric diffuse intrinsic pontine gliomas highlights platelet-derived growth factor receptor α and poly (ADP-ribose) polymerase as potential therapeutic targets. J Clin Oncol 28(8):1337–1344
41. Cummings MC, Simpson PT, Reid LE, Jayanthan J, Skerman J, Song S et al (2014) Metastatic progression of breast cancer: insights from 50 years of autopsies. J Pathol 232(1):23–31
42. Van Poznak C, Somerfield MR, Bast RC, Cristofanilli M, Goetz MP, Gonzalez-Angulo AM, Hicks DG, Hill EG, Li MC, Lucas W, Mayer IA, Mennel RG, Symmans WF, Hayes DF, Harris LN (2015) Use of biomarkers to guide decisions on systemic therapy for women with metastatic breast cancer: American Society of Clinical Oncology Clinical Practice Guideline. J Clin Oncol 33(24):2695–2704
43. Iacobuzio-Donahue CA, Fu B, Yachida S, Luo M, Abe H, Henderson CM, Vilardell F, Wang Z, Keller JW, Banerjee P, Herman JM, Cameron JL, Yeo CJ, Halushka MK, Eshleman JR, Raben M, Klein AP, Hruban RH, Hidalgo M, Laheru D (2009) DPC4 gene status of the primary carcinoma correlates with patterns of failure in patients with pancreatic cancer. J Clin Oncol 27(11):1806–1813

44. Udager AM, Shi Y, Tomlins SA, Alva A, Siddiqui J, Cao X, Pienta KJ, Jiang H, Chinnaiyan AM, Mehra R (2014) Frequent discordance between *ERG* gene rearrangement and ERG protein expression in a rapid autopsy cohort of patients with lethal, metastatic, castration-resistant prostate cancer. Prostate 74(12):1199–1208
45. Grasso CS, Wu YM, Robinson DR, Cao X, Dhanasekaran SM, Khan AP, Quist MJ, Jing X, Lonigro RJ, Brenner JC, Asangani IA, Ateeq B, Chun SY, Siddiqui J, Sam L, Anstett M, Mehra R, Prensner JR, Palanisamy N, Ryslik GA, Vandin F, Raphael BJ, Kunju LP, Rhodes DR, Pienta KJ, Chinnaiyan AM, Tomlins SA (2012) The mutational landscape of lethal castration-resistant prostate cancer. Nature 487(7406):239–243
46. Gavrilovic IT, Posner JB (2005) Brain metastases: epidemiology and pathophysiology. J Neurooncol 75(1):5–14
47. Allan DS, Belanger R, Busque L, Cohen S, Fish D, Roy DC et al (2005) Maintaining high autopsy rates in a Canadian blood and marrow transplant program: preserving a diagnostic and research tool. Bone Marrow Transplant 35(8):781–785
48. Tanaka H, Takamori H, Kanemitsu K, Chikamoto A, Beppu T, Baba H (2012) An autopsy study to clarify characteristics of local recurrence after extended pancreatectomy with intraoperative radiation therapy in patients with pancreatic cancer. Langenbecks Arch Surg 397(6):927–932
49. Finn RS, Brims FJ, Gandhi A, Olsen N, Musk AW, Maskell NA et al (2012) Post mortem findings of malignant pleural mesothelioma: a two-centre study of 318 patients. Chest 142:1267–1273
50. Nakagawa K, Aoki Y, Tago M, Terahara A, Ohtomo K (2000) Megavoltage CT-assisted stereotactic radiosurgery for thoracic tumors: original research in the treatment of thoracic neoplasms. Int J Radiat Oncol Biol Phys 48(2):449–457
51. Collins M (1998) The ultimate gift: Mortui vivos docebunt. Med Pediatr Oncol 30(5):267–268
52. HIll RB, Anderson RE (1992) The autopsy in oncology. CA Cancer J Clin 42(1):47–56
53. Penninckx B, Van de Voorde WM, Casado A, Reed N, Moulin C, Karrasch M (2012) A systemic review of toxic death in clinical oncology trials: an Achilles' heel in safety reporting revisited. Br J Cancer 107(1):1–6
54. Start RD, Bury JP, Strachan AG, Cross SS, Underwood JC (1997) Evaluating the reliability of causes of death in published clinical research. BMJ 314(7076):271
55. Shojania KG, Burton EC, McDonald KM, Goldman L (2003) Changes in rates of autopsy-detected diagnostic errors over time: a systematic review. JAMA 289(21):2849–2856
56. Burton EC, Trocxlair DA, Newman WP (1998) Autopsy diagnoses of malignant neoplasms. JAMA 280(14):1245–1248
57. Tavora F, Crowder CD, Sun CC, Burke AP (2008) Discrepancies between clinical and autopsy diagnoses: a comparison of university, community, and private autopsy practices. Am J Clin Pathol 129(1):102–109
58. Lott Limbach AA, Prayson RA (2012) Utility of autopsy in uncovering unexpected neuropathology. Ann Diagn Pathol 16(5):350–353
59. Seftel MD, Ho M, Pruthi D, Orbanski S, Rubinger M, Schacter B et al (2007) High rate of discordance between clinical and autopsy diagnoses in blood and marrow transplantation. Bone Marrow Transplant 40(11):1049–1053
60. Twigg SJ, McCrirrick A, Sanderson PM (2001) A comparison of post mortem findings with post hoc estimated clinical diagnoses of patients who die in a United Kingdom intensive care unit. Intensive Care Med 27(4):706–710
61. Tejerina E, Esteban A, Fernandez-Segoviano P, Maria Rodriguez-Barbero J, Gordo F, Frutos-Vivar F et al (2012) Clinical diagnoses and autopsy findings: discrepancies in critically ill patients. Crit Care Med 40(3):842–846
62. Silfvast T, Takkunen O, Kolho E, Andersson LC, Rosenberg P (2003) Characteristics of discrepancies between clinical and autopsy diagnoses in the intensive care unit: a 5-year review. Intensive Care Med 29(2):321–324
63. Pastores SM, Dulu A, Voigt L, Raoof N, Alicea M, Halpern NA (2007) Premortem clinical diagnoses and postmortem autopsy findings: discrepancies in critically ill cancer patients. Crit Care 11(2):R48

64. Sonderegger-Iseli K, Burger S, Muntwyler J, Salomon F (2000) Diagnostic errors in three medical eras: a necropsy study. Lancet 355(9220):2027–2031
65. Schwanda-Burger S, Moch H, Muntwyler J, Salomon F (2012) Diagnostic errors in the new millennium: a follow-up autopsy study. Mod Pathol 25(6):777–783
66. Thurnheer R, Hoess C, Doenecke C, Moll C, Muntwyler J, Krause M (2009) Diagnostic performance in a primary referral hospital assessed by autopsy: evolution over a ten-year period. Eur J Intern Med 20(8):784–787
67. Baillargeon J, Kuo YF, Lin YL, Raji MA, Singh A, Goodwin JS (2011) Effect of mental disorders on diagnosis, treatment, and survival of older adults with colon cancer. J Am Geriatr Soc 59(7):1268–1273
68. Berg J, Downing A, Lukes RJ (1970) Prevalence of undiagnosed cancer of the large bowel found at autopsy in different races. Cancer 25(5):1076–1080
69. Sinko J, Csomor J, Nikolova R, Lueff S, Krivan G, Remenyi P et al (2008) Invasive fungal disease in allogeneic hematopoietic stem cell transplant recipients: an autopsy-driven survey. Transpl Infect Dis 10(2):106–109
70. Jahn JL, Giovannucci EL, Stampfer MJ (2015) The high prevalence of undiagnosed prostate cancer at autopsy: implications for epidemiology and treatment of prostate cancer in the prostate-specific antigen-era. Int J Cancer 137:2795–2802
71. Combes A, Mokhtari M, Couvelard A, Trouillet JL, Baudot J, Henin D et al (2004) Clinical and autopsy diagnoses in the intensive care unit: a prospective study. Arch Intern Med 164(4):389–392
72. Nomori H, Ohba Y, Yoshimoto K, Shibata H, Shiraishi K, Mori T (2009) Positron emission tomography in lung cancer. Gen Thorac Cardiovasc Surg 57(4):184–191
73. Tu DG, Yao WJ, Chang TW, Chiu NT, Chen YH (2009) Flare phenomenon in positron emission tomography in a case of breast cancer--a pitfall of positron emission tomography imaging interpretation. Clin Imaging 33(6):468–470
74. Underwood JC, Cotton DW, Stephenson TJ (1989) Audit and necropsy. Lancet 1(8635):442
75. Vonsattel JP, Del Amaya MP, Keller CE (2008) Twenty-first century brain banking. Processing brains for research: the Columbia University methods. Acta Neuropathol 115(5):509–532
76. Medicine SUSo. http://lokey.stanford.edu/cores/shared-services.html - tissuebank 2009 [cited 2012]
77. Mucci NR, Moore HM, Brigham LE, Goldthwaite CA, Little AR, Lockhart NC et al (2013) Meeting research needs with postmortem biospecimen donation: summary of recommendations for postmortem recovery of normal human biospecimens for research. Biopreserv Biobank 11(2):77–82
78. Thayyil S, Robertson NJ, Scales A, Weber MA, Jacques TS, Sebire NJ et al (2009) Prospective parental consent for autopsy research following sudden unexpected childhood deaths: a successful model. Arch Dis Child 94(5):354–358
79. Underwood JC (2006) The impact on histopathology practice of new human tissue legislation in the UK. Histopathology 49(3):221–228
80. Thayyil S, Robertson NJ, Scales A, Sebire NJ, Taylor AM (2008) Parental consent for research and sudden infant death. Lancet 372(9640):715
81. Wiener L, Sweeney C, Baird K, Merchant MS, Warren KE, Corner GW et al (2014) What do parents want to know when considering autopsy for their child with cancer? J Pediatr Hematol Oncol 36:464–470
82. Womack C, Jack AL (2003) Family attitudes to research using samples taken at coroner's postmortem examinations: review of records. BMJ 327(7418):781–782
83. Lambe S, Cantwell N, Islam F, Horvath K, Jefferson AL (2011) Perceptions, knowledge, incentives, and barriers of brain donation among African American elders enrolled in an Alzheimer's research program. Gerontologist 51(1):28–38
84. Majores M, Schoch S, Lie A, Becker AJ (2007) Molecular neuropathology of temporal lobe epilepsy: complementary approaches in animal models and human disease tissue. Epilepsia 48(Suppl 2):4–12

The Future of Molecular Pathology

John S. Mattick

This is a time of unprecedented change, challenge, and opportunity in molecular pathology, which will fundamentally alter its nature, structure, and business models. While there will remain a place for traditional tests, and perhaps some opportunities for new metabolomic, proteomic, and immunological assays, the emergent field of genomics, driven by the extraordinary technical advances and attendant cost reductions in DNA sequencing, will transform genetic diagnoses and lead to the demise of individual genetic tests, cytogenetics analyses and, to a large extent, both cellular pathology and traditional microbiology. Moreover, molecular genetic analyses will change from being case-by-case diagnostic to prognostic at population scale, reducing the incidence of diseases and making preventative, mitigating, or ameliorative strategies personal and precise. Genome analysis will become standard medical practice, not simply in molecular 'pathology' but in health management, and transform the healthcare system. It will drive the amalgamation of molecular pathology with clinical genetics. It will also require integration of genomic data with phenotypic information, both traditional and non-traditional, made available in national and global databases that will be used by all parts of the research and healthcare systems for biomedical discovery and economic management.

J.S. Mattick (✉)
Garvan Institute of Medical Research,
384 Victoria Street, Darlinghurst, NSW 2010, Australia

St Vincent's Clinical School, UNSW Australia, Sydney, NSW 2052, Australia
e-mail: j.mattick@garvan.org.au

S.R. Lakhani, S.B. Fox (eds.), *Molecular Pathology in Cancer Research*,
DOI 10.1007/978-1-4939-6643-1_16

Genetic Disability

Until recently, and still substantially, diagnoses of simple ('Mendelian') genetic diseases (such as muscular dystrophy, cystic fibrosis, motor neuron disease, Huntingdon's disease, and thalassemia) have been informed by clinical symptoms and/or family history, and confirmed by specific, usually PCR-based, DNA tests limited to those loci which are known to be involved.

However, while specific genetic disabilities are rare and familial inheritance rendered almost invisible by the recessive nature of most mutations, cumulatively at least 2 % of all babies suffer a serious physiological, developmental, and/or intellectual disability due to damaging mutations in conventional (i.e., protein-coding) genes [1–3], of which there are ~19,000–20,000 [4–7]. Indeed it seems that most individuals carry a significant burden of such mutations [8], and it has been estimated that 20–30 % of infant deaths and ~50 % of all admissions to pediatric hospitals are due to genetically determined disorders [9–12], and that 12 % of all individuals will suffer a consequential hospitalization event at some point in their lives [13].

This also includes the finding that cerebral palsy, thought to be mainly caused by perinatal environmental factors such as pregnancy or birth trauma, is substantially due to genetic abnormalities [14–16].

Hitherto the vast majority of disease causative mutations were impossible to identify and therefore to treat in an informed way, but the system, more for psychosocial than medical reasons, has been obliged to undertake a litany of phenotypic tests—the so-called diagnostic odyssey—which are not expected to and usually do not lead to any productive insight about the *cause* of the condition, as opposed to obtaining finer detail of the problem, and reassuring families that "we are doing all we can." This is a lot of (otherwise) useless expenditure.

This situation is changing rapidly with the advent of cost-effective DNA capture and sequencing protocols, leading to use of targeted gene panels, for example, in cardiac or retinal conditions, and more recently relatively agnostic whole exome sequencing, which has less depth but more breadth than panel-based tests, of particular value in developmental and intellectual disorders of unknown etiology.

Exome sequencing delivers something approaching 25–40 % diagnostic yield beyond the obvious, but I predict this phase to be transitory, and be quickly overtaken by whole genome sequencing (WGS), at least in advanced jurisdictions.

WGS, for technical reasons, has much more even, as well as more comprehensive, exome coverage, and therefore higher diagnostic yield for protein-coding mutations. Indeed in our facility, we are obtaining an average of 50 % diagnostic yield for undiagnosed genetic disability by WGS (Tony Roscioli, personal communication).

It also provides additional information on other sources of genetic variation (deletions, insertions, inversions, translocations, etc.) and a lifetime reservoir of information that can be interrogated repeatedly in other contexts, including variations in regulatory sequences that may be important in other conditions or for determining risk for complex diseases.

I predict that WGS will replace all others and become the universal genetic test, notwithstanding some current technical blind spots, which will be rectified soon enough. Despite the drag from marginal cost considerations and infrastructure requirements, including the analytics, the value proposition is so high that delays in moving to WGS are largely due to lack of vision and inertial difficulties of changing established systems in commercial laboratories, and the activation costs of moving to new protocols and SOPs. Reciprocally, this means that established pathology organizations, and those working within them, are more vulnerable to disruptive change by more new and less constrained players, a self-destructive cycle that is increasingly repeated across the innovation landscape.

Cancer

High throughput DNA sequencing, to poll the status of the cancer genome, epigenome, and transcriptome, will also transform the diagnosis, treatment and prevention of cancer. First, population-scale sequencing of the genomes of children and adolescents with cancer will likely identify the inherited components of cancer risk, leading to better identification of individuals and more active screening of populations at risk, and early intervention.

Second, for spasmodic cancers that largely arise later in life, despite systemic reluctance, the current well-worn practice of sending biopsies to a cellular pathologist is likely to obligatorily accompanied and possibly completely supplanted by genomic sequencing, which is far more informative in terms of identifying the so-called driver mutations that are amenable to therapeutic inhibition. "Grade 4" will not cut it. Cancer genome sequencing will become the expected standard of care, driven by patient demand, especially with the rise of targeted drugs.

Third, deep sequencing can identify cancer-derived DNA in circulating blood, and be used to monitor its incidence, burden, and progression (drug resistance and metastasis) [17–20]. In this and other cancer contexts panel- or exome-based capture tests may persist, because of the greater depth of coverage, important where biopsies may have mixed normal and cancer cells. Innovative DNA capture protocols will provide far more precision in leukemia, making it possible, for example, to diagnose many if not most causative translocations in a single test (Tim Mercer, personal communication).

Preconception and Prenatal Screening

One of the surprising findings of recent years has been that 6–8 % of the DNA circulating in the blood of pregnant women is derived from the fetus. Cellular testing for trisomies by chorionic villus sampling (CVS) and amniocentesis, the major source of demand for cytogenetics is being rapidly supplanted by deep sequencing, which can detect trisomies with greater accuracy and no risk to mother or child [21, 22]. The word on the street is that even young mothers who are at low risk are accessing the test at personal expense. They appreciate the value proposition, and it won't be long before health systems do too. The primary *raison d'être* and demand for cytogenetics is being wiped out overnight.

Furthermore, research papers have appeared reporting WGS of the fetus from maternal circulation, which I predict will become common in obstetrics [23–25]. Prenatal genomic testing of embryos, however, has its problematic and for many unpalatable sequelae. I expect that, especially as population-scale genomic profiling becomes more common and accessible, prenatal screening will be rendered largely redundant by preconception WGS analysis of the parents, to identify the incidence or otherwise of damaged alleles in common, and therefore the high (25 %) risk of a homozygous or compound heterozygous child, which can be avoided by pre-implantation testing of IVF-derived embryos. The value proposition for the health system is enormous, and for the families inestimable.

Adverse Drug Reactions and Optimization of Drug Treatments

Adverse reactions to prescription drugs account for up to 7.6 % of hospital admissions in Australia [26, 27], with similar incidences in Europe and the USA [28], some with lifelong consequences. Genomic testing can predict and avoid a large fraction of such toxic drug buildups, with a recent trial in Melbourne indicating national savings of $480 m per annum in the area of psychiatric drugs alone [29]. In addition, many drugs, such as beta-blockers, antidepressants, and the anticoagulation drug clopidogrel, only work in a fraction of those individuals for whom they are prescribed [30–32], which therefore constitutes, at the other end of the spectrum, useless expenditure.

The main thing that is required, following a decision to implement by health agencies, is to put in place the infrastructure, which is not complicated, to obtain the relevant genetic information in a timely manner at the point of prescription. Of course, once population-scale sequencing is commonplace, such information will be immediately available by querying the database.

Infection and Immunity

Microbiological testing will also be revolutionized by WGS, which can not only rapidly and agnostically identify pathogenic species, but also their biotype, including virulence determinants and antibiotic resistance, as well as complex mixtures thereof. This will render eliminate the need for clumsy, incomplete, and expensive culturing and make antibody-based tests largely unnecessary. The latter will be further made redundant by innovative sequence capture protocols of B-cell immunoglobulin and T-cell receptor loci that will be able to poll the full repertoire and history of pathogen, autoantigen, and tumor neoantigen exposure.

One Test to Rule Them All

There will of course continue to be many situation-specific molecular pathology analyses, in cancer, infections, neurological conditions, and immunological diseases, among others, where targeted genomic and non-genomic methods will have high value and where the choice will be dependent on the combination of cost and specificity. Horses for courses.

Nonetheless, a significant proportion of existing tests and many new ones will be replaced or introduced, respectively, by WGS. At present the demand for and utility of such tests is driven primarily by disease (pathology), but I predict that within the next 5–20 years every member of our community who consents will have their genome sequenced and incorporated into their medical records and health management plans.

The initial uptake of WGS may be driven by disease diagnosis in molecular pathology and by preconception screening in clinical genetics, but will rapidly expand throughout the community as the costs continue to decline, as they assuredly will, and the value proposition grows with more sophisticated and comprehensive genotype–phenotype databases.

The major diseases afflicting society, and the major cost burden on the public and private healthcare systems—such as heart disease, cancer, diabetes, autoimmune, and inflammatory diseases such as arthritis, osteoporosis, stroke, dementias, and neuropsychiatric disorders, among others, as well as viral and bacterial infections—have major genetic factors.

While it is the case that the genetic risk factors for complex diseases are not well understood, it is expected that with the flood of genomic data coming down the pipeline that these factors will become increasingly well defined, driving further research and incorporated into personalized advice and preventative programs to reduce the incidence or severity of chronic disease in an aging population, with huge benefits for quality of life and health economics.

There are very few diseases whose impact cannot be mitigated by anticipatory action—by lifestyle modification, early detection, and preventative therapies, including pharmaceutical intervention, such as regular monitoring of individuals at high risk for colon cancer or prescription of anticoagulant drugs for individuals at heightened risk for blood clotting disorders like deep vein thrombosis and stroke. In cardiovascular disease, a risk-focused approach (as opposed to managing the consequent illnesses) has had a major impact on system burden over the past 30 years. The same may be achieved through genomics for anticipating risk for cancer and most other diseases that impose whole of life or later life chronic burdens. The distinction between molecular pathologists and clinical geneticists will disappear.

Infrastructure for a New World

No hospital or molecular pathology laboratory, let alone individual clinician, will have the capability to analyze genomic data or translate it into medical advice. The consequence is that nature of the molecular pathology industry and professional practice will fundamentally change.

A number of things are required. First, most molecular medicine and much of the healthcare system will need to become genomically literate and receptive. Second, genomic medicine clinics staffed by clinical geneticists and genetic counselors will need to be established, as a portal for referral entry into genomic testing.

Third, and most importantly, central genotype–phenotype correlation databases will need to be built to provide well-curated, evidence-based, continuously-updated information on the clinical significance of genetic variants and recommended actions (treatments and/or avoidance strategies) across the many domains of medicine and health (e.g., intellectual disability, dementia, diabetes, osteoporosis, cancers, cardiac, autoimmune, and neuropsychiatric diseases, to name a few), with appropriate consent and privacy provisions and protections, to accredited clinicians, primary healthcare providers, health agencies, and researchers.

My view is that these databases, while globally linked, will need to be pulled together and run by national governments, because of the scale of the infrastructure (including the data storage and computational capacity, software development, and domain-specific genotype–phenotype correlation specialists) required the jurisdictional idiosyncrasies in legal and regulatory frameworks and financial rebate systems, and because of the public good and privacy imperatives associated with the use of such databases.

From Molecular Pathology to Data Analytics

The integration of genomic data with information from electronic medical records and other sources, including smart devices and patient input, which is essential, will allow both directed and agnostic interrogation by pattern analysis of the data and metadata to identify, for example, patient responses and unexpected co-morbidities to inform patient stratification for more effective treatment and more efficient use of healthcare and insurance resources.

This will also become a central reservoir, indeed a goldmine, for medical and health research, for example, in determining why some people do not respond to a particular therapy, and going forward to close the gap. An illustrative example is the discovery of the fact that certain people of East Asian origin do not respond to the anticancer drug Imantinib, due to a secondary genetic variant in one of the apoptosis genes required for Imantinib action, leading to the development of a supplementary strategy to bypass the problem and rescue those individuals [33].

One can only imagine the savings by public and private healthcare system from being able to accumulate, access, analyze, and act on such information. However there are also many questions and challenges that are and will be raised by the transition to genomic medicine.

Incorporating genomic information into routine healthcare will require changes to public policy and public health management. These include identification of at-need communities where genomic information can have an immediate and/or lifelong benefit for the individual and their family, as well as the health system; assessment of the costs and the net health, social and economic benefits of genomic information, and their trajectory; consideration of the approach for equitable population-wide introduction of genomic sequencing; definition of the clinical support systems and interfaces, databases, analytical systems, educational infrastructure, and consent and privacy provisions, among others, that will be required to obtain maximum benefit from the widespread usage of genomic information in healthcare; identification of public policy considerations that will require regulatory or legislative change or health system framework development; and provision of guidance on public policy and private sector implications of genome sequencing on personal insurance, private health and employment disclosures, and introduction of protections to be put in place.

As Bill Gates said, we overestimate the change that will occur in the next 2 years and underestimate the change that will occur in the next ten. The twentieth century was just the warm up—this is the century of molecular medicine.

References

1. Robinson A, Linden MG (1993) Clinical genetic handbook. Blackwell Scientific Publications, Boston
2. Sankaranarayanan K (2001) Estimation of the hereditary risks of exposure to ionizing radiation: history, current status, and emerging perspectives. Health Phys 80:363–369
3. Ropers HH, Hamel BC (2005) X-linked mental retardation. Nat Rev Genet 6:46–57
4. Goodstadt L, Ponting CP (2006) Phylogenetic reconstruction of orthology, paralogy, and conserved synteny for dog and human. PLoS Comput Biol 2, e133
5. Clamp M, Fry B, Kamal M, Xie X, Cuff J, Lin MF, Kellis M, Lindblad-Toh K, Lander ES (2007) Distinguishing protein-coding and noncoding genes in the human genome. Proc Natl Acad Sci U S A 104:19428–19433
6. Church DM, Goodstadt L, Hillier LW, Zody MC, Goldstein S, She X, Bult CJ, Agarwala R, Cherry JL, DiCuccio M, Hlavina W, Kapustin Y, Meric P, Maglott D, Birtle Z, Marques AC, Graves T, Zhou S, Teague B, Potamousis K, Churas C, Place M, Herschleb J, Runnheim R, Forrest D, Amos-Landgraf J, Schwartz DC, Cheng Z, Lindblad-Toh K, Eichler EE, Ponting CP, Mouse Genome Sequencing Consortium (2009) Lineage-specific biology revealed by a finished genome assembly of the mouse. PLoS Biol 7, e1000112
7. Ezkurdia I, Juan D, Rodriguez JM, Frankish A, Diekhans M, Harrow J, Vazquez J, Valencia A, Tress ML (2014) Multiple evidence strands suggest that there may be as few as 19,000 human protein-coding genes. Hum Mol Genet 23:5866–5878

8. Genomes Project Consortium, Abecasis GR, Altshuler D, Auton A, Brooks LD, Durbin RM, Gibbs RA, Hurles ME, McVean GA (2010) A map of human genome variation from population-scale sequencing. Nature 467:1061–1073
9. Berry RJ, Buehler JW, Strauss LT, Hogue CJ, Smith JC (1987) Birth weight-specific infant mortality due to congenital anomalies, 1960 and 1980. Public Health Rep 102:171–181
10. Hoekelman RA, Pless IB (1988) Decline in mortality among young Americans during the 20th century: prospects for reaching national mortality reduction goals for 1990. Pediatrics 82:582–595
11. Hoyert DL, Freedman MA, Strobino DM, Guyer B (2001) Annual summary of vital statistics: 2000. Pediatrics 108:1241–1255
12. McCandless SE, Brunger JW, Cassidy SB (2004) The burden of genetic disease on inpatient care in a children's hospital. Am J Hum Genet 74:121–127
13. Emery AEH, Rimoin DL (1990) Principles and practice of medical genetics, 2nd edn. Churchill Livingstone, New York
14. Moreno-De-Luca A, Ledbetter DH, Martin CL (2012) Genetic insights into the causes and classification of the cerebral palsies. Lancet Neurol 11:283–292
15. Oskoui M, Gazzellone MJ, Thiruvahindrapuram B, Zarrei M, Andersen J, Wei J, Wang Z, Wintle RF, Marshall CR, Cohn RD, Weksberg R, Stavropoulos DJ, Fehlings D, Shevell MI, Scherer SW (2015) Clinically relevant copy number variations detected in cerebral palsy. Nat Commun 6:7949
16. McMichael G, Bainbridge MN, Haan E, Corbett M, Gardner A, Thompson S, van Bon BW, van Eyk CL, Broadbent J, Reynolds C, O'Callaghan ME, Nguyen LS, Adelson DL, Russo R, Jhangiani S, Doddapaneni H, Muzny DM, Gibbs RA, Gecz J, MacLennan AH (2015) Whole-exome sequencing points to considerable genetic heterogeneity of cerebral palsy. Mol Psychiatry 20:176–182
17. Dawson SJ, Tsui DW, Murtaza M, Biggs H, Rueda OM, Chin SF, Dunning MJ, Gale D, Forshew T, Mahler-Araujo B, Rajan S, Humphray S, Becq J, Halsall D, Wallis M, Bentley D, Caldas C, Rosenfeld N (2013) Analysis of circulating tumor DNA to monitor metastatic breast cancer. N Engl J Med 368:1199–1209
18. Murtaza M, Dawson SJ, Tsui DW, Gale D, Forshew T, Piskorz AM, Parkinson C, Chin SF, Kingsbury Z, Wong AS, Marass F, Humphray S, Hadfield J, Bentley D, Chin TM, Brenton JD, Caldas C, Rosenfeld N (2013) Non-invasive analysis of acquired resistance to cancer therapy by sequencing of plasma DNA. Nature 497:108–112
19. Ignatiadis M, Dawson SJ (2014) Circulating tumor cells and circulating tumor DNA for precision medicine: dream or reality? Ann Oncol 25:2304–2313
20. Murtaza M, Dawson SJ, Pogrebniak K, Rueda OM, Provenzano E, Grant J, Chin SF, Tsui DW, Marass F, Gale D, Ali HR, Shah P, Contente-Cuomo T, Farahani H, Shumansky K, Kingsbury Z, Humphray S, Bentley D, Shah SP, Wallis M, Rosenfeld N, Caldas C (2015) Multifocal clonal evolution characterized using circulating tumour DNA in a case of metastatic breast cancer. Nat Commun 6:8760
21. Fan HC, Blumenfeld YJ, Chitkara U, Hudgins L, Quake SR (2008) Noninvasive diagnosis of fetal aneuploidy by shotgun sequencing DNA from maternal blood. Proc Natl Acad Sci U S A 105:16266–16271
22. Chiu RW, Chan KC, Gao Y, Lau VY, Zheng W, Leung TY, Foo CH, Xie B, Tsui NB, Lun FM, Zee BC, Lau TK, Cantor CR, Lo YM (2008) Noninvasive prenatal diagnosis of fetal chromosomal aneuploidy by massively parallel genomic sequencing of DNA in maternal plasma. Proc Natl Acad Sci U S A 105:20458–20463
23. Lun FM, Tsui NB, Chan KC, Leung TY, Lau TK, Charoenkwan P, Chow KC, Lo WY, Wanapirak C, Sanguansermsri T, Cantor CR, Chiu RW, Lo YM (2008) Noninvasive prenatal diagnosis of monogenic diseases by digital size selection and relative mutation dosage on DNA in maternal plasma. Proc Natl Acad Sci U S A 105:19920–19925
24. Kitzman JO, Snyder MW, Ventura M, Lewis AP, Qiu R, Simmons LE, Gammill HS, Rubens CE, Santillan DA, Murray JC, Tabor HK, Bamshad MJ, Eichler EE, Shendure J (2012) Noninvasive whole-genome sequencing of a human fetus. Sci Transl Med 4:137ra176

25. Chitty LS, Bianchi DW (2015) Next generation sequencing and the next generation: how genomics is revolutionizing reproduction. Prenat Diagn 35:929–930
26. Miller GC, Britth HC, Valenti L (2006) Adverse drug events in general practice patients in Australia. Med J Aust 184:321–324
27. Roughead L, Semple S, Rosenfeld E (2013) Australian Commission on Safety and Quality in Health Care (2013). Medication Safety in Australia. ACSQHC, Sydney, Literature Review
28. https://www.health.gov.au/internet/main/publishing.nsf/Content/D341DA146481106ACA257BF00020A7CD/$File/Phase1Literature review.pdf
29. Kleinman R. Personalised prescribing promises to save mental health millions. The Age
30. Thomas FJ, McLeod HL, Watters JW (2004) Pharmacogenomics: the influence of genomic variation on drug response. Curr Top Med Chem 4:1399–1409
31. Wiita AP, Schrijver I (2011) Clinical application of high throughput molecular screening techniques for pharmacogenomics. Pharmgenomics Pers Med 4:109–121
32. Storelli F, Daali Y, Desmeules J, Reny JL, Fontana P (2015) Pharmacogenomics of oral antithrombotic drugs. Curr Pharm Des 22:1933–1949
33. Ng KP, Hillmer AM, Chuah CT, Juan WC, Ko TK, Teo AS, Ariyaratne PN, Takahashi N, Sawada K, Fei Y, Soh S, Lee WH, Huang JW, Allen JC Jr, Woo XY, Nagarajan N, Kumar V, Thalamuthu A, Poh WT, Ang AL, Mya HT, How GF, Yang LY, Koh LP, Chowbay B, Chang CT, Nadarajan VS, Chng WJ, Than H, Lim LC, Goh YT, Zhang S, Poh D, Tan P, Seet JE, Ang MK, Chau NM, Ng QS, Tan DS, Soda M, Isobe K, Nothen MM, Wong TY, Shahab A, Ruan X, Cacheux-Rataboul V, Sung WK, Tan EH, Yatabe Y, Mano H, Soo RA, Chin TM, Lim WT, Ruan Y, Ong ST (2012) A common BIM deletion polymorphism mediates intrinsic resistance and inferior responses to tyrosine kinase inhibitors in cancer. Nat Med 18:521–528

Index

S.R. Lakhani, S.B. Fox (eds.), *Molecular Pathology in Cancer Research*,
DOI 10.1007/978-1-4939-6643-1

N

MIX
Papier aus verantwortungsvollen Quellen
Paper from responsible sources
FSC® C105338
FSC
www.fsc.org

If you have any concerns about our products,
you can contact us on
ProductSafety@springernature.com

In case Publisher is established outside the EU,
the EU authorized representative is:
Springer Nature Customer Service Center GmbH
Europaplatz 3, 69115 Heidelberg, Germany

Printed by Libri Plureos GmbH
in Hamburg, Germany